# Managing Enterprise Information Technology

by CJ Rhoads

Library of Congress Cataloging-in-Publication Data
Rhoads, CJ 1960–
Managing Enterprise Information Technology/ CJ Rhoads.
Includes bibliographical references and index.
1. Information Technology. 2. Microsoft Office. 3. Computers. 4. Managing Information Systems.



Student textbook:
Print Version: ISBN 978-1-61305-023-1
Electronic Version: ISBN 978-1-61305-024-8

HPL Publications, PO Box 564, Douglassville, PA 19518
www.HPLConsortium.com
Printed in the United States of America
The paper used in this book complies with the
Permanent Paper Standard issued by the National
Information Standards Organization (Z39.48–1984).
10 9 8 7 6 5 4 3 2 1



HPL Consortium
http://HPLConsortium.com

# Table of Contents

# Illustrations

## Figures

# Tables

# Acknowledgements

As first noted by Bernard of Chartres, we all stand on the shoulders of giants. Any author will tell you that a book may have only one name on the cover, but they are always the product of many individuals. My support is legion; it would take another book to name them all. But a chosen few, at least, will be named here. They are no less responsible for its birth than I.

There are many on this list who helped me years ago, and some who have passed away. I leave them on this list to honor them for their legacy lives on in my writing and current activities.

There are those who specifically helped me by supplementing my knowledge with their expertise: Jack Bradt, Vanessa DiMauro, Francois Dumas, Pamela Gockley, Scott Herman, Brenda Moretti, Joe Puglisi, and Patrick Schmid. My graduate students when I wrote the first version of this book (my first book) also helped a great deal; they are: Anil Aras, Michael Axman, Robert Gourley, Julie Gower, Brett Harbach, Tammy Hein, Brian Meares, Susan Miller, Dave Musante, Mariya Papazova, Bryan Rathman, Andrew Sims, and Todd Dierksheide. More recently, some of my graduate students were especially helpful in my review and update of this book: Kimberly Duhamel-Murray, Scott Matthews, Tea Vid, Margaret Ramsden, Daniel Held, and Shana Hashagen.

Additionally, tops in my book would be Jeff Olson, initially my agent, and then my series editor, and then my acquisitions editor, who believed in me enough to ask me to step in as series editor when he moved up.

Several people contributed by allowing me to use something they developed: T. Max Devlin, Bill Jelen, and Ned Lilly.

The following people did the difficult task of reviewing various chapters of the book: Bill Bauerle, Fred Beste, Polly Beste, David Bonner, Jack Bradt, Jeff Benzel, Joel Cardis, Greig Clark, Laura Donovan, Roger Ganser, Roger Hibbs, Con Karlson, Jan Lipper, George Lipper, John McNamara, Skip Torresson, and Molly Tremblay. The following people reviewed the final revisions of various chapters: Reuel Launey, Thomas Penn, Sandra Walker, David Freschman, Steve Meng, Lila Bellando, Dan Conley, Leigh Wood, Jay Coen Gilbert, Robert Clemens, Doug A. Bloom, Dan McKinney, Sean Mallon, Laura Lorber, Anne Marie Smith, Michael Reddy, Fred Hafer, Jenn Mossholder, Maria Koba, John Moore, John R. Moore, Jon Reed, Anne Marie Smith, Gergely Tapolyai, and Carol Tunnessen.

Many of my former clients appear in these pages, though few with their real names. In any case, my own knowledge is always expanded greatly by those for whom I serve, for I learn much from them in the course of our activities together. I'd especially like to extend my heartfelt thanks to Lee Baker, Peter Baker, David Bosler, David Dries, Bob Goodman, Michael Guido, Joanne Just, Conrad Karlson, Andy Klee, Pat Krick, Harriet Layton, Jane Palmer, Samantha Reimert, LuAnn Seyler, Patrick Sullivan, Anthony Triano, and Linda Wade.

Additionally, many CEOs and senior executives have taken time out of their busy schedules to appear on panels or speak on the topics presented here. I learned from each of them as well: Bill Alberte, Sandy Becker, Ed Barrett, Steve Bobick, Tom Baumann, Greg Baxter, Larry Bergman, Terry Carpenter, Dennis Cichelli, Andrew Downer, Don Gould, Dennis Hague, Keith Hill, John Kruglinski, Bob Lewis, Rob McCord, Ray Melcher, Brenda Moretti, Sean Moretti, Michael Mullen, Dianah Neff, Chris Miller, Jayashree Raman, Kevin Raudenbush, Edie Ritter, Harry Roberts, Bill Ruhl, Laura Scott, Shawn Shirk, Charles Sullivan, Rick Sutton, Brian Von Stetten, Phil Theis, Chet Winters, and John Weidenhammer.

Furthermore, even more CEOs and business owners took time to answer survey questions and speak with me at length about the challenges they face as they manage their information technology: James Adams, Vernon Babilon, Ed Barrett, Carolyn Bazik, Christopher Bernard,

Chand Bhutra, Steve Bobick, Dale Bolger, Larry Bonfante, Ann Borza, Michael Boylan, William Buckholz, John Bungert, Rick Burkey, James Chan, Chris Del Valle, Joseph DiAngelo, Mark Dillon, Ann Earon, Jill Edwards, Bob Ehlinger, Suzanne Fairlie, William Flippin, Diane Galbraith, John Gantz, Paul Gentile, Matthew Gilbert, John Glaser, Lisa Guzzardo, David Haas, Scott Hafer, Robert Harrop, Michael Hund, Fidel Ikem, Werner John, Paul Jones, Charles Kelly, Andy Klee, Chris Kraras, Ginger Kunkel, Michael Lawrence, George Lipper, Daniel Little, Timothy Mack, Donald Martin, Laura Matrisciano, Terry Maurer, Dave Meckley, Charles Miller, Tom Murphy, Joseph Narieka, Stephanie Olexa, Steve Olshevski, Joe Ostraner, Robert Page, Melvin Pankuch, Gregory Payne, Michael Pellegrino, William Peterson, Susan Phillips Speece, Rick Principato, Joseph Puglisi, Ronald Rhode, Mark Rimler, Peter Rittenhouse, Jeff Salvino, Joseph Schaeffer, Lisa Scheller, Robin Schroeder, Grace Schuler, Anne Searle, Michael Shea, Ryan Smith, Helane Stein, Richard Stichter, Tom Stine, Camille Stock, Steven Talarico, Barry Unger, Jody Wagner, Peter Webster, Steve Willems, Charles Willis, Stacey Zbyszinski.

I would also like to thank some of the former and current faculty and staff of the College of Business (in alphabetical order): Okan Akcay, Dan Benson, Pat Blatt, Gary Chao, Henry Check, Donna DeLong, Arifeen Daneshyar, Mark Dinger, Ken Ehrensal, Tom Grant, Keshav Gupta, David Haas, John Hamrick, Ray Heimbach, Roger Hibbs, Eileen Hogan, Arian Hungaski, Fidel Ikem, Kim Kilgus, Sue Kong, Jonathan Kramer, John Kruglinski, Rajeev Kumar, James "Doc" Ogden, Patricia Patrick, Elisabeth Rogol, George Rogol, Paul Sable, Norman Sigmond, David Wagaman, and John Walker. Their encouragement and support over the years enabled me to focus on writing when I needed to do so. Additionally, I thank the employees of my business: Dan Chen See, Jade Alatishe, Elle Hargrave, Stella Deeble, George Deeble, and Bonnie Taylor, who all had to put up with my split focus at times while I was writing.

Furthermore, several people picked up the slack for me in the Pacem In Vita program I developed—a leadership program for kids based on Taijiquan. I must thank Jack Solchurch, Judy Butler, Vicky Lee Levy, Luke Jih, Heidi Markwalder, Victoria Mehl, Judd Meinhart, Sandy Wise, and Bill Wheeler.

Also helpful was Siobhan Hutchinson, and Jacqui Blatt who put up with my missing activities while we were on vacation together (but I still had writing to do!). Furthermore, I have to think my integrative healthcare team; people who keep me painfree since my 2002 accident with massages, therapy, and/or classes: Polly Truex, J. Michael Clark, Michelle Wurtz, Ramesh Wallen, Betsy Chapman, John Loupos, Niharika Agrawal. Without their help I would not be able to work.

Of course none of this would be possible without my many mentors over the years (in alphabetical order): Fred Beste, Mike Bolton, David Bosler, Jack Bradt, Tom Casey, Betsy Chapman, Martin Cheatle, Effie Chow, Jim Collins, Vanessa DiMauro, Jonathan Dreazen, Francois Dumas, Marsha Egan, Dale Falcinelli, Keshav Gupta, Steve Higgins, John Lucht, Nancy Magee, John McNamara, Ray Melcher, Pete Musser, Maggie Newman, Josephine Painter, Joe Puglisi, Leo Robb, Robert Rubin, Paul Sable, Lee Scheele, Steve Sperling, Alan Weiss, and Kevin Wren. Though the amount of time I spent with each varied, each one has given me a gift that has turned out to be of extreme value in my never-ending quest to improve myself.

No list would be complete without my fabulous family: my mother, Judith Liffick, and my father, Martin Devlin, and his wife, Jeri, along with my brothers and sisters, Jeffrey Devlin (and his wife Melissa), T. Max Devlin, Eileen Piccolo, and Denise Rankin. I also received a great deal of help and advice from several of my accomplished aunts and uncles, most of whom are authors in their own right: Tom and Barb Liffick, Blaise and Alana Liffick, Kathy Liffick, Mike Lillich, Charles and Elsie Jane Lorber, and Anita and Gary Young.

I also wish to thank my former husband Bob, who was patient and loving for thirty-seven short years. I was lucky to have found my soul mate so early in my life.

And I can't forget Marianne Pawlikowski, Sister Mary Morley, and Sister Patricia Dotzauer as well — all who provided the various LBI getaways that formed the backdrop of many hours of my writing.

For anyone who I have inadvertently left off this list, I am heartily sorry. My own feeble-minded brain does the best that it can, but it is far from perfect. I can only thank those people who I remember to write a note about at the time – I'm sure there were many who also helped, but I had forgotten to make the note. Feel free to let me know if you feel that you should be included in the acknowledgement section of next edition of this book. ☺

# Introduction

With so many textbooks on Managing Enterprise Information Technology over the years, why do we need another one?

## Why You Should Read This Book

As an entrepreneur, business owner or manager, do you ever make decisions that involve information technology? In the fullness of time, did your decisions turn out to be great, good, or perhaps not so good?

Have you ever made a technology decision that, in retrospect, turned out to be a mistake? You're not alone. The information in this book will improve your track record in making decisions involving information technology. The information described here will help you assess past decisions and will provide guidelines as you make decisions in the future. This information will enable your organization to experience smooth computing, the *holy grail* of effective organizations and profitable businesses.

In brief; the purpose of this book is to help those who are making these technology decisions to improve their ratio of great-decisions/not-so-good-decisions.

Some guidelines you may recognize immediately as functional. You might find that you've been using them without realizing it. Others may strike you as counterintuitive or will go against what others have told you. But you will find that, in practice, following these guidelines will work, for both you and your organization.

## Few Know the Secrets or Understand Their Importance

Making the right information technology decision has gotten easier over the past twenty-five years. In 1995 The Standish Group published the "CHAOS" report, noting that only 9 percent of the technology projects in large companies were successful[*]. That same study reported that even successful technology projects fulfilled only an average of 42 percent of their original requirements. According to Henny Portman's review of the most recent CHAOS report (which has been published every two years since the first one), project managers have gotten much better at technology projects[†]. Success rate shot up to a whopping 36% (admittedly better than 9%).

My experience mirrors these grim statistics. Decades ago, most technology projects ended up achieving no more than a mere shadow of their original goals—even, or especially, the ones that are paraded as successes by those who initiated or paid for the projects. People who worked on the project know the truth—the project was a failure by all objective standards of measure. We've gotten much better at managing technology projects for a variety of reasons. We stopped treating them as if they were construction projects (they are not). We made them smaller and stopped trying to plan for a three-year technology project. (By the time those projects were finished, the technology was always obsolete!) We added real project management to the process and got much smarter about what we could estimate and what we couldn't estimate. But given the poor track record of technology projects, how are managers and business owners expected to be able to make good technology decisions? Well, they generally don't.

---

[*] Crear, J., L. Gesmer, J. Johnson, J. Lynch, H. Mulder, T. Mulder and V. Lou (1995). The CHAOS Report, The Standish Group International, Inc.: 9.

[†] Portman, H. (2020). "Review CHAOS Report 2018." On Portfolio, Programme and Project Management https://hennyportman.wordpress.com/2020/01/03/review-chaos-report-2018/ 2020.

The fact is, there are many more technology losers (leaders who make the wrong decision) than technology winners (leaders who make the right decision). The difference between the two is not whether they make mistakes—they all make mistakes. The difference is whether they learn from their mistakes or keep making the same mistakes repeatedly.

This book is designed to set you on the path to becoming a technology winner—someone who makes the right decisions and understands how to get the best value for your money. Whether you are just choosing a bunch of smart phones for your employees or designing a completely new system from the ground up; the lessons to be learned from this book can help.

People who made poor technology decisions were not stupid. They were probably intelligent, dedicated executives making what they thought at the time was the best decision. But an experienced technology decision maker like me can usually tell from the very beginning whether a technology project is doomed to fail. If it was so easy for me to spot, then why did these executives decide to do these projects? Even more important, given the track record of most decision makers, why do business leaders continue to make the same mistakes over and over? We will explore not just the answer to those questions, but what they could have done differently that would have saved them from the failed projects.

This book is especially important for entrepreneurs, CEOs, business owners, managers and executives—leaders all—whose organizations rely upon technology, but who have a difficult time understanding what makes a good decision and a bad decision. Sometimes I talk with entrepreneurs and business leaders who tell me that they don't make any technology decisions. But research shows that leaders make between one and five significant technology decisions each day, and most leaders dedicate up to 12 hours per week gathering information to make decisions. Just the fact that business leaders think that they are not making technology decisions shows how large the gap is. Here is the fact: There are no businesses that aren't impacted by information technology. All businesses are influenced by information technology either directly or indirectly, and all business decisions are technology decisions.

## Book Conventions

Throughout this book you may note several conventions designed to make learning easier. First, at the beginning of each chapter is an overview that lists the objectives. The wording is "In this chapter, readers will learn:", followed by a list of topics. The section could just as easily be stated "Upon completion of this chapter, the reader will be able to define, differentiate, understand, discuss, and analyze the following:" which is the more typical wording of an educational objective, (but very wordy).

At the end of each chapter is an interactive section, which may include discussion questions, class activities, or individual activities. In a classroom environment, these may be assigned as homework. If you are reading this independently, you are encouraged to do the hands-on activities as they are designed to be enlightening. Occasionally these hands-on assignments refer to a data file which can be downloaded from D2L or the textbook website (https://hplconsortium.com/publications/meit). Following the activities, there are a set of self-study questions to help assure a reader that they've mastered the material.

It is the author's intent to make sure that every single acronym is defined in either the text, a footnote, or in a table. The exception will be the technology map; space does not allow for expansion of many of the acronyms. However, all acronyms will be explained within the chapter.

You may notice that sometimes the acronym is given first, and the words they stand for follow in parentheses. Other times the words are first written out, and the acronym is in parentheses. For example, you might see Operating System (OS), but NIC (network interface card). Whichever appears first, the acronym or its defining terms, will reflect whichever is **more commonly used** among IT professionals. A person familiar with technology usually refers to the "operating system" rather than the "OS." The "network interface card" is normally referred to as the "NIC" card (even though the "C" is redundant). Some acronyms are almost always referred to without reference to the words they stand for, e.g., "IP," and almost never "internet protocol" (which may be confusing to some, because "Intellectual Property" is also known as "IP").

You will also notice many stories interspersed. These stories are all based on real live experiences, but the names (and genders) of those involved have been changed. Additionally, sometimes the details have been modified or simplified to better highlight the lesson the story is designed to support. Stories are often set in a grayed textbox to differentiate them from the normal text.

This book will contain both footnotes (at the bottom of the page) and endnotes (in the Reference section at the end of the book). Endnotes are references in Chicago $17^{th}$ (numbered order) format, and will typically be the books and articles which are the sources of the information endnoted. Footnotes are generally personal notes from the author that don't belong directly in the text.

## The Impact of Information Technology on Business

Jim Collins reported the following conversation in his best-selling book:

> *Taking a short break from the rigors of writing this book, I traveled to Minnesota to teach sessions at the Masters Forum. The Masters Forum has held executive seminars for the past fifteen years, and I was curious to know which themes appeared repeatedly over those years. "One of the consistent themes," said Jim Ericson and Patty Griffin Jensen, the program directors, "is technology, change, and the connection between the two."*
>
> *"Why do you suppose that is?" I asked.*
>
> *"People don't know what they don't know," they said. "And they're always afraid that some new technology is going to sneak up on them from behind and knock them on the head. They don't understand technology, and many fear it. All they know for sure is that technology is an important force of change, and that they'd better pay attention to it.*

The impact of information technology over the past two decades cannot be overemphasized. The World Economic Outlook (published by the International Monetary Fund) identified the information technology as a "revolution" that had a major impact on the world economy. The impact of information technology on economic growth went from 17 percent (from 1974 to 1990) to 25 percent (from 1991 to 1995) to 50 percent (from 1996 to 2000). Now, twenty years later, they don't even bother to try and quantify the impact because, due to the proliferation of the Internet and other world-wide technologies, the impact is 100%.

The comparative investment businesses make in technology has skyrocketed in the past two decades. During the 1970s and early 1980s it was typical to see from 3 percent to 5 percent of business expenses spent on technology. Since the late 1980s, it is much more typical to see between 7 percent and 45 percent (depending upon the type of business).

### *How Do We Manage this Technology In Business?*

How can we ensure that we are making the right technology decision at the right time? How can we be sure that we are getting the value we expected? How can we avoid the mistakes of failed projects? What can we do to maximize our limited resources? That's what this book is all about.

The solution to technology problems is a combination of experience and research. Research shows that top-performing companies are very careful about the money they spend on information technology. They work hard at ensuring that they are making the right information technology decision at the right time. The harder task is to clarify exactly how those top-performing companies made the right information technology decision at the exact right time. That's where experience—one thing this book offers—can step in to get the job done.

The consequences of a failed information technology project can be dreadful and unexpected. A bad decision can cost money, careers, companies, and lives. With a little more knowledge, however, bad decisions can be avoided.

## It Ain't Rocket Science

Just because making the right decision at the right time is rare doesn't mean that it is difficult. It may seem like enchanted magic to those who haven't done it yet, but it is easier than it looks—once you know how. This book will tell you what the vendors and technology people won't—the secret to making the right decision at the right time.

You might be asking: If it is so easy, why don't those who are knowledgeable share this information? The most accurate answer is that they might not know that others don't know what they know. Or they don't even know what they know because the knowledge has become intuitive and implicit. But there might be another reason. For most vendors and consultants, sharing the inside secrets would likely hit them in the pocketbook.

Nicholas Carr and Paul Strassmann are right on target when they claim that business leaders spend too much money on information technology and don't get enough value. There is a huge knowledge gap between those who know, and those who don't, and those who don't know are paying the price to those who do know {Carr, 2004, Does IT Matter? Information Technology and the Corrosion of Competitive Advantage;Strassmann, 1997, The squandered computer : evaluating the business alignment of information technologies}.

These esteemed researchers aren't saying that information technology is not important (despite the provocative names of their books). What they are saying is that IT no longer guarantees a competitive advantage all by itself. **The alignment between IT and the organization is the determining factor between winners and losers in business**. Many businesses spend a lot of money, but don't know how to achieve that magical alignment, so their IT is a waste.

## You Can't Learn to Ride a Bike from a Book

Alas, you may read this entire book and still not benefit from the lessons right away. Books, seminars, workshops—they all can introduce the concepts and set the stage, but they are just a beginning. To learn something, really learn it, we need to experience success doing it multiple times. We need constant reinforcement, encouragement, and guidance. David Sandler* understood that when he wrote the book *You Can't Teach a Kid to Ride a Bike at a Seminar*. He likens the process of learning to ride a bike with learning how to do other things (in his case, selling).We can't learn anything overnight, and we can't apply knowledge if we don't get the right kind of circumstances to practice with. Consider this book an ongoing reference as you work to manage technology better in the coming years. The technology terms may well become out of date, but the guidelines discussed here have stood the test of three decades of time. They will not go out of date. The more often you recognize the situations discussed here, and apply the lessons learned, the more you will get out of this book.

* Sandler, D. H. (1995). You can't teach a kid to ride a bike at a seminar. New York, Dutton.

## A Smorgasbord Of Common Sense

It may sound cliché, but I stand on the shoulders of giants. If you're an avid reader of Drucker, Austin and Peters, Deming, Covey, Collins, Hammer, Strassmann, Carr, and Weiss, you will recognize the underlying current of their commonsense words flowing throughout this book*:

- Pay attention to the customer.
- Focus on what you're good at.
- Begin with the end in mind.
- Lead, don't manage.
- Establish good, principled relationships.
- Distinguish between that which can be changed from that over which you have no control.
- The process, not the technology, is important.
- Be passionate in your search for excellence.

The truth is common sense. And common sense sells. So, if you've been reading business books over the past four decades (as I have), you will recognize all the common sense.

Despite having eaten at the table of others' labors, the full responsibility for this particular meal is mine. Like a cook who can take the raw materials and knows just what to add to serve up a delicious meal, my unique experiences have led me to observe trends and connections that seem to be hidden to many others. I hope you find this book to be a smorgasbord of chewable advice and succulent knowledge to which you return for sustenance again and again. If this book helps you, it will be for two reasons: (1) the quality of the ingredients provided by the masters, as well as (2) the support, encouragement, and feedback of those listed in the acknowledgment pages. There are literally hundreds of people who, over the years, have sparked an inspiration, contributed ideas, or helped with logistics. They deserve all the credit.

Read on as we share the secret recipe for successfully planning our information technology strategy.

---

* (in order of introduction)
Drucker, P. F. (2001). The essential Drucker : selections from the management works of Peter F. Drucker. New York, HarperBusiness.
Peters, T. J. and N. Austin (1986). A passion for excellence : the leadership difference. New York, N.Y, Warner Books.
Aguayo, R. (1990). Dr. Deming : the American who taught the Japanese about quality. Secaucus, NJ, Carol Pub. Group.
Covey, S. R. (2005). The 8th habit : from effectiveness to greatness. New York, Free Press.
Collins, J. and M. Hansen (2011). Great by Choice: Uncertainty, Chaos, and Luck--Why Some Thrive Despite Them All. New York, HarperBusiness.
Hammer, M. (1996). Beyond reengineering : how the process-centered organization is changing our work and our lives. New York, HarperBusiness.
Strassmann, P. A. (1997). The squandered computer : evaluating the business alignment of information technologies. New Canaan, Conn, Information Economics Press.
Carr, N. G. (2004). Does IT Matter? Information Technology and the Corrosion of Competitive Advantage. Boston Massachusetts, Harvard Business School Press.
Weiss, A. (2009). Million Dollar Consulting, Mcgraw-hill.

1

# Chapter 1: What Is Information Technology and Why We Need to Map It

Overview: In this chapter, readers will learn:

- ⇨ Why Information Technology Decision-making is so important.
- ⇨ Terms Used in Information Technology
- ⇨ Technology Layers and Information Technology Maps
- ⇨ Hardware, Software, Firmware, and Network Configuration Systems

Before we can manage information technology, we must learn enough about it so that we do not appear to be philistines in the digital age. This knowledge will make it easier to direct people adequately and make the right decisions. Do not fear; this book will not be a dull recitation of buzzwords. But it will give you enough of an understanding of the secrets of the technology map that you will be better at making decisions about technology.

This first chapter focuses on defining information technology, calculating the risk of not understanding it, and explaining the concept of the technology map and why you need it.

## What Is Information Technology

What comes to mind when someone says, "information technology" or "IT"? Business people over the decades would not even recognize the term. Throughout its short history, IT has been known by many names. IT terminology keeps changing so it is difficult to keep up*.

Other terms that people have used over the years for the same thing as what we mean by Information Technology are: information systems (IS), computer systems, personal computers (PC), microcomputers, mid-range computer systems, mainframe computer systems, transaction processing systems, decision support systems, enterprise collaboration systems, process control systems, executive information systems, operations support systems, telecommunications, data processing (DP), electronic data processing (EDP), information processing, data systems, document management, information management, systems management, communication networks, electronic communications, big data, analytics, enterprise resource planning (ERP), the Internet of Things (IoT), e-commerce, Internet technologies, networking technologies, artificial intelligence, world wide web, mobile devices, digital world, social media, cyberspace; and it goes on and on.

The scope of IT has changed too. Over time, IT has evolved from a back-end business process, controlled by experts with dedicated equipment designed to automate simple repetitive tasks, to a world-wide network of all-purpose communication devices used by everyone.

* Even here at Kutztown University, I've waged a 15 year effort to get them to change the name of the Information Systems undergraduate course to Information Technology, but the reticence of bureaucracy knows no bounds.

For this book, information technology is a bit broader than any of these individual components. We define information technology as "the people, processes, software, and hardware that make up the information flow in the operations of an organization."

Notice that people is listed first, and hardware last—purposefully. A common point of confusion is thinking that information technology is just the hardware and software. People and processes are much more critical components. We can't discuss any one without discussing all the others.

Furthermore, the connections between these components are more important than any one of the individual components. A common root cause for failed technology systems is an incompatibility between the people, the processes, and the hardware/software chosen. It is often the case that, if evaluated individually, each component works just fine. People may be dedicated and hardworking. Processes may be well defined and top-notch. Hardware may be the best that money could buy. Software may be well designed and bug-free. Network bandwidth and security can't be beat. But together, they just don't work. We can't talk about technology without talking about the relationships between all the components.

Having said that, this book is not really about information technology itself. This is a guide on how to manage information technology, which means how to manage the people and processes, as well as the software and hardware. This is a book on making decisions about information technology. This book can help ensure good leadership when it comes to information technology.

Technology leadership is as important for the managing executives of a large Fortune 500 company as it is for the new business leader. There is a difference, however. Executives in large corporations have resources; they can rely upon other people in the technology division within their company to actually do the managing (although as you read on, you'll learn why they shouldn't abdicate to them). Entrepreneurs, managers, and small business owners, on the other hand, often don't have the same level of resources, and their level of risk rises with their lack of knowledge about the technology. While everyone can use this book, for entrepreneurs, managers, and small business owners, the information is absolutely vital and can be found nowhere else. But both levels need the information in this book.

The first step in understanding the importance of information technology is to calculate the risk of making the wrong decision regarding it.

## Calculating the Risk

Very few people go to the trouble of calculating the cost of ignorance in any field, much less information technology. Often it seems that not knowing something is not our fault, or a petty issue, or simply not important. Because each problem by itself seems so small, we falsely believe we can ignore it. Many businesspeople think that learning the details of technology is beneath them—like the unwillingness of senior leaders of a bygone era to learn how to type (i.e., "That's what secretaries are for.").

I had one client, Jackie, who insisted that she had a few minor problems, but they were "nothing she couldn't handle" and it "wasn't causing that much of an issue." After doing an assessment (see Appendix B), we narrowed down her problems to two very common ones. Her company was: (1) spending money on software that didn't work as promised; and (2) unsure that the computer systems were going to be safe in the event of a disaster. After a little thinking, she realized that in addition to the $12,000 she had spent in one year on a particular software application that still wasn't working, there was the $4,000 she had spent for a week of a consultant's time and the $3,000 salary she had paid the other two people in her company who were working on the problem. Then she realized that she had to add the missing $80,000 in increased income that she had expected from the working software, and the $10,000 that she had expected to save by retiring the old computer system the software was supposed to replace. Just one small failed project, caused by the fact that she didn't know something she should have known, cost the company over

$105,000. That's a big chunk for a company with only $2 million in revenues and losses that year of $75,000. And Jackie told me she could think of three different technology projects that had similar outcomes. The key is that these problems would have disappeared had she just taken the time to understand the technical issues instead of ignoring them.

Like many business leaders, Jackie viewed the technical issues as "not under her purview." Information technology issues are the director of information system's problem, aren't they? Getting the Enterprise to hit warp speed was Scotty's problem, not Captain Kirk's, wasn't it?*

Well, here's the question: Did Captain Kirk expect Scotty to hit warp 20? Of course not—everyone knows that the best the Enterprise could do was warp 10. Did Kirk expect Scotty to run the ship without dilithium crystals? Of course not. Everyone knows that a starship needs dilithium crystals. Business leaders need to have enough general knowledge and understanding of IT issues so that they don't have expectations beyond what's possible. If business leaders expect that a new system can be implemented without first spending six months cleaning up the data, they are demonstrating a lack of knowledge about IT. If business leaders expect people to be just as productive on a new system after only one week of training, they have unrealistic expectations. These kinds of information technology failures are costing businesses billions of dollars every year. So the answer is no: IT issues cannot be separated from business issues, and therefore must be the manager's problem, the business owner's problem, and not the IT director's problem.

It is difficult for managers to get their heads around the cost of ignorance. Consider the question of how much a business should spend to keep the systems up and running in the event of a disaster. When I ask the question "What is your information technology worth to you?" most managers have a hard time answering. The whole thing is too big for most people to get their minds around. But if a business owner spends too much money on disaster recovery to keep the business "safe," he or she risks losing the business, because the costs can exceed the profits needed to keep the business going. The business's competitors might only spend 10 percent on disaster recovery. If there really were a disaster, the business that invested in disaster recovery technology might continue while the competitor would not. But if there isn't a disaster, the business owner would go out of business and the competitor would continue!

We can calculate properly the value of our IT by taking it just one system at a time. What if there were a fire in the room with the primary computer system? Well, we'd lose the computer. Let's say that the cost to replace the hardware and software was $15,000. The data is even more important. If we have to hire people to re-enter just one month's worth of data (assuming you have paper backup at all, of course), it might cost—say, three people at $15 per hour for two weeks—$3,600. Multiply that by 12 months, and the cost would be $43,200. Suddenly the cost of a special computer-safe fire extinguisher isn't that much (but the cost of a completely outfitted backup site might be too much).

Let's look at another example of risk calculation, this time dealing with the cost of technical support incompetence. I have a friend, Bob, who is a CEO of a relatively small company, and he complained to me about having to send his laptop back to the technical support people three times before they finally replaced the whole thing. He considered it an inconvenience, but not worth the effort to fix.

But then I walked Bob through the real cost of the incompetent technicians who took three tries to get it right. Bob's time is valuable. Imagine that, at his salary, packing up and shipping the laptop three times would cost about $450. Add to that $30 a piece for shipping, and $225 for the techs' time. And, of course, the final cost of replacing the laptop at $4,000. Now we're up to $4,765. If

* Okay, yes – I'm probably aging myself here as I was a fan of the original Star Trek. Perhaps it would make more sense to say that it was Raffi's or Cristóbal's job to run the ship, and Jean Luc Picard simply expected them to *make it so*.

this is an average cost of a problem, and there are 50 problems like that a year—that's $238,250. Not chicken feed.

One of the most important parts of the equation, however, is the cost of lost opportunity. Bob was highly involved in the sales process. He was just about out of commission the entire time his laptop was gone. How much more could he have made for the company? Which opportunities were lost because he was trying to get his laptop working?

The high cost of lost opportunity was brought home to me when I was speaking at an event run by a chapter of the Association of Information Technology Professionals (AITP). One of the members, Chuck, a VP of sales, talked at length about his knowledge management system, an electronic library of information made available to those who need it. The knowledge system aggregated information about the products of his company as well as facts about their competitor's offering. The knowledge system cost about $300,000 to put together. Chuck's bosses were constantly howling about the high cost.

Then one day Chuck was on the phone with the company's top customer—a company worth well over $15 million in annual sales. The customer was unhappy and threatening to take his business to the company's competitor. Chuck was desperate. Thinking quickly, while he was on the phone with the customer he pulled up everything he could on the competitor from the knowledge base—and he came up with a fact about the competitor that, once revealed to the customer, made the customer stay with his company. He relayed this story to his boss, and complaints about the cost of the system were never heard again. Cost: $300,000. Savings: $15 million in five minutes. That's quite a return on investment (ROI).

Generally, to calculate the total potential loss from information technology, we look at the following:

- Replacement value of existing hardware/software
- Value of data lost
- Value of our time
- Value of others' time
- Cost of lost opportunity

Then, to calculate the total gain from information technology, we look at the potential increase in income, as well as the future savings in expenses. Sometimes those savings can come from odd and indirect ways, such as not having to pay for maintenance of future systems.

Multiply your results by the number of typical projects in a year. This is just a rough estimate, but once all the amounts are added together, you might have a better idea of the value of learning more about information technology. This is a lot of risk.

Actual calculating the potential income and savings from technology over the costs is an essential exercise. Checking back a year later to see if the income and savings were realized is an even more essential exercise (albeit rarely done).

## Technology Map and Layers

Information technology is "layered," like a pyramid or a building. Knowing which layer is which is essential to solving IT problems. This chapter will introduce the concept of a *technology map* and provide an overview of each layer. The layers of the technology map, starting from the bottom and moving upward, are *networking, hardware, software, process, and people.* (Figure 1).

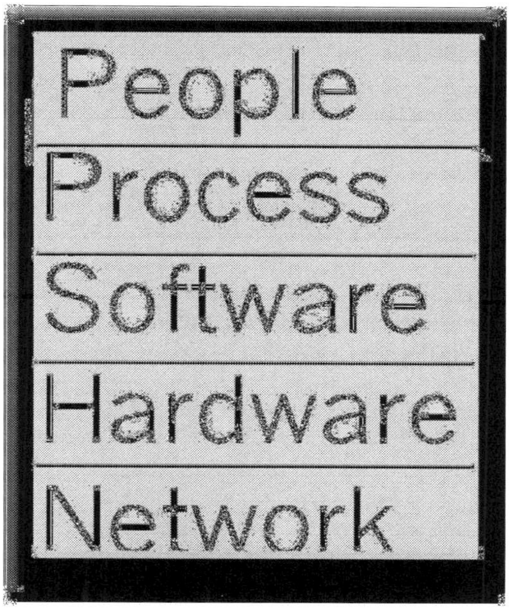


**Figure 1. Layers of Technology Map**

### *Why We Need to Understand the Layers*

Businesses started processing information electronically in the 1960s using large, multi-million dollar computers housed in special rooms and requiring a number of people to maintain the hardware and program the software. Only technical people understood the process; business people were kept out of the "*glass house*." They could see what was going on and found the results useful, but they weren't invited into the electronic data processing center, which was often under the control of the company's finance division.

Everything was owned and controlled at one level; there weren't any layers. When a business invested in a computer, it bought everything from one company for one price: the hardware, operating system software, specialized software, development software, programming, networking devices, and maintenance services were all included. It was an expensive undertaking, manageable by only the largest of companies. Banks were among the first to realize a return on investment in utilizing computers, because the cost difference between processing transactions manually and computerizing the process was huge.

However, when personal computers became popular and networking became feasible, it was a different world. It was possible for products and services to be purchased separately, *unbundled.* Each piece was available from a specialized vendor, who worked independently. Large technology firms, including IBM { XE "International Business Machines (IBM)" }and AT&T (formerly American Telegraph & Telephone) were broken up through government intervention designed to promote innovation and increase fairness in competition. However, coordinating a compendium of interrelated products not specifically designed to function well together sometimes created problems, which could result in unhappy end users. The design of IT systems has since become more and more complex. To get the most value from IT, we must understand the different components and how they fit together. To traverse this difficult terrain, we need a *technology map.*

## Why We Need A Technology Map

Whether you are a technology person who has learned the business side, or a businessperson who has learned the technology side, you probably already have, inside your head, a technology map.

A technology map is a mental map of how technology works—how the different components of the systems in your company fit together. The map gives us a good understanding of how technology interacts and fits together. The map demonstrates the layering effect of IT.

What does that mean? Technology layers fit together like the layers of an onion, and each one can be peeled, one by one. Managers make better technology decisions when they understand how to peel the onion, when they can envision the map and all the related layers. If we have a map of all the different layers and the way they are connected to each other, we can see where new concepts, buzzwords, and descriptions should be placed on the map. We can see where some items inherit certain characteristics because of their location or their relationship to the other items on the map. We can more easily see what portions (if any) can be replaced. We can also more easily tell when something is missing.

Think of it this way: A translator knows that there must be a verb somewhere in the sentence even if she doesn't know which word is the verb. If a translator is reading a sentence and can't find the verb, she knows that something is missing, and that the sentence is incomplete. Similarly, a technical person knows that there must be an operating system in the technology even if she doesn't know which operating system is there. If it's not obviously apparent, the technical person would ask the vendor about the operating system and would know by the answer which hardware the software runs on.

There are several reasons why we need a map. By envisioning the layers in a mental map, we can translate between the specific examples in our own systems and the general components that exist in the industry. We know which vendors with whom we should be talking about problems or replacements. We will know what we can, and cannot, buy.

One of my clients was a large regional bank. At this bank, the department heads made many of the technology decisions. The loan department was headed up by Sam, an experienced executive in banking, but not so experienced in IT. Sam saw a demonstration of a loan-decision support application at a conference—and he was dazzled. Sam knew that his existing system (which was PC based) could not do all the marvelous things that this software could do. He figured that he could make more loans of higher quality and would need fewer loan officers with this new software. His ROI calculation showed that it would pay for itself within a few months. So he adjusted his department budget and wrote out a check for $100,000 to buy the software. When the installation tapes arrived, he sent them down to the system administrators in the information technology department to install.

There was only one problem: The software ran only on a mainframe computer.

This particular department of the bank ran only small microcomputers. Sam had never even thought to ask on which kind of computer the software ran. He didn't realize that even if the systems people made arrangements to use the software, it would have taken four times the cost of the software to install a mainframe and hire the expertise necessary to run it (so much for the ROI).

You wouldn't buy new tires for your car without first checking to see if they fit the wheels, would you? And yet managers and business leaders purchase technology all the time without first checking to see if it fits into the existing environment.

Because a technology map gives us a starting point for questions, we can use it to help us detect the difference between hype and reality. Furthermore, the technology map will help managers understand how to tell when the best time is to make purchases.

Probably the most important reason we need the technology map is so that we can talk sensibly to those who are implementing the technology. Having a well-formed map will gain the respect of the information technology folk.

Sam wasted $100,000 simply because he didn't have a technology map. But that wasn't his biggest problem. Even more important, Sam lost the respect of the people under him and he became a laughingstock. The technical folks still circulate the story, years later, with glee. They can't wait to tell new people joining the company about Sam's "great" software purchase. I was there less than a week before I heard the story (to many guffaws), but I wasn't surprised when I checked it out and found that it was true. To avoid the same fate, we can work to improve our own technology maps.

There are hundreds of stories like this one—you can probably think of a few yourself. But how do we avoid these mistakes?

## *Getting over the Hump*

Due to the complexity and newness of much of technology, it is easy to understand why business leaders commonly respond, "I don't need to know that," when someone tries to explain it. In such cases, they let other people make the critical decisions for them. The problem is that the people making the decision often don't have the same vested interest in the business, or they have different criteria for success than the business leader. They won't necessarily make the right decision.

Is it any wonder the technology rarely accomplishes the desired return? In an even worse scenario (because the technology leaders don't understand the underlying technology but can't admit it), the decision makers can be "wowed" by polished salespeople striving to make a sale. Since the salespeople often don't understand the technology themselves, they make promises they just can't keep, and the buyer is always the loser.

Even if business leaders do try to attempt to understand their own information technology, they may become frustrated by the confusing technical jargon and contemptuous attitudes of some technical people.

In my role as a technology educator, I have heard thousands of complaints about how difficult technology concepts are to learn, and how impatient technology people are in explaining them. The funny part is, I have also listened to countless CIOs and technical people who complain about the overwhelming challenge and complexity of business and corporate culture. I agree—corporate culture and business acumen are extremely hard to learn. I remember in 1986, as a school teacher starting my first business, I encountered the term "marketing" and had no idea what it meant. I tried to get my brother-in-law (who had a business degree) to explain it to me, and it seemed like it took me forever to understand the concept. My brother-in-law became impatient with my inability to understand what, to him, was a simple business concept. But I was a teacher. I had not encountered the term during my education.

Good business leaders learned all those business concepts over many, many years and within the context of running businesses. Business leaders immediately know the following terms: finance, sales, marketing, operations, process flow, capitalization, expenses, receivables, payables, etc. These terms are easy for businesspeople to understand but, believe it or not, hard to explain outside the context of a business. Similarly, technology people immediately understand terms like operating systems, network interface cards, Internet protocols, printer drivers, application servers, and Web monitors without even thinking about it. But they are difficult to explain outside of the context of information technology.

IT terms are not difficult; they just haven't had time to permeate our culture yet. The next time you feel frustrated because an IT person seems to be talking Greek, consider this analogy. Imagine that it is your job to help some people who were magically transported here from the 1840s. Now give them a road map* and explain to them how to drive to New York.

Today, we can count on people having a certain amount of general knowledge about cars, roads, maps, and so on. For example, when we look at a map, we know immediately that the big thick lines are interstate highways, the smaller red lines are two-lane roads, the big blue splotches are

* Okay – dating myself again. But even if you gave them a GPS, the analogy works.

bodies of water. We already know that we are going to get into a car and drive on those roads, and we intuitively understand the relationship between the car we drive and the representation of the roads on the paper (or on the GPS screen). We also know different brands of cars (Porsche, Cadillac, Mercedes, Toyota, etc.), and we can picture them in our minds. We even know a bit about the social status of the people driving them based upon the brands. We know that if we go on a long trip, we will stop for gas on the way. We already know how to put gas in our car. We know that eventually we will need to change the oil and put in new brake pads. We grew up with the knowledge. We don't have to explain any of that to someone today, so giving them a map and telling them how to drive to New York would be simple.

But the people who grew up in the 1840s don't have the benefit of that background knowledge. They still live in a world filled with horses and buggies. They've never been in a car. They don't know what a paved road is. If we were tasked with explaining not only the map, but all the background information that goes into the context of what the road map is and how it works, we might show a bit of impatience and frustration, too.

But the difficulty just increases the importance. Don't you, as a manager and business leader, really appreciate when the technology folks learn a bit about the profit-and-loss nature of business, or work harder at getting along with others? Similarly, technology folks generally have much more respect for business leaders who learn a bit about the categories of information technology. To speak with them, we need a technology map.

## Technology Map Documents

A technology map is a series of documents that include a written description, lists (like an inventory list) categorized by layer along with illustrations (usually network designs and process flows), which is used in a business to help the user create a mental map of how the technology works. It also includes the financial status of vendors and the current/obsolete/next versions of the technology in use. It is designed to give a good understanding of the different components of the systems and how they fit together for that organization. One you look at the entire map, you can see how the map demonstrates the layering effect of IT. The basic categories that might be included in a technology map are shown in Table 1.

**Table 1. Technology Map Basic Categories**

| MAJOR | MINOR | CATEGORIES |
|---|---|---|
| People | Ability | High Ability, Low Experience |
| | | Low Ability , High Experience |
| | Experience | High Ability, High Experience |
| | | Low Ability , Low Experience |
| Process | Who (Role) | Flow charts and process documentation – often obtained by watching people and what they do as well as by asking them. |
| | What (Task) | |
| | When (Prompt or Time) | |

| MAJOR | MINOR | CATEGORIES |
|---|---|---|
| Software | Collaboration | Email |
| | | Cloud |
| | | Groupware |
| | | Social Media |
| | | Search |
| | Financial | ERP, EDI, XBRL |
| | Media Management and Design | Graphics and Publishing |
| | | Music Managers and Editors |
| | | Video Editors |
| | Specialized | CAD (Computer Aided Design)/CAM (Computer Aided Design/Manufacturing)/CAE (Computer Aided Engineering) |
| | | Logistics Software |
| | | Vertical Market |
| | Office Application | Word Processing |
| | | Spreadsheet |
| | | Database |
| | | Presentation |
| | Development | Languages |
| | | Web Development Tools |
| | Utilities | Backup; Antivirus; Enterprise Network Management |
| | Operating System | Single User |
| | | Multi User |
| Hardware | Types | Smartphone, Phablet, Tablet, Personal Computer, Workstation, Server, Mainframe, Supercomputer |
| | Main Components | CPU, HD, SSD, DASD, RAM, ROM, BIOS, Motherboard, Ports, Heat Sink, Fan, Power Supply |
| | Peripherals | Mouse, Keyboard, Monitor, Scanner, Camera, Printer, Flash Drive, DVD, Smartphone, Touchpad, Game Controller, Video Camera, Speaker, Headphones, Earphones, etc. |

| MAJOR | MINOR | CATEGORIES |
|---|---|---|
| Network | Network Software | Browser (or Web Browser) |
| | | Application Servers; Web Servers; Database Servers |
| | Network Hardware Protocol | Connection Type: Packet Switched (Ethernet) Point to point leased line (T1,T3), Frame Relay |
| | Network | Satellite, PBX, NID, NTU, ADM, ATM Switch, SONET Ring), etc.<br>Hubs, (Ethernet Switch, FiOS Switch, DSL Switch, Cable Modem, WIFI transceiver)<br>Device connection hardware: Dial-up Modem, or NIC, or transceiver.<br>Cabling; Coaxial, Unshielded Twisted Pair, Fiber Optic, etc. (Also on this level is transmission via radio waves to Satellite or Wifi hub) |

While a summary of these layers are briefly explained here, the ensuing chapters of this book will delve much more deeply into each of the layers.

### *Network*

*Networking* is a term used to describe the sharing of data between multiple devices. In the technology map, the network is the bottom layer, and is comprised of both hardware and software. When multiple networks connect, it is called an *internetwork*.

Networking itself has multiple layers, often referred to as the *Open System Interconnection* (OSI) model (Figure 2).

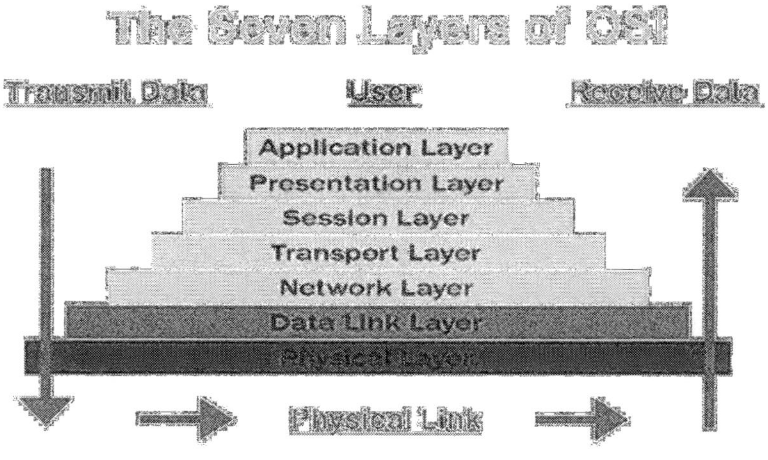


**Figure 2. OSI Model of Networking**

The networking devices are all on the Physical layer. The software that controls those networking devices are on the other layers. Ethernet and WIFI are on the Datalink layer. Internet connections are on the Network and Transport layers*. The higher level layers, Application, Presentation, and Session, are all the software that actually controls data moving across the network.

### Hardware

Hardware is what you can *see* when you look at a computer. Hardware can be categorized as a main component or a peripheral (a device connected to the main component such as a mouse or a monitor). Here are some of the different types of main component hardware systems:

- Mobile Devices such as smartphones, phablets, tablets, netbooks, etc.
- Personal Computers (formerly called microcomputers) such as laptops, game consoles, desktops, etc.
- Workstations a more powerful device than a personal computer, generally used by software developers or graphics designers.
- Servers (formerly called minicomputers or midrange computers) a multi-user computer that provides services such as files, applications, web, database, etc. Generally they run on a flavor of the Unix operating system such as Linux. They can be used in parallel to form extremely powerful systems.
- Mainframes and Supercomputers, also multi-user system, generally much larger and more powerful than individual servers. Most are dedicated to a single focus or task such as processing check transactions for a bank or predicting the weather or computer generated graphics for a movie.

### Software

Software is a set of *instructions* for computers, written in a language and presented in a format the computer can understand. Software is always a layer above the hardware layer. Software is ephemeral. You can look at a paper or a screen filled with the written letters and characters of software code, but that isn't really the software. Software instructions are stored in digital format, such as is found on a DVD (digital video disc), a CD (compact disc), a flash drive, or a hard drive. While a physical device is being used, software resides in its RAM (random access memory). Software can be stored locally in the physical machine a person is using, or remotely on a server located somewhere else. Software, whether residing locally or remotely on a server, can't really be seen, felt, or touched: it consists of binary code stored electronically. Software can be categorized as System software, Development software, Utility software, or Applications software. Most software falls into the Applications software category.

### Process and People

In the technology map, above the networking, hardware, and software layers are the process and people layers. *Process* refers to what people actually do with computers. When you read an email, you are following a process. When you download a video, you are following a process. When you order a book online, you are following a process.

The most important aspect of the technology map is the *people*. It is people who use IT (and not the other way around), and so understanding the productivity of people as they use technology is critical to understanding IT.

* Later we will discuss TCP/IP, which stands for Transport Control Protocol/Internet Protocol. The Transport Control Protocol is the part of the internet software which is on the Transport layer, and the Internet Protocol is the part of the internet software which is on the Network layer. In the early days of networking, before the Internet, there were many competing networking protocols such as Netware and NetBUI. Back then, the distinction was important. Now – not so much.

## Points of Confusion: Network, Hardware, and Software

As explained, the term hardware is used to describe the physical components of the technology map, e.g., the desktop computer or the smartphone device. Software is the name given to the set of instructions that tell the hardware what to do. Networks, however, are always comprised of both hardware and software, so splitting the network out as a separate category can lead to some confusion. The Hardware and Software layers of the map only refer to hardware and software that is NOT associated with the Network itself. Hopefully, the details in the chapter will enable the reader to make the distinction so that they can understand their technology maps.

Furthermore, although there is no category for it in the technology map, mention must be made of a device that comes in between hardware and software called *firmware*. Sometimes instructions get saved, or "burned," into the circuitry of a component, such as a chip. Then it's not simply hardware, but it's not only software either, so we call it firmware. Your computer has a form of software called a BIOS (Basic Input Output System), but instead of being stored on the hard drive like all the other system software, the BIOS software is stored inside your computer on a ROM* (Read Only Memory) chip called a CMOS (complementary metal-oxide semiconductor), which is a special kind of memory chip that doesn't require much power. Attached to the ROM chip is a tiny battery, so that it retains its instructions even when the power is off, and doesn't forget how to boot up (turn itself back on when needed). Since the battery lasts for years, most computers will become obsolete long before the battery runs out.

Most networking devices have some sort of firmware inside them. Furthermore, everyday items that aren't typically thought of as computers may have an embedded component that provides some kind of processing capability. Today, almost any item you might purchase may contain firmware: automobiles, medical equipment, washing machines, dishwashers, ovens, vacuum cleaners, and even toasters often contain an embedded chip to control or measure some aspect of the function of the device. Every cell phone, every camera, and every device that connects to a computer contains a firmware chip.

## Categories, Brands, and Providers of a Technology Map

To make sense of the map, it must be put into the context of the categories, the brands, and the providers of the information technology within our businesses. We need to understand what we have, as well as the characteristics of each brand and provider within each category for each layer of the map.

To give a nontechnical example, if we purchase a Porsche, we know that we are going to get an expensive, fast car that others envy. That tells us something about where we should store it. We wouldn't put it unattended in a crime-infested center city lot. But if we buy a Hyundai, we know that we will get an inexpensive, affordable ride. We wouldn't build a beautiful garage for it, and we wouldn't expect to open it up by going 120 mph on the highway, either.

Now for a technical example. If the layer we are replacing is a database and we are considering DB2, Oracle, mySQL, SQL, or Access, we would need to know that DB2 and Oracle are robust and expensive, SQL is cheaper and easier, mySQL is an open source (i.e. free) software, and Access is a small single-user database that should not be relied upon for any critical business functions. And if our programmer tells us to use Sybase, we need to know that despite being entrenched as the database of choice for Wall Street systems for over 30 years, Sybase is now obsolete and no longer sold. We also need to know that DB2 needs a mainframe computer from

* ROM (Read Only Memory) is different from RAM (Random Access Memory). RAM memory is volatile, and information stored in RAM disappears as soon as the power is disconnected, unless there is battery backup. ROM instructions, on the other hand, do not disappear when the power is turned off. RAM, on the other hand, is volatile and information stored in RAM does disappear as soon as the power is turned off.

IBM, Oracle does not work well in any Windows environment, and SQL does not work well in a UNIX environment. Putting SQL on a mainframe would be like putting that Hyundai in a huge custom-built marble garage; it wouldn't make any sense. Putting Oracle on a Windows-based machine is like trying to drive the Porsche on a dirt road. Furthermore, if we only need to support 20 people accessing the database, we would know that Oracle is overkill (like buying a 747). Nor would we try to squeeze 200 people into a tiny plane, which is what putting 200 people on SQL would be like. This is the type of information we gain by understanding the technology map and what brands are in which layers and what other layers they depend upon.*

### Understanding New Technology

Having a technology map enables us to quickly understand new technologies even if we've never heard of them before. Waaay back in 1998, I was asked to take over a mainframe conversion project in my role as a vice president at a large credit card company. I had a deep understanding of some of the branches of the map (applications, databases, and networking), but I didn't know anything about mainframes or development tools. Because of the map, I was able to quickly relate new terms and concepts to what I already knew. The words were different, but all the layers were there in one way or another. Keep in mind that *I* had to make these translations. Often programmers and systems analysts within a single "branch" only understand their own maps where the layers differ. Since they've only worked on one system that is not separated into components the same way, they may not understand other systems the same way.

For example, in a PC, two of the layers in the main hardware are the central processing unit (CPU, the brains) and the hard drive (HD, the storage). On a mainframe, the equivalents are the processor (the brains) and the direct access storage disk (DASD, pronounced "dazzdy"). The first time someone used the term DASD and I asked what that was, the person gave me a long drawn-out explanation that I didn't quite understand. But I sensed a similarity, so we kept circling the concept until I figured out that DASD is like a hard drive.

I used the same trick when I first heard the term IPL. I quickly figured out that to IPL a mainframe is the same as booting up a PC. Same concept, different terminology. Piece by piece I found all the mainframe terms for the already familiar PC terms I knew.

The best place to use the technology map is when working with vendors. Imagine that a vendor is telling us about a new product called Malugas. We have never heard of it, so we ask what it is and the vendor tells us that Malugas is a web server. If we understand the technology map, knowing that Malugas is a web server tells us what it does (monitors the network card for requests to see web pages across the Internet). We would also immediately be able to name four or five competitors (like Apache, IIS, Nginx, Lightspeed or GWS). Furthermore, without even asking, we will understand what kind of hardware it sits on (multiuser server, either Windows Server 2016 or 2019, or Linux or some other flavor of Unix, but not Windows 10 or MacOS or any other desktop system). We would also know what kind of software interacts with it (browser programs such as Edge, Firefox, Chrome, or Safari).

But what if we were told that Malugas is an enterprise database? We would immediately know what it does (stores related tables of information), be able to name four or five competitors (like Oracle, SQL, DB2, or MySQL), know what kind of hardware it sits on (probably a multiuser server, definitely not a desktop), and what kind of software it interacts with (application programs coded in many different languages like PHP, Java, Python or C++). This immediate recall of all the different characteristics of each layer or branch enables technology people to understand the strengths and weaknesses of each new product that is introduced by placing it on the map. The

* If you are overwhelmed by all these terms, don't worry. They are explained in subsequent chapters.

new product gains all the characteristics of the current brand and provider already in the category on the map.

## Other Influences in Learning New Technology

Learning how to use information technology is different than learning a subject like history or English because it involves a combination of acquiring knowledge and practicing a series of actions (e.g., typing and moving a mouse or touching a screen). Being able to learn how to use a computer is closer to learning how to play a musical instrument than memorizing facts. To play the guitar, you read chord notation *and* physically manipulate your fingers to simultaneously press the strings and strum the chord. How long does it take a person to learn to play a musical instrument? A day? A month? A year? Longer? Chances are you said "longer." Though anyone can be taught to play a single series of notes very quickly, it takes time to establish the habits and gain the knowledge necessary to play an instrument well. Similarly, although almost anyone can be taught how to use a single program relatively quickly, it takes time to learn how to use a technology well.

But experience is not the only factor leading to success. There are people who can pick up a guitar and just play it. They have never taken a lesson in their life, and yet they can play songs that they've heard, "by ear". They are naturally talented at playing that instrument. Similarly, some people seem to be naturally talented at information technology. They don't need to be taught every single type of program; they use their general knowledge of different applications and "figure out" the new ones without being taught. If you've ever played a video game, you can easily understand that the more video games you play, the more quickly you can learn how to play a new video game. The knowledge you gain is generalizable and can apply other uses of the computer.

Therefore, the two characteristics that most influence how quickly a person can learn new technology are: 1. natural ability, and 2. the amount of prior experience with using other technology devices.

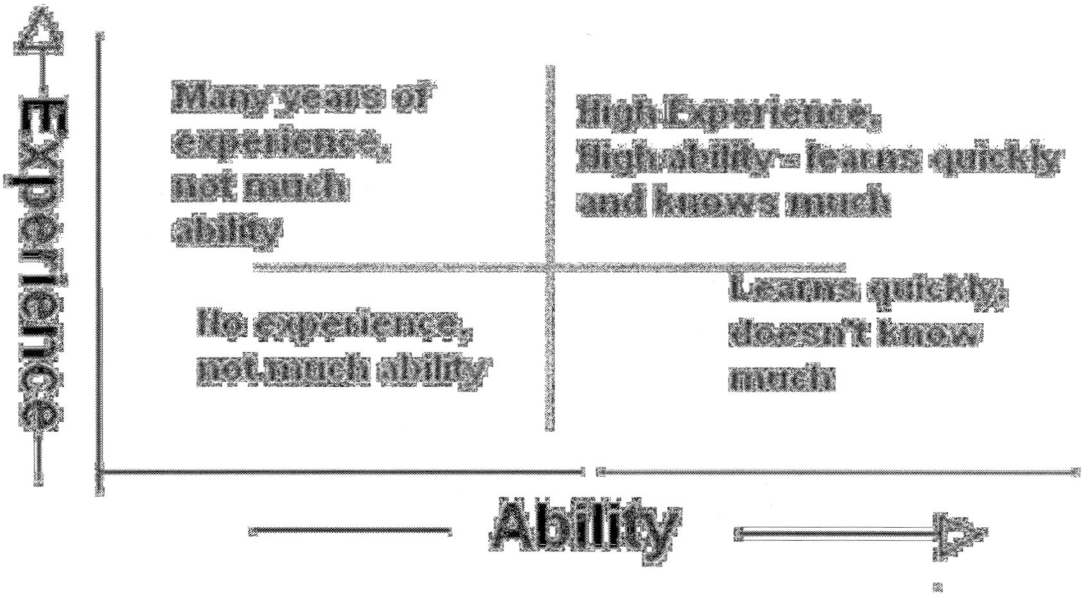


**Figure 3. Experience and Ability.**

It is helpful for anyone who is attempting to learn new applications on the computer to assess their own experience and ability. The amount of time that needs to be invested to learn a new program decreases if the experience and ability of the person learning is high. People who don't have a lot of experience, however, or who don't seem to have any talent with information technology, will need much more time to learn any new application or system.

When creating a technology map, it is important to know the experience and ability level of the employees and customers and vendors of a computer. One way to assess how much talent people have with information technology is how much they already use IT for personal purposes. Thirty years ago, most computers were found in businesses. Today, especially with the advent of tablet computers (e.g., Apple iPad, Amazon Kindle Fire, Barnes and Noble Nook, Samsung Galaxy Tab, Google Nexus 9, Verizon Elipsis, LG G-pad, and Microsoft Surface) and smart phones (e.g., LG G5, Galaxy S7, Apple's iPhone 6S, Google Nexus 6p), consumer use of information technology is more widespread than business use. People use computer devices to communicate with friends and family, take pictures, draw and design art, compose and listen to music, record and share videos, play games, invest in stocks, organize their collections, watch movies, and much more. While this textbook will emphasize business rather than consumer uses of the computer, when individuals utilize new information technologies for personal use, they gain experience that is helpful in business applications.

Learning to work with technology is akin to learning a whole new language; it will take a while for you to absorb it, and if you are completely new to it, initially nothing will make sense. Eventually, however, you will be able to "think" in the "technology language", and it will be easier for you to understand what is going on. Whenever you are having trouble grasping an IT concept, turn around your perspective and think like a computer. What is the computer expecting from you? Where are you in the navigation of the application? On which layer are you having difficulty? Learn the terminology, recognize the menus, and try to remember the shortcuts. When in doubt, google it. The amount of help available online for learning hands-on technology is amazing. There are thousands of YouTube instructional videos on using Excel and other Office software, and blogs and social media sites where users answer one another's questions.

## Difficulty of Sharing Maps

Over the years as I've taught people the importance of this concept, I have tried numerous times to illustrate a single all-encompassing technology map. Every time I tried to diagram the technology map, however, I realized that the technology map:

- Represented only the last environment I was in and was often inappropriate as an example for the person to whom I was showing it.
- Was too shallow in some areas and too deep in others for the purposes of the person to whom I was showing it.
- Was often already out of date and incorrect, and it often represented categories that no longer existed or was missing new categories that were now necessary.

The upshot is that each technology map is unique, and one person's map may simply confuse other people (especially if it is the only example they see). I might show you an example of one company's systems which are highly virtual and web-based, but if your company relies upon mainframes, it would just be confusing.

In Appendix A, you will find several example maps. Hopefully the variety of maps will help you understand the potential scope and size of your own information technology map. But remember—each company's map is unique. The important activity for you is to develop your own map for your own business.

We don't need a map of Utah if we are driving around in New York. But if we are driving in New York, we'd better have a map of New York. We don't need to understand every single type of information technology to make good decisions, we just need to understand our own technology map.

A technology map starts with the inventory of the networking hardware and software. Then it moves up to the non-network hardware both on and off the network.* For each layer we should know the name of the hardware/software, the category of it, the names of the vendor and the version or brand. Additionally, for each vendor we should know where they fit in the industry among their competitors. We need to know if we are using the Lexus of the industry, or if we are riding in a Chevy.

You may find that it is harder to get a simple inventory than you expect it to be. Leslie was one of my clients, the vice president of information systems (IS) at a $100 million manufacturer. Leslie was new to her role in IS. When she asked her staff to provide an inventory of hardware and software, she was given a short list that included one minicomputer, 27 terminals, three routers, and five software programs. She immediately realized that this could not be the complete list (after all, there was a PC on her desk and it was not on the list). In this company, since the departments themselves purchased their own PCs and servers, the IS department did not keep an inventory. Her staff offered to walk through the data center and count what was there—but there was no way to tell what else was in use at other locations. This is not a minor matter. In this case, not having an up-to-date inventory cost this company big bucks.

Leslie was told by the CEO that she had only two weeks to make a decision regarding the new purchasing system, which had to be installed by the end of the year. Two weeks was not enough time to get a complete inventory for 200 employees spread out over three locations. Leslie was forced to make a decision without knowing what hardware and software the company already owned.

Leslie chose a Unix-based purchasing system. She thought she had made a great choice, and that everyone would love the neat new features of the sophisticated system.

The problems started right away.

First, many of the PC systems didn't have a new enough operating system to connect to the Unix-based system. Additionally, the new Unix-based system was very text-based and used function keys to access and manipulate the menus. The employees were used to GUI systems†. Employees expected that a new system would include movable windows, menus across the top, and submenus available on objects—all of which could be clicked on with a mouse. The written requirements that had been given to Leslie did not describe any of this; it was just assumed that a new system would be graphical. (A common problem in requirements documents is that they don't list features and functions already included in existing systems. Requirements documents often focus on new features the users want but don't have.)

As you can imagine, the employees resented using this purchasing system, and many simply refused to learn it. After several years, the seriousness of the bad decision became clear. The new software never achieved the hoped-for gains because the people doing the purchasing either wouldn't use it because it was too difficult to use, or couldn't use it because their own systems didn't have a new enough operating system to connect. Even after the CEO issued an edict that all orders must go through the new system, most people continued to order items using email and independent websites.

---

* Again, to point out how much technology has changed, at the time I started my career and I wrote the first version of this book, most technology was independent of networks. Computers were just starting to get connected through local area networks, and few were connected to others across buildings or towns. Now, you'd be hard pressed to find a single device that is NOT connected to the network. It's a good thing (perhaps even prescient) I named the *network* as its own layer back then when it was just a tiny portion and often not even in the map.

† A GUI is a Graphical User Interface like you would find on a Windows machine, a Macintosh, or a mobile system. While popular technology is all GUI, there are still many business systems out there that are text-based. Putting a "mobile" or "GUI" front end on those old text-based systems is one of the top income producers in the technology industry.

The CFO had calculated the ROI at the beginning of the project. The CEO was expecting an annual savings of $4 million through volume discounts. But because few buyers used it, there was limited opportunity for grouping the orders of different departments for volume cost savings. The result of running the new system was that the company was spending an extra $2 million each year and saving nothing. Instead of the calculated 80 percent of the orders going through purchasing, or even 35 percent, which was the percentage that went through the previous purchasing system, only 10 percent of the orders went through the new purchasing system. If Leslie had held off on deciding about the purchasing system until she knew what already existed, there would not have been a problem.

In addition to having an up-to-date inventory, we also must know the capabilities of our existing technology. Again, this is difficult if you don't understand the technology at all.

I have a good friend, Jon. I admire Jon—a highly intelligent person and at the top of the field. However, Jon is not very knowledgeable about computers. Jon called me up one day to ask me where to buy software or hire someone to send out an e-mail to about 30 people every month. I tried to explain to Jon that you can use a contact management system to merge an e-mail to a distribution list, which was called an address book in the e-mail system. Jon could send out up to 50 emails at a time that way (though to send more, an email marketing system like MailChimp or Constant Contact would be needed). But Jon was convinced that specialized software and a technical expert was needed to accomplish this task. Jon ended up spending $1,500 on specialized software before discovering that, indeed, all that was needed was to learn how to merge to an address book using any client email system.

For a small or new company, keeping a hardware or software inventory is as simple as having someone create a spreadsheet with basic information on each system including the vendor, brand, version, etc. Once we get beyond a few dozen employees, we would want to keep the information in a database rather than a spreadsheet, but the columns would be the same. With thousands of employees, however, you would want to employ an asset manager which will automate the process of listing all of the hardware and software connected to the network.

A common problem is phantom inventory. If you ask most IT managers if there is an inventory, they will say yes. The problem occurs when you ask to see it. That's when we find out that the list either doesn't exist or is extremely out of date. It is not enough to simply ask if there is an inventory. The manager or business owner should ask to see it and check it for accuracy. Since keeping track of this information is tedious and boring, no one will do it unless a leader asks to see it.

One of my clients assured me that all the company's hardware and software inventory was completely up to date. When we finally got the list, we learned that it had been updated - four years earlier. There had been many upgrades and new employees since then.

One thing that makes doing a hardware and software inventory easier is standards (which we will talk about in detail in later chapters). For now, suffice to say that if you have 500 employees and each one has a slightly different configuration of hardware and software, the inventory is now a major months-long project. If the company had adopted standards and everyone has one of 3 or 4 different configurations, the job of inventory becomes much easier.

## Start at the Inventory and Build

Often, managers and business owners recognize the need to understand the full technology map, but they get overwhelmed in the beginning because there are so many items in the inventory that they don't recognize. Don't let the sheer volume dissuade you. Start by just looking at it for a few days or a few weeks. Use the maps in Appendix A and compare them. Try to envision the categories that each piece of hardware and software fits into. Ask technology people questions about how different items in the inventory relate to each other. By doing so, you've already started creating your technology map.

Keep in mind that you don't need to get it right the first time. Accept also that there is no such thing as getting it exactly right because the technology itself changes, and the categories may change each time you buy something new. There is no single right or wrong answer. The important thing is to get something down on paper, and to have a mental understanding of the layers. After a while, you won't even be aware of how the map becomes the foundation that enables you to add new information, constantly updating and staying knowledgeable about information technology.

# 1-1. Interactive

1. Here's a short exercise. Calculate the value of increasing your knowledge about information technology. Choose a typical initiative or project in your company, then put a value on each of the items listed in Table 1. Then compare the costs of the project (including ongoing maintenance costs) to the cost of not doing the project. What do you find?

**Table 2. Calculating the Risk/Cost Benefit of IT Knowledge**

| Category | $ Cost |
|---|---|
| Replacement value of existing hardware/software | |
| Value of data lost | |
| Value of our time | |
| Value of others' time | |
| Cost of lost opportunity | |
| Total potential loss from Technology | |
| Potential increase in income | |
| Future Savings in expenses | |
| Total potential gain from Technology | |

2. **Discussion**: How experienced are you with computers, compared to your classmates? What kinds of programs have you already learned to use? How many hours would you estimate that you have spent using a computer since the day you were born?
3. **Discussion**: How much natural ability towards computers were you born with? How quickly, compared to your classmates, do you pick up a new program? Do you prefer for someone to show you how to use a new program step-by-step, or would do you prefer to be pointed in the right direction and allowed to "play" by looking at all the menus and trying different things?
4. **Discussion**: Do consumer technologies such as tablets and cell phones make it easier or more difficult to learn the more powerful capabilities of computers? What are the commonalities? What are the differences?
5. **Classroom Activity:** Imagine yourself standing in line with all of your classmates, with the student least experienced with computers at the back of the line and the most experienced at the front. Where do you stand?
6. **Individual Activity:** What kinds of technology do you currently use? Do you have a smartphone, a tablet, and/or a personal computer? How often do you use them? What do you use them for?

## 1-2. Self Study Test Questions

7. What is information technology (IT)?
8. What other names has IT been called over the past 5 decades?
9. In Business, what are the risks of ignorance when it comes to IT? How might you quantify the risks?
10. What is a technology map? Why do we need one?
11. What are the layers of IT?
12. Why is the order of the layers of components of IT important?
13. Why do we need to understand the layers?
14. Is a technology map the same for all businesses? Is it static or dynamic?
15. What is the difference between hardware and software?
16. Why is hardware and software and networking confusing?
17. What is firmware?
18. What are the two characteristics that influence how long it takes a person to learn a new computer program?

2

# Chapter 2: Networking Hardware

Overview: In this chapter, readers will learn about the following topics:

- ➯ The Bottom Layer: Networking Hardware
- ➯ Local Area Networks (LAN) (including HAN & CAN)
- ➯ Wide Area Networks (WAN) (including MAN)
- ➯ Physical Cabling
- ➯ Physical Wireless Protocols

## The Term: Networking

Most people have a general understanding about personal computers because they use them regularly. Terms like mouse and menu and icon are common vocabulary.

The same cannot be said for networking. For many people, how information gets from the personal computers in our homes and businesses to other people's homes and businesses is a complete mystery—a black box. It seems to work like magic—until it breaks. Then suddenly we are left without the basic information we need to figure out if we should call the phone company, the device manufacturer, or the PC help desk.

A *network* is an interconnection of three or more communicating entities. Never has a term been so utterly confusing to everyone who uses it. In business, when people boast about their networks, they are usually referring to the number and level of people they know. Entrepreneurs and business leaders use their network or hone their network or build their network, usually one node at a time.

In the world of IT, however, the term network has a very different meaning. It refers to how computers are connected. There are personal area networks and local area networks and campus area networks and metropolitan area networks and wide area networks and global area networks. There is the public Internet as well as private Internets and intranets and extranets. Networks can be wireless through the air, or directly connected through copper wires or fiber optics (and usually a combination of all three).

## Network: The Bottom of the Layers

At the very bottom of the layers is the physical equipment and the types of connections that are part of the hardware of networking and internetworking (connecting networks to each other). Table 3 shows the hardware (Modems, Hubs, Routers , etc.) and connection types at this lowest layer.

**Table 3. Technology Map Network Hardware Categories**

| MAJOR | MINOR | CATEGORIES |
|---|---|---|
| | Network Protocol | Connection Type: Packet Switched (Ethernet) Point to point leased line (T1,T3), Frame Relay |

| MAJOR | MINOR | CATEGORIES |
|---|---|---|
| Network | Network Hardware | Satellite, PBX, NID, NTU, ADM, ATM Switch, SONET Ring etc.<br>Hubs, (Ethernet Switch, FiOS Switch, DSL Switch, Cable Modem, Router)<br>Device connection hardware: Dial-up Modem, or NIC, or transceiver.<br>Cabling; Coaxial, Unshielded Twisted Pair, Fiber Optic, etc. (Also on this level is transmission via radio waves to Satellite or Wi-fi hub) |

## Local Area Networks and Wide Area Networks

Years ago, physical networks were separated into two types: local area networks (LANs) and wide area networks (WANs). Wide area networks originally ran at speeds of under 2 megabits per second (not very fast), whereas local area networks were faster than 10 megabits per second, and often up to 100 megabits per second. Today there are a few more options. The various levels of networking today are often categorized this way:

- *Personal Area Network (PAN)* which resides around a particular person. It includes things like headphones or earphones connected to a cellphone, or a speaker connected to a laptop computer, or a tablet connected to a monitor, television set or projection system. PANs usually involve only Bluetooth, which is a very short-distance way of connecting devices within close proximity to each other.
- *Local Area Network*, which usually resides within a building or nearby buildings in a campus environment. Sometimes when multiple buildings are included, the LAN is referred to as a CAN (*Campus Area Network*). If the network is in a home, it is often referred to as a HAN (*Home Area Network*). The technology, however, is exactly the same between a LAN and a HAN and a CAN (usually wired Ethernet and wireless WIFI.)
- *Wide Area Network*, connecting devices in other parts of the city, country, or world. WANs involve telecommunications devices and equipment owned and managed by phone and cable companies. When a wide area network is available for a whole city or country, it is referred to as a *Metropolitan Area Network (MAN)* which is a publicly financed network, usually delivered through a WIFI technology, within a certain town or city.

Today, typical local area networks run at speeds of 100 *megabits per second*[*] to 1 gigabit per second, eclipsing older local area networks which ran between 2 and 16 megabits per second. For the right kind of money, you could even get a 100 gigabit per second *wide* area network (though never through cellular or mobile technology, only wired. 5G tops out at 20 gigabits per second). In the past, the phone company provided access to the wide area network from a home or business, and the home or business accessed their own devices through a local area network. While that is still the most common configuration, there are currently many situations these days where a single provider services both the wide area and local area portions of the network. It used to be that all local area networks were wired, whereas wide area networks could be either wired or wireless. Today, both local area networks and wide area networks can be either wired or wireless.

[*] Notice the fact that data is rated at *megabits per second* or *gigabits per second.* Since there are 8 bits in a byte, this is very different from *megabytes* or *gigabytes.* When people just say *gigs* or *megs* they are not telling you which it is – bits or bytes. Furthermore, *speeds* cannot be stated without the timeframe: *per second.* Whereas a file stored on a drive is measured in *megabytes,* the speed of the connection that got it there would be in *megabits per second.* More about this is discussed in the section ***Speed and Size of Connections.***

Additionally, in the past, the types of cables used for local area networks and wide area networks were different, but advances in cabling technology have eliminated this distinction; often the same cables are used for both. And phone companies are no longer the sole provider of WAN services; cable companies and mobile phone companies also provide those services (and now even other companies like Google).

## Physical Cabling

At the bottom of the bottom of the network portion of the technology map is the physical connection: the cabling. There are several types of physical wires used in the cables of data networks. In general, the type of cable for data needs to be of a much higher quality than telephone cable. The signals sent over telephone wires are analog in nature, and travel at an extremely low frequency (approximately 5–15000 Kilohertz). Telephone wire cannot send 10 or 100 Megahertz digital signals, and high-speed electronic equipment is far less forgiving of the inferior wire's minor signal fluctuations than is the human ear.

There are several *common transmission problems* that are caused by poor cabling or low quality wireless transceivers. The common standard for installing wire is known as *structured wiring.* Wiring is installed following a *physical topology* (though the type of network that runs over the wire may also follow a *logical topology*). Finally, we will go over several *physical cable types.*

### Common Transmission Problems

Some problems involved in transmitting data bit streams across a connection are universal, regardless of whether the information is transmitted wirelessly, through "open air," or sent through cables. Copper wire cables carry electromagnetic transmissions (electrons), whereas fiber optic cables employ laser emitting diodes to send light (photons). Any type of transmission (electromagnetic, laser pulses, or radio waves) may be interrupted, which interferes with communication of data. Breaks in the data stream may be caused by attenuation, cross-talk (including near-end cross-talk), jitter, signal interference, or environmental factors.

**Attenuation** is a loss of signal strength due to distance. Attenuation can be a result of poor signal strength, impedance* mismatch, poor quality conductors, or other problems with the physical configuration of the network. All data media are subject to attenuation, but some more than others.

**Crosstalk** is a result of the signal "leaking" outside of the wire carrying the transmission or outside of the frequency carrying the radio wave. In some respects, all wires can act as antennae, both transmitting signals to and receiving signals from other cables in physical proximity. This is especially true in the case of cables which run side-by-side for some distance. Fiber optic cable is generally immune to crosstalk because the laser-emitting diode stays within the cable.

**Near-End-Crosstalk** (often abbreviated as NEXT) is a transmission difficulty associated with systems which use separate conductors for transmission and reception. When the signal is sent (at "full strength") on the transmission line, an incoming signal on the receiver line (which has been become weaker because of attenuation as it traveled some distance on the wire) can "get clobbered," or be drowned out, by the more powerful, outgoing signal.

One of the techniques for controlling crosstalk and near-end crosstalk is the use of shielded cable, but shielded cable is very expensive. Another way to control crosstalk is by twisting the paired wires sending the same signal around each other. Electrically, twisting the conductors around each other is very similar to using a separate shield element. Higher-quality cable (through tighter

* Impedance is similar to *resistance.* Impedance is measured in Ohms and refers to the relationship between the current and the voltage of a cable. An excellent non-technical explanation of impedance can be found at http://forum.ecoustics.com/bbs/messages/34579/131527.html, written on April 12, 2005 and sponsored by ecoustics.com, a Hi-fi online retailer. Accessed initially May, 2008, last accessed September 3, 2020.

twists) and higher quality connectors decrease near-end crosstalk, while precision circuitry can decrease its effects.

**Jitter** is the result of minor inaccuracies in the transmission of a signal, frequency attenuation, and signal clocking (inaccurate timing of the carrier signal). When the timing is out of whack, it has a cumulative destabilizing effect that can result in a lost signal. All data media are subject to jitter, but some more than others.

**Signal Interference** (also known as EMI or electromagnetic interference) is caused by something from an outside system. Simply switching on the lights in an office, for instance, usually creates enough of a radiated electromagnetic "spike" to interrupt network transmissions and cause interference. Again, only fiber optic cabling, because it does not use electrical signals, is immune to electromagnetic interference.

**Environmental factors** and **Physical objects** generally only impact wireless communication. Tall buildings, concrete walls, mountains – any kind of physical structure that makes it difficult for radio signals to get through will interfere in wireless signals. Additionally, weather can interfere with wireless signals. For example, lightning or sunspots can interfere and prevent wireless communication.

You may think that if you are using Wi-Fi instead of wired, you can avoid worrying about interference, but that would not be correct. In fact, Wi-Fi has a lot more potential for interference. Wi-Fi generally cannot get through metal or concrete walls, for example. One Wi-Fi network can interfere with another Wi-Fi network, so you can't have two of them share the same physical space without a lot of interference (or at least cooperation, putting them on different bands).

## Structured Wiring & Wireless

*Structured wiring* is the name given to the practice of using specific methods of placing wires in a building. It was developed by AT&T in the early days, when they wired buildings for telephones. It describes the general layout and terminology to use. When data wiring began to be installed in office buildings, it only made sense to use structured wiring as well. Structured wiring was used in most buildings for both voice and data cabling.

Structure wiring calls for a centralized wiring closet on each floor that is connected to wall plates in each office or room. This required the use of different kinds of *distribution frames*. A *Main Distribution Frame* (MDF) refers to cabling between floors, buildings, or areas. This is sometimes called the *backbone wiring*. On each floor of the building, a wiring closet or *Intermediate Distribution Frames* (IDF) is used to distribute the cable to individual computers and printers and servers. The distribution frame is illustrated in Figure 4.

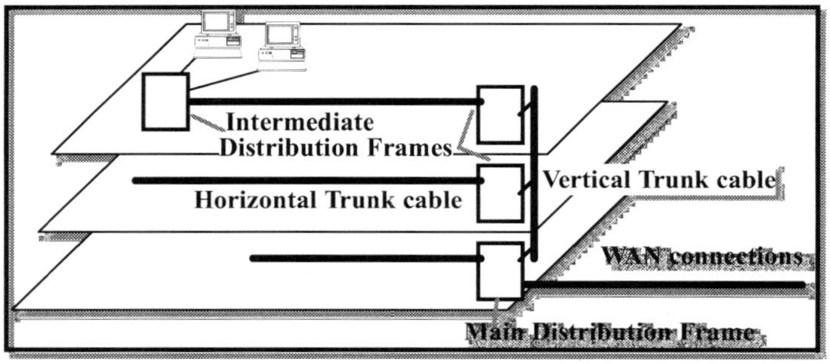


**Figure 4. Structured Wiring Distribution frame.**

While wireless systems do not use the structured wiring methods or terminology, they do have a similar structure for the distribution frame.

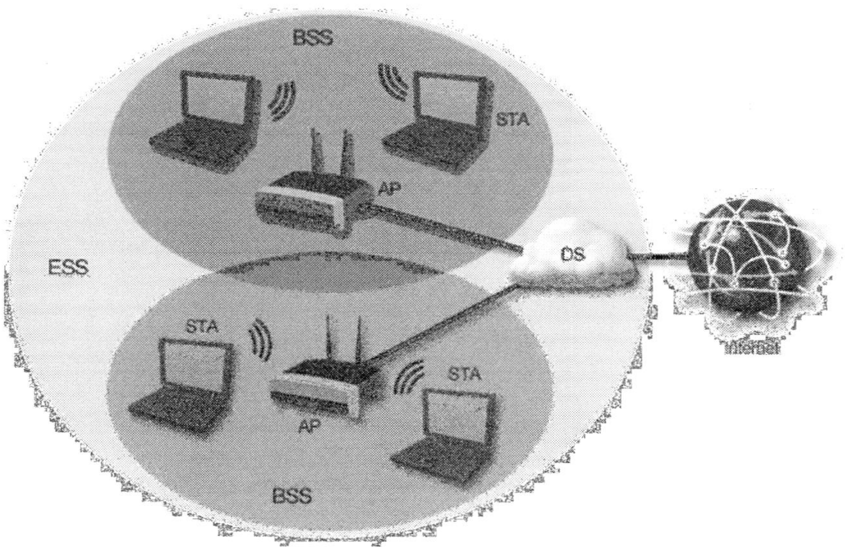


**Figure 5. Wireless Distribution Frame**

The components for a wireless distribution frame are listed in Table 4*. They are similar to the wired distribution frames, but use different nomenclature.

**Table 4. Components of Wireless Network Structure**

| **Component** | **Description** |
|---|---|
| Station (STA) | A STA is a wireless network card (NIC) in an end device such as a laptop or wireless PDA (personal digital assistant or smartphone). STA often refers to the device itself, not just the network card. |
| Access Point (AP) | An *Access Point* (AP), sometimes called a *Wireless Access Poin*t (WAP), is the device that coordinates all communications between wireless devices as well as the connection to the wired network. It acts as a hub on the wireless side and a bridge on the wired side. It also synchronizes the stations within a network to minimize collisions. The least secure location for a wireless cell access point is against a perimeter wall. |
| Ad Hoc | An *ad hoc* network works in peer-to-peer mode without an AP. Instead, the wireless NICs in each host communicate directly with one another. An ad hoc network:<br>Uses a physical mesh topology.<br>Is cheap and easy to set up.<br>Cannot handle more than four hosts.<br>Requires special modifications to reach wired networks.<br>You will typically only use an ad hoc network to create a direct, temporary connection between two hosts or a tiny home network. |
| Infrastructure | An *infrastructure* wireless network employs an AP that functions like a hub on an Ethernet network. With an infrastructure network:<br>The network uses a physical star topology.<br>You can easily add hosts without increasing administrative efforts (scalable).<br>The AP can be easily connected to a wired network, allowing clients to access both wired and wireless hosts.<br>The placement and configuration of APs require planning to implement effectively.<br>You should implement an infrastructure network for all but the smallest of wireless networks.<br>**Note:** In the graphic above, the wireless network is configured to use infrastructure mode. |
| Basic Service Set (BSS) | A BSS, also called a *cell*, is the smallest unit of a wireless network. All devices in the BSS can communicate with each other. The devices in the BSS depend on the operating mode:<br>In an ad hoc implementation, each BSS contains two devices that communicate directly with each other. |

* Much thanks to Nguyen Hoang Anh Tuan who gave permission to use the graphic and table from https://sites.google.com/a/pccare.vn/it/security-pages/wireless-network-structure.

| Component | Description |
|---|---|
| | In an infrastructure implementation, the BSS consists of one AP and all STAs associated with the AP.<br>The graphic above shows two BSSs. |
| Independent Basic Service Set (IBSS) | An IBSS is a set of STAs configured in ad hoc mode. |
| Distribution System (DS) | The distribution system (DS) is the backbone or LAN that connects multiple APs (and BSSs) together. The DS allows wireless clients to communicate with the wired network and with wireless clients in other cells. |
| Extended Service Set (ESS) | An ESS consists of multiple BSSs with a distribution system (DS). The graphic above is an example of an ESS. |
| Service Set Identifier (SSID) | The Service Set Identifier (SSID), also called the network name, groups wireless devices together into the same logical network.<br>All devices on the same network (within the BSS and ESS) must have the same SSID.<br>The SSID is a 32-bit value that is inserted into each frame. The SSID is case-sensitive.<br>The SSID is sometimes called the BSS ID (Basic Service Set ID) or the ESS ID (Extended Service Set ID). In practice, each term means the same thing. Note: Using BSS ID to describe the SSID of a BSS is technically incorrect. |
| Basic Service Set Identifier (BSSID) | The BSSID is a 48-bit value that identifies an AP in an infrastructure network or a STP in an ad hoc network. The BSSID allows devices to find a specific AP within an ESS that has multiple access points, and is used by STAs to keep track of APs when roaming between BSSs.<br>Note: Do not confuse the BSSID with the SSID. They are not the same thing. |

## *Physical and Logical Topology*

The word *topology* refers to the cable configuration. There are physical topologies and logical topologies which are analogous to hardware (the cables) and software (the protocol). Star topology is the preferred and predominant physical wiring scheme. In the star topology, as shown in Figure 7, all nodes are connected to a single point.

Ethernet, one of the network types discussed later in this chapter, has a logical topology of a **bus** (which is just a straight line with two ends), but the physical topology is a star. To see how this is done, imagine starting with a long string and laying it out so that each segment meets in the center, as seen in Figure 6[*]. While requiring longer cable lengths than any other standard topology, this geometry enables the easiest management, maintenance, and reconfiguration possibilities.

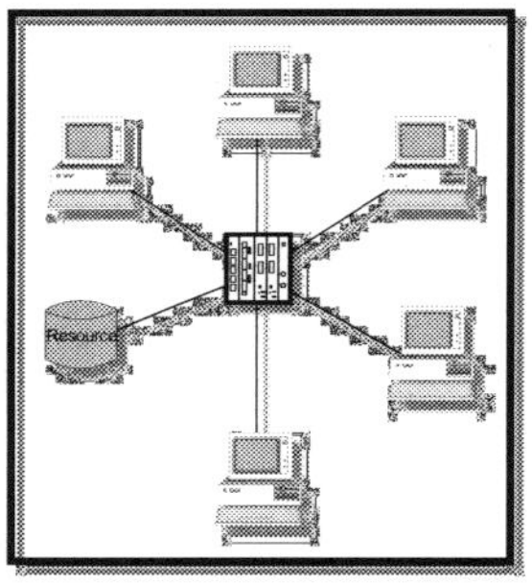

**Figure 7. Star topology - physical wiring.**

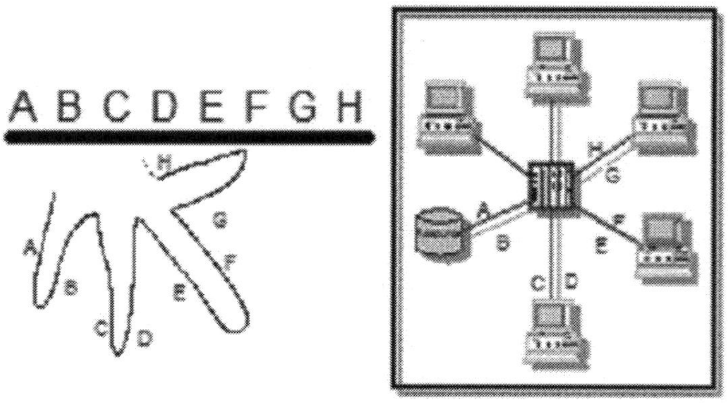


**Figure 6. Logical Bus with Physical Star topology**

---

[*] Much thanks to T. Max Devlin who provided illustrations for Figures 6, 7, 14, 15, 16, and 17 as well as much of the discussion on the technical aspects of networking.

## Types of Network Cables

The cable used by the trunk and the user connections (nodes) are usually two different types. Trunk cable is generally designed for heavy-duty use, while the node connections use lighter cabling to wire individual offices.

In a building with standard structure wiring, the hub is connected to a patch panel. The patch panel, simply an array of connectors, is connected, in turn, to the installed wiring running through the floors, ceilings, wiring ducts, walls, and/or elevator shafts and risers of the building. This cabling connects at the wall plate in the offices. The trunk cables connect the different hubs on the different floors together. The cables that connect from the hub to the patch panel are called patch cables. Additionally, patch cables connect the wall connector to the user's equipment (NIC).

There are three different types of cable used in most networks; coaxial, unshielded twisted pair, and fiber optic.

> All wire is not equal. Even two lengths of cable which appear identical may have slightly (or widely!) differing electrical performance. Specifically, with UTP (unshielded twisted pair) cabling, the quality of the cable varies tremendously, and can be understood by the number after the CAT. CAT3 cable is much less resistant to interference than CAT5 or CAT6 (which have more twists per inch).
>
> Coaxial cable is far more resistant to the environment (including physical variations in temperature, etc., as well as electrical interference) since it has shielding. However, though electrically superior to UTP, coaxial cable is usually far more expensive.

### *Coaxial Cable*

Coaxial cable consists of a center conductor within an insulation core. Metal braiding is wrapped around the core and covered by a heavy-duty jacket. Coaxial cable comes in a variety of specifications with differing electrical characteristics.

Coaxial cable is rather expensive and difficult to work with, especially the thicker (1/2") "Thicknet" cable. Cable companies, however, sell network connectivity along with television services, and this has led to a rise in the use of coaxial cable for networks.

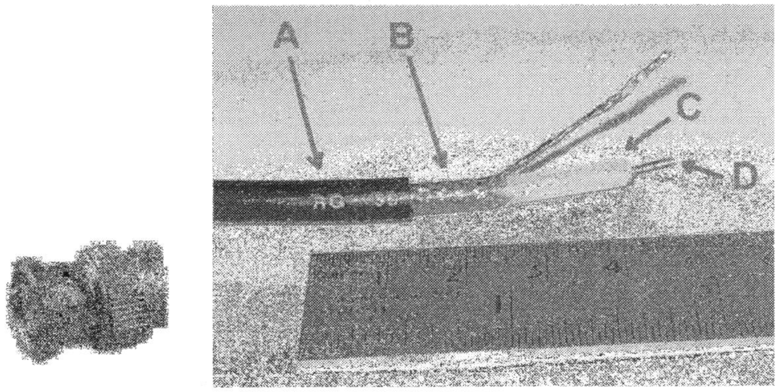


**Figure 8. Coaxial cable and a BNC connector for coaxial cable.**

Coaxial cable typically uses BNC connectors. The coaxial cable and BNC connectors can be seen in Figure 8. To preserve the electrical characteristics of the cable system, unconnected coaxial cable ends generally require a terminator, which looks like a little metal cap on the BNC connector.

## Unshielded Twisted Pair (UTP)

Unshielded twisted pair is currently the most popular type of cabling*. Four pairs of wire are inside the plastic sheath. Each pair of wires is twisted around itself (see Figure 9). While UTP is the easiest to install and the cheapest of the standard cable types used for data networks, the quality of unshielded twisted pair cabling also varies greatly.

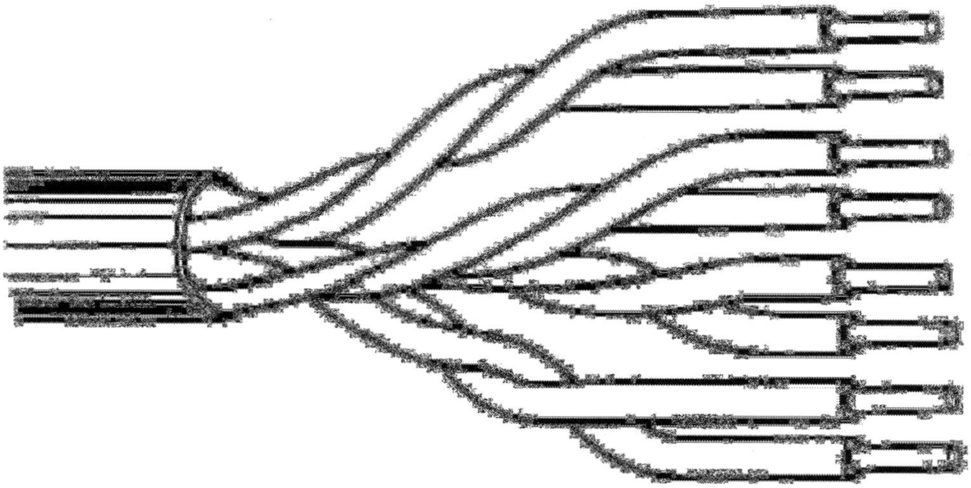

**Figure 9. Inside the UTP cable (four-pair).**

The quality of unshielded twisted pair cabling is determined by the size of the conductor, the thickness and type of insulating jackets, and most importantly (from a data network viewpoint) the number of twists per foot. Twists per foot affect the electrical characteristics of the wire. In effect, the "tighter" the twists, the less high frequency signal energy can "leak out." That is, the more twists per foot, the greater the bandwidth of the cable. This vital statistic, twists per foot, will vary between cables from different manufacturers, and even varies between the individual pairs in a single cable sheath. Because the outside twisted pairs are usually wrapped around the inner twisted pairs, giving them a larger circumference in proportion to the inner pairs, which are twisted around each other, the inner pairs may have more twists per foot. This varies, of course, based on the manufacturing methods and the number of pairs in the cable.

The quality of the UTP cable is categorized as 1, 2, 3, 4, 5, 6 or 7. "Regular" telephone wiring is low-grade unshielded twisted pair and is category 1 or 2. The Ethernet specification (also called 10BaseT) specifies UTP category 3 or higher, although most installations today use category 5 or 6. Technical people will shorten this and refer to the cable as "CAT5" or "CAT6". Notice the markings on the UTP cable shown in Figure 10.

* Because UTP is the most common cabling and Ethernet is the most common networking protocol, stores and salespeople and even sometimes technical people will call the UTP cable an Ethernet cable. But it can't be. Ethernet is a protocol, not a type of cable. The Ethernet protocol can run on many different types of cables – or even over the air without a cable.

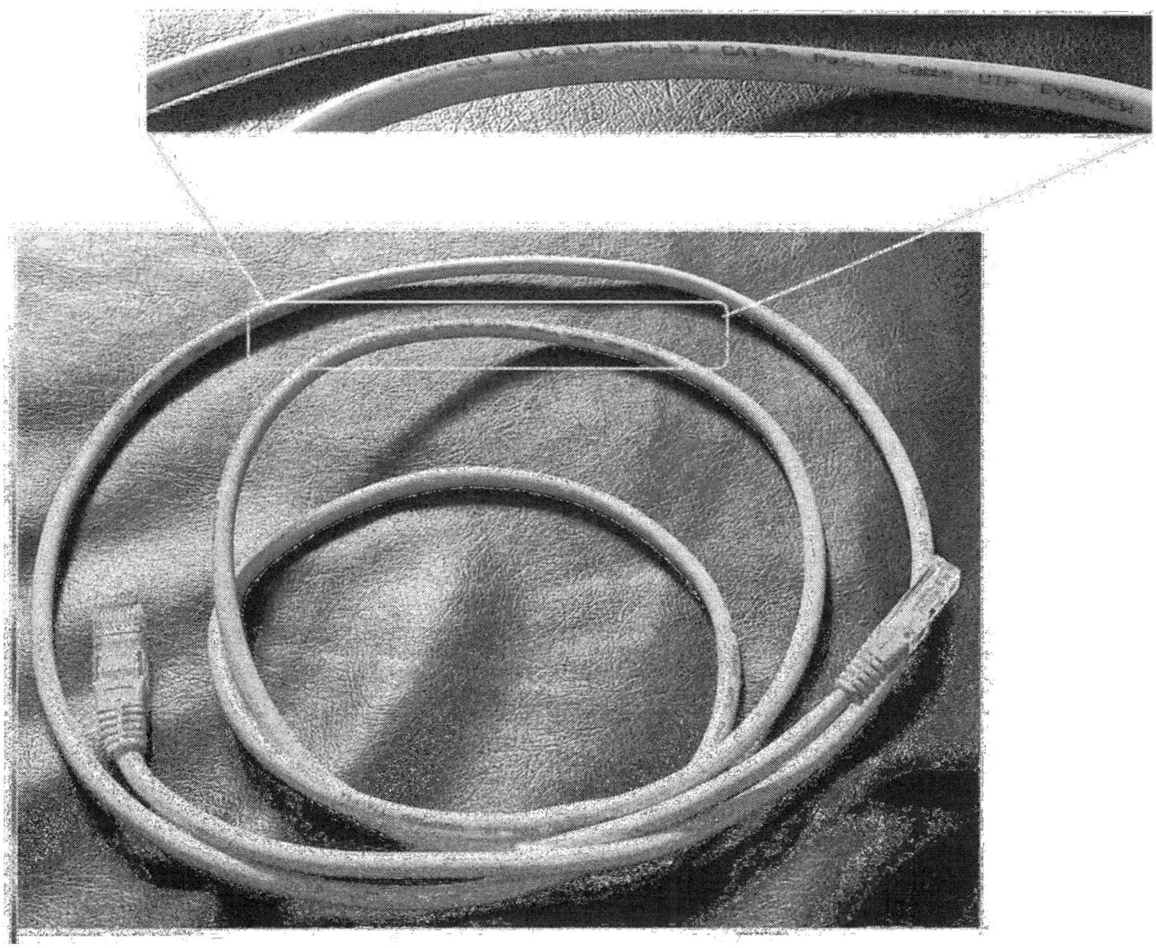

**Figure 10. Category 5e Unshielded Twisted Pair (UTP) patch cable.**

The type of connector used for landline telephone cable is RJ-11. It is not often used for data network connections. The RJ-11 connector has four contacts for 2 pairs of wires and is relatively narrow. The connector widely used for data network UTP connections is RJ-45. RJ-45 connectors are slightly wider and usually have eight contacts. Both can be seen in Figure 9.

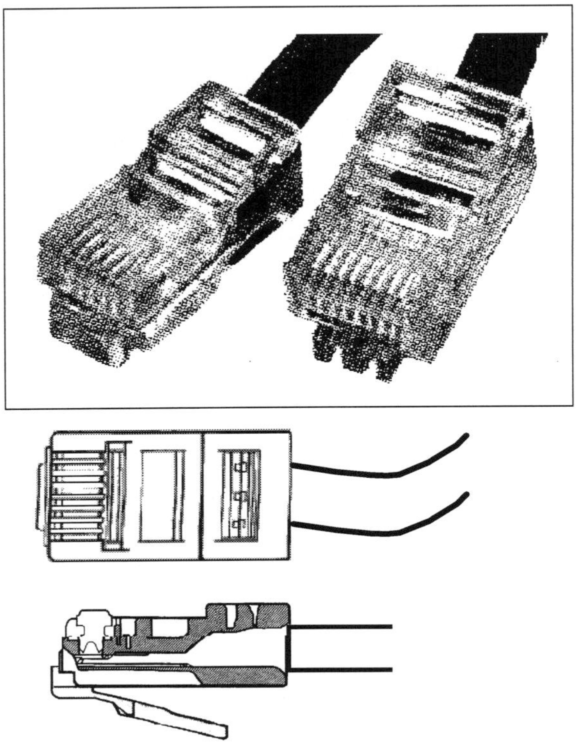

**Figure 11. RJ-11 & RJ-45 UTP connectors.**

## *Fiber Optic Cable*

Fiber optic cable is a glass or plastic thread surrounded by a cladding. A protective plastic coating then surrounds the cladding (see Figure 12).

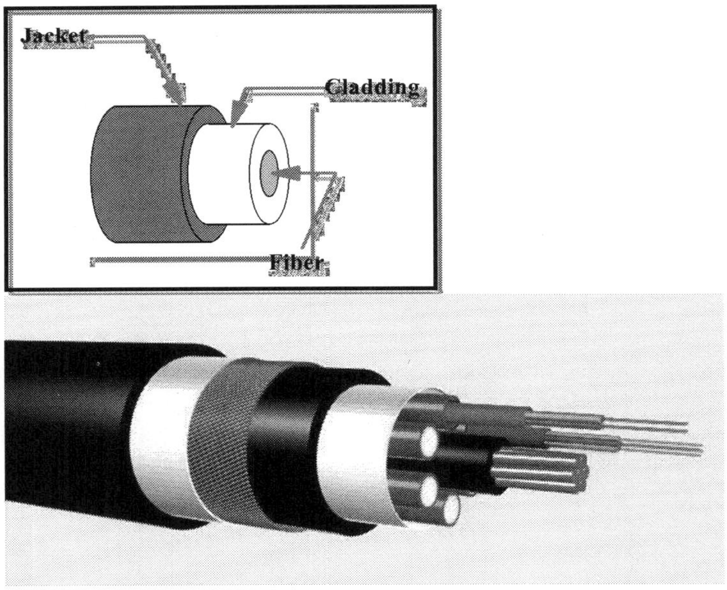


**Figure 12. Inside Fiber Optic cabling.**

Unlike any other cable, fiber optic cable does not carry data via electrical impulses. Instead, the signal is converted to light using a laser or light emitting diode (LED). The light impulses can travel immense distances through the fiber with minimal attenuation. Because the energy is not electrical, the signal is completely immune to noise and electromagnetic interference (including crosstalk and voltage spikes). Since the light impulses are contained within the cladding (as shown in Figure 13.) and do not "radiate" as electrical impulses do, fiber optic transmissions are more secure than electrical transmission methods.

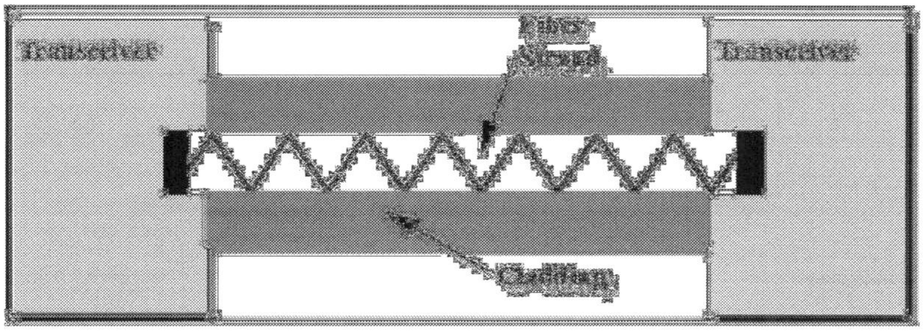

**Figure 13. Fiber optic cable is resistant to interference.**

Fiber optic cable is the most expensive type of standard networking cable, so it is rarely used for internal house or building wiring. Most connectors used with fiber optic cable are ST (Straight Tip) bayonet-style connectors. Similar to, but smaller than, BNC connectors, ST (or just S) connectors are "push-on, twist-to-lock" connections. The installation of fiber optic cable is costly and more difficult than other types of cable, since special skills and tools are required, and the connections must be much more precise than with electrical wire. Additionally, fiber optic cable is very sensitive to temperature; in cold weather the glass or plastic can shatter if bent.

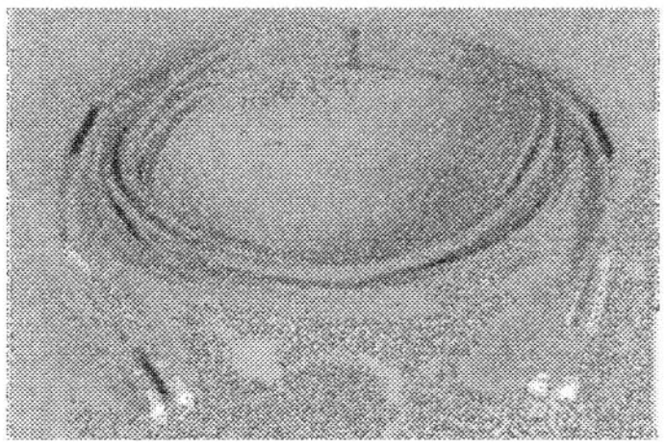

**Figure 14. Fiber optic cable with ST connectors.**

## Wireless Media

The media for wireless technology cannot be seen or touched, but it is still "real" and are different frequencies of radio waves. WAN satellites generally use microwaves (from 1 gigahertz [GHz] to 300 GHz, with a wavelength between 30 centimeters and 1 millimeter) to communicate and require a parabolic dish transceiver to send the signal. LAN networks use radio waves usually using **UHF** (ultrahigh frequency, which is at 2.4 GHz with a 12 centimeter wavelength) or **SHF** (superhigh frequency, which is 5 GHz with a 6 centimeter wavelength). Signals at these frequencies are too high for humans to hear and too low for humans to see, but the computer transceiver has no problem receiving them.

## Ethernet Hardware

In order to communicate, devices need to use the same *protocol* on every layer of the technology map. At the bottom layer, the physical layer, the most popular choice is Ethernet*. In the ‘80s there were many others (Token Ring, ArcNet, FDDI, LocalTalk, AppleTalk, etc.), but Ethernet has proven to be the most inexpensive with reasonable reliability, so it has become the worldwide

* Ethernet is a protocol at the physical networking layer. TCP/IP is the protocol at the logical internetworking layer. Technically the Ethernet protocol itself is 'software", though in actuality Ethernet "exists" in the cabling, hub, and NIC cards in the form of firmware (which is software burned onto a hardware chip). That's why we cover Ethernet under hardware, while we cover TCP/IP under software. You could quibble and say that TCP/IP actually exists as firmware on the router, but we assess TCP/IP to exist more in the TCP/IP client software that is included with almost all operating systems, therefore, more software than hardware.

standard for the physical layer of the network. The official name for wired Ethernet is 802.3*, which was established by the Institute for Electrical and Electronic Engineers (IEEE) in January 1980. By 1995 it was the de facto standard.

The official name for wireless Ethernet is 802.11, but everyone just calls it *Wifi,* though officially is it either Wi-Fi or WLAN (for Wireless Local Area Network). The first wireless Ethernet protocol was adopted in 1997, and there have been over a dozen variations to the protocol (generally designated with one or two letters, such as 802.11a, 802.11g, 802.11n, 802.11ac, 802.11ax, etc.)

To understand Ethernet, we need to understand the difference between packet switching and circuits. Additionally, to understand Ethernet, we need to understand what CSMA/CD stands for. Finally, we need to recognize what kind of data is in an Ethernet packet.

## Packet Switching vs. Circuits

Packet switching is a different approach to transmitting data. In the old days, when one computer wanted to communicate with another computer, they had to establish a *circuit.* A circuit is a semi-permanent connection that allows two computers to transmit messages back and forth. It doesn't matter which medium (fiber, copper wire, satellite, radio waves, microwave, etc.) connects the two computers. The connection starts up and stays connected until the computers no longer need the circuit to transmit to each other. The model was based upon the type of connection that was used by the phone company for a telephone call. When two people are talking to each other on analog phone lines, the connection between them is established upon the first ring and stays on until the last person hangs up. That connection is called a circuit. Another name for a circuit is a *point to point* connection.

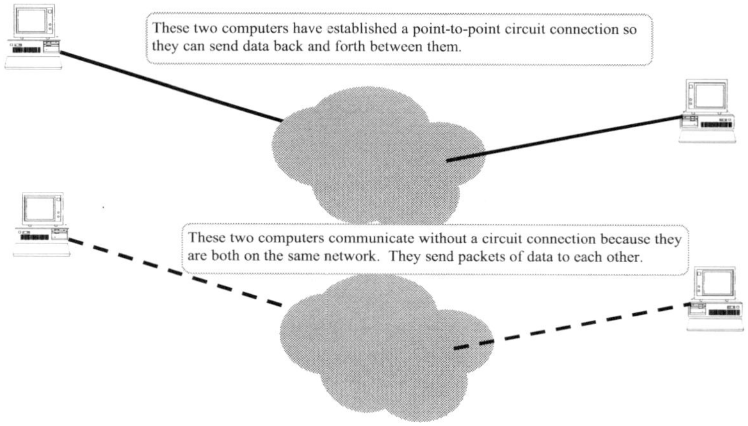


**Figure 15. Point-to-point circuit vs. packet-switching network.**

With packet switching, however, there is no permanent connection between two computers (see Figure 15.). Each message to be transmitted is broken up into millions of datagrams, and each datagram is encapsulated inside a data packet, as illustrated in Figure 16. The packet gets transmitted at nano speeds–at around the speed of light.† Ethernet was developed based on the concept of packet switching.

* Ethernet 802.3 is similar to, but not exactly the same as, Ethernet II, which was the LAN protocol developed by Bob Metcalfe and David Boggs a few years before.

† Don't confuse the speed of a single packet on a network with the speed of the data flow of a network.

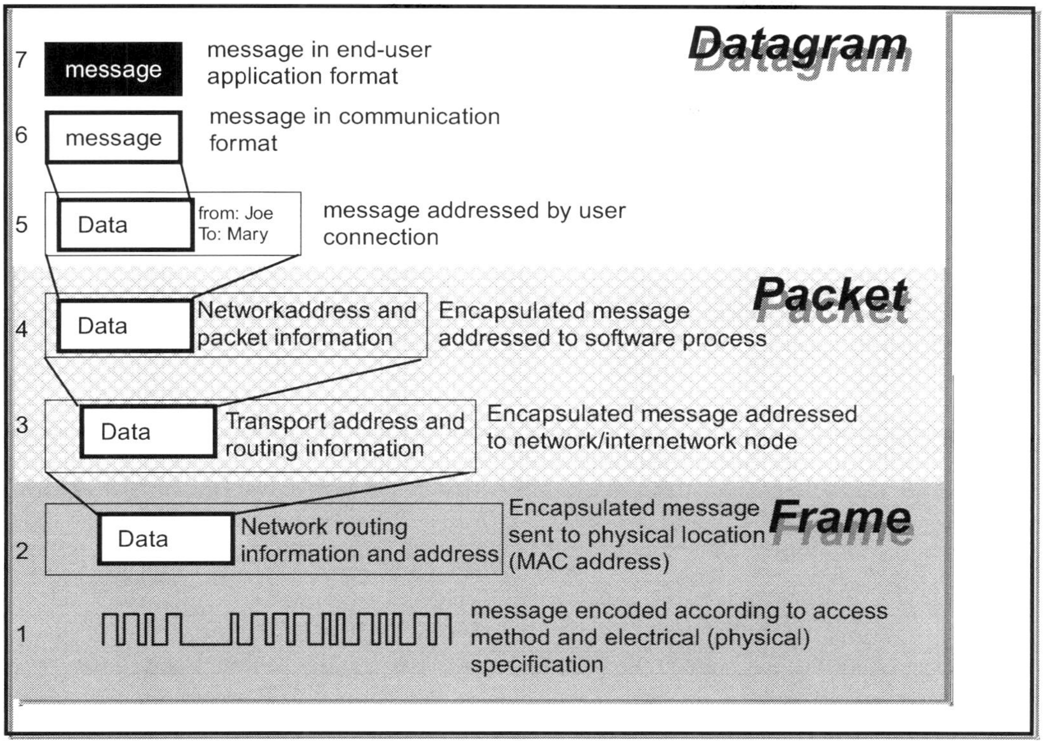


**Figure 16. Datagrams encapsulated into a packet.**

## CSMA/CD

There are different types of packet switching. Ethernet uses a method called Carrier Sense Multiple Access with Collision Detection (CSMA/CD)*. A computer is connected to the network line through an Ethernet NIC card. There are many computers with NIC cards on the same line, referred to as *multiple access*. To send a packet of information, the NIC "listens" to the wire to see if any other device is transmitting. The NIC senses if there is a *carrier* (a message already transmitting on the wire). If so, the NIC reads the message in order to check the destination address of the packet. If the destination address turns out to be itself, the NIC reads the message and hands it over to the application in memory (which may be waiting for the message). If the NIC doesn't sense any carrier on the line, the NIC knows the line is clear of all transmissions. When the line is clear, if the NIC has a message to send, it begins transmitting. The Carrier Sense Multiple Access part of the transmission is completed.

* CSMA/CD is often called "shout on a wire" technology, because when one computer transmits a message, every computer on the segment (i.e. within "shouting" range) can "hear" the message. Of course, now that wireless is so prevalent, it might also be called "shout in the air" technology.

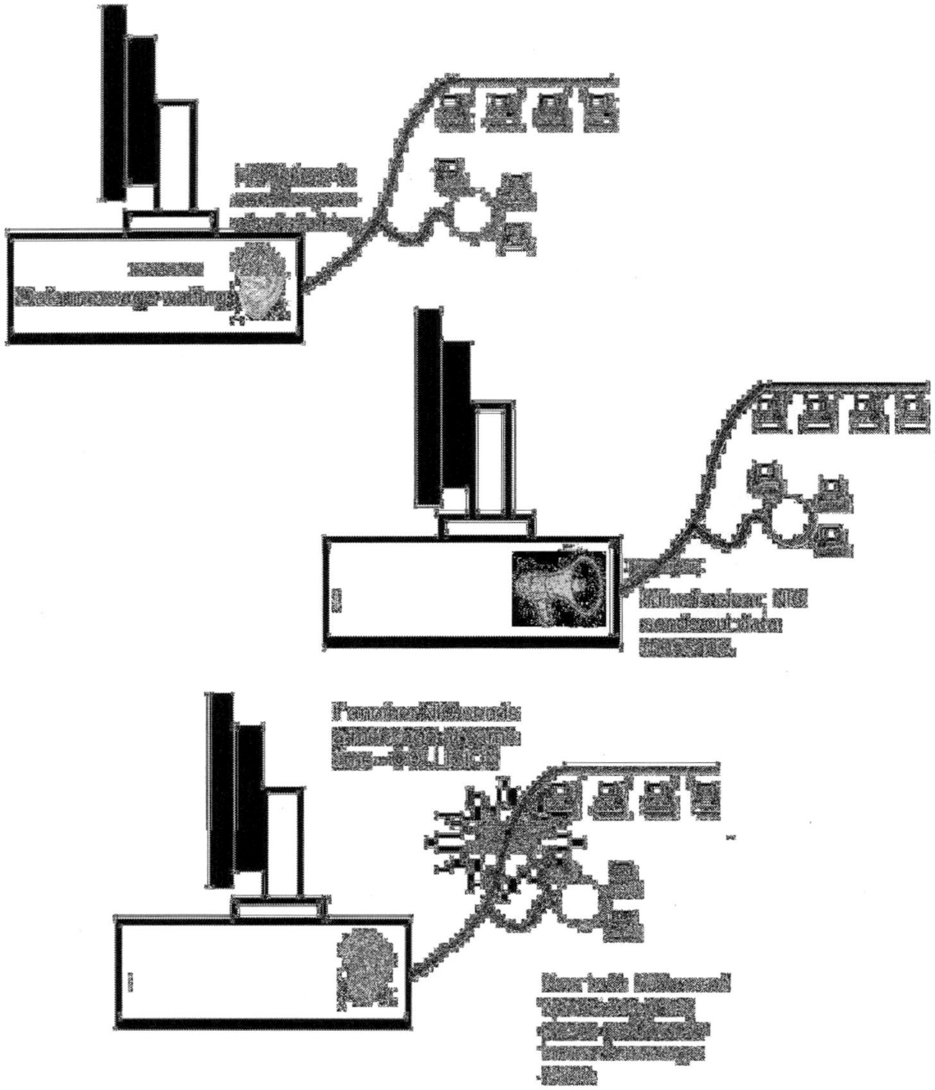

**Figure 17. Illustration of Carrier Sense Multiple Access / Collision Detection (CSMA/CD).**

However, if another device transmits at the exact same time as the NIC, there's trouble. In that case, the NIC senses the *collision* of the two messages and stops transmitting. (Alternatively, if interference causes a collision, the result is the same.) Usually, in a collision, the messages from both NICs get corrupted, so neither message gets through. The NIC waits a set period of time until the line is clear of all transmissions, and then tries again. This part of the cycle is called *collision detection*. Together it means that Ethernet is a CSMA/CD communication method. This process is illustrated in Figure 17.

Keep in mind that this is happening at nano speeds. A hundred computers can share the line without any problems as long as none of them tries to "hog" the line and transmit continuously (which would prevent all the other computers from being able to transmit at all). Also keep in mind that the information being transmitted is seen by all the computers on the network. The NIC card reads the destination address within the packet to find out for which NIC it was meant. It only reads and conveys the datagram inside the packet if it matches the destination address.

Each NIC has a unique identifying number known as a MAC ( Media Access Control) address that was burned into it when it was manufactured at the factory. The source and destination MAC addresses are found in the outer layers of every packet placed there by the sending NIC. That's how the NIC card knows to whom to address the packet.

This same methodology is used for Wi-fi, but instead of transmitting on a wire, the message is transmitted through the air at a certain frequency.

## Ethernet Frame Format

The Ethernet frame's makeup is illustrated in Figure 18. The length of the encapsulated data (application/presentation data, along with session, transport, and network protocol data) can range from 64 to 1500 bytes long, and must be a multiple of eight.

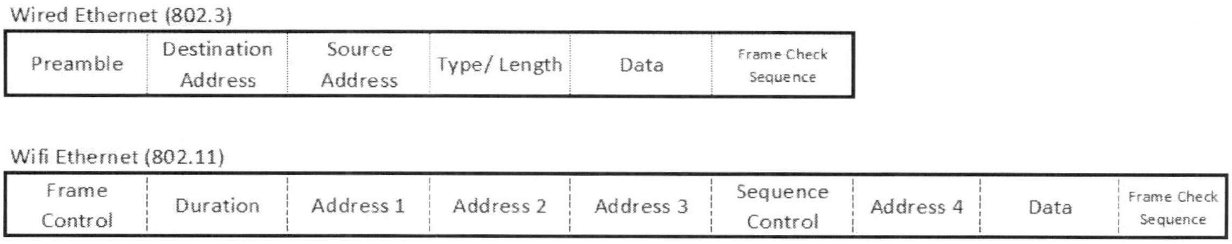


**Figure 18. Ethernet frame.**

The different sections of the 802.3 Ethernet frame are described in Table 5. The 802.11 Ethernet frame is a bit more complicated but follows the same general pattern.

**Table 5. Ethernet 802.3 Frame Breakdown**

| Section | Purpose | Notes |
|---|---|---|
| Preamble / Starting Delimiter | Sets start of frame and synchronizes timing | The preamble is "removed" at the physical layer. |
| Destination address | MAC (media access control) address of the NIC the frame is sent to. | MAC address is burned into the NIC card by the manufacturer |
| Source address | MAC address of the NIC generating the frame | Also comes from the NIC card. |
| Length | Length of data field | how many bytes of data |
| Data | Data, including encapsulated higher-level protocols | Varies in length. Must be a multiple of eight, padded if required. |
| Frame Check Sequence | Error checking | Uses Cyclic Redundancy Check (*CRC*) |

## Ethernet Facts

There are several other bits of information about ethernet which are important to know.

**Other Ethernet devices:** The NIC in a personal computer is usually connected through a cable to a *hub*, which is the generic name for either the *switch* or the *router/switch* combination*. Alternatively, the wireless receiver in a phone, tablet, or computer is connected through the air to a Wi-fi hub. In terms of layers, the *switch* is the hardware device needed to send and receive the signals between the devices.

These days, many people call the router/switch combination a *modem*, which is technically incorrect but very prevalent. Back when the data was transmitted across the media in analog (using the frequencies normally used for either telephone conversations or television transmissions), the digital data signal needed to be *modulated* into analog and then *demodulated* back into digital. The term *modem* is a combination of the beginning of the words for two processes *mod-dem = modem.* In 2012 broadcasting companies stopped using analog frequencies for television shows, and few people still use modems to access the internet through dialup technology. Nonetheless, because many people were used to the term *modem* to refer to *that box they connected to the line,* it was difficult to get them to use newer terminology. Especially since some cable companies were still using a modem process (though phone companies had switched to Digital Subscriber Lines which used switches). So now even the ISPs (Internet Services Providers) use the term modem to refer to the box that houses both the router and the switch or hub.

**Speed of connection:** Originally, Ethernet was designed to transmit information at 10 megabits per second (10 Mbits/s), but newer versions transmit at 100 Mbits/s (Fast Ethernet) or 1000 Mbits/s (GigaEthernet)†. The most common configuration today is 100 Mbits/s from the hub to the desktop, and 1000 Mbits/s between hubs or from the hubs to the server.

**Reliability and expense:** Although the description of the traffic patterns of an Ethernet network sounds like chaos, they are generally pretty reliable. Because Ethernet is so common, the equipment and cabling are relatively inexpensive.

**Length of connection:** One issue with Ethernet is the length of the connection. An Ethernet transmission traveling on a cable wire can only travel about 100 feet‡. An Ethernet baseband connection traveling over the air is even shorter. If you have a large house, for example, you will need a Wi-fi extender to be able to get a strong Wi-fi connection in the rooms that are far from your Wi-fi access point.

**Number of nodes:** The maximum number of nodes (computers and servers) on a typical Ethernet network segment would be about eighty (unless the users are power users and transmit a lot of data to the server and back – in which case fewer nodes can coexist on the same segment).

**Contention:** Using Ethernet protocol, only one computer can transmit data at any one time. If too many devices on the network are trying to transmit simultaneously, collisions become very frequent, causing the computers to continuously try to send the same data over and over again. The consequence is a traffic jam that can result in gridlock; no network traffic can get through. Contention is the word applied when there is too much traffic on the network.

---

* The *router* is at the logical layer, where TCP/IP is the protocol. Generally, Ethernet hubs are *switches*, but the same physical device can be both a switch for Ethernet and a router for TCP/IP. If you have a wireless router in your home provided by your Internet Service Provider (ISP), it is probably a combination device, either a modem/router or a switch/router.

† There is a 10 Gigabit per second Ethernet switch in use, but very expensive and rare, and a 1000 Gigabit per second (or Terabit per second) switch under development, but not yet in use. I recently read an article that said that 100 Gig Ethernet was common, but the author simply didn't know the difference, I believe. Admittedly, it's a bit much to keep track of the M, G, and Ts.

‡ The maximum length for a cable is dependent upon the quality of the cable and the quality of the hub.

**Segmenting:** A switch can relieve contention by setting up separate segments for the Ethernet devices. Instead of 80 computers sharing a single segment, the switch can be set up to separate out 20 computers on four different wires, thereby decreasing contention and speeding up transmissions.

The typical enterprise network, therefore, is actually a collection of many individual Ethernet segments. In a typical home, however, there will only be one Ethernet segment controlling 3 – 5 devices. Unless more than one of those devices decides to stream a movie at the same time, there usually would not be a problem with contention.

## Long Distance Physical Network Hardware

When many different devices are connected over a wide geographic area, then Ethernet is no longer sufficient (in most cases). Ethernet can go at most a few hundred feet. A typical multiple-connection network illustration can be seen in Figure 19.*

The term *broadband* is often used when discussing long distance physical networks, especially when at least one segment of the journey is wireless. Broadband, with its multiple frequencies and separate carrier wave was originally used to distinguish from baseband, which had only one frequency, that doubled as a carrier wave. Ethernet local area networks within a building or campus were baseband, while communications between buildings were broadband. However, because Ethernet is now often the protocol over long distances (especially for systems provided by cable companies), the technical distinction has been lost. The term broadband is applied whenever long distances are involved, regardless of network protocol.

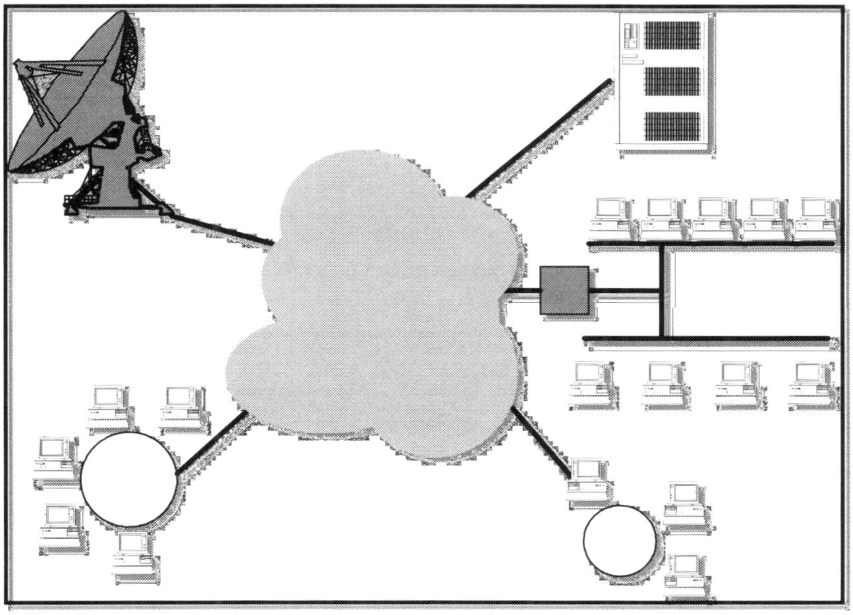

**Figure 19. An enterprise network showing the "cloud".**

* Note the picture of the cloud in the figure. This is not the internet "cloud" discussed in more detail in *Cloud* in *Chapter 9:* Process of Managing Information Systems, but rather represents the illustration of a cloud as was formerly used in documentation of networks. The cloud represented the components of the network that were owned and controlled by the data service provider, typically the phone company, that were connected to the business' in-house equipment.

## Hardware for Wireless Communication

It may seem strange to see wireless communication in a networking section, because, like software, you can't see wireless communication. But, like radio and television, which have broadcasted wirelessly since their invention, the hardware device at the ends of a connection determine the type of wireless connection.

The analog voice and television broadcasts of the past used bandwidth differently than data communications of today. Even wireless cell phone networks were originally analog, but now all phone transmissions are digital. Analog transmissions use a point-to-point method and a frequency modulation, while digital transmissions use a packet-switching method with a binary (on/off) code. Wireless data communications these days are mostly digital.

Wireless capability has grown considerably over the past ten years. More than a decade ago, wireless grew faster in Europe than in the United States because the U.S. was already heavily invested in wired solutions. The gap is getting smaller each year, as wireless devices, such as smartphones, tablets, and laptops that cannot use wired connections, become more common.

Wireless options are complicated by the fact that governments are starting to regulate the use of wireless networking frequencies, along with the phone, radio, and television frequencies they have regulated in the past.

In 2016, a landmark ruling regarding "net neutrality" put the broadband services that ISPs offer under the purview of the FCC (Federal Communications Commission), the same way that telephone communication is regulated. It also proposed that broadband services could not be "throttled" or given special traffic parameters. In other words, neither a company nor an individual can pay a broadband provider to allow them to have a "faster lane" for their data. But neither can the broadband provider prevent a company from using the connection to provide any content they wish, at the same speed as everyone else. This type of regulation has been typical in European countries, but until recently, how data communication broadband was regulated in the United States had not yet been determined.

Prior to 2016, the FCC forced all television stations to broadcast their television content digitally. In 2012, the FCC outlawed analog television transmissions. This switchover, from analog to digital, was required because digital and analog are not compatible technologies. Compared to digital, analog was a "frequency hog", and weak signals resulted in snowy, shadowy video. Using the same amount of space required by a single analog station, up to six digital stations could be transmitted. Digital transmission is compatible with modern technology and is less prone to problems due to a weak signal. While the picture might occasionally pixilate (freeze or degenerate into large blocks of a single color), the problem is much less prevalent than the often snowy, shadowy transmissions of previous generations of television.

## Types of Wireless Technologies

There are several major types of wireless networks. As can be seen in Figure 20*, the protocols are focused either on Wireless Personal Area Network (WPAN), Wireless Local Area Network (WLAN), or Wireless Wide Area Network (WWAN).

Wi-fi has already been explained; it is the wireless local area network (WLAN) that uses the Ethernet protocol. *Bluetooth* is the WPAN, and is used mostly to connect devices such as earphones, speakers, and monitors to computers and smartphones. *Cellular networks* is another name for the WWAN.

---

* Verizon Wireless, 1xEV-DO Rev. A Technology (a white paper) http://b2b.vzw.com/assets/files/Rev_A_WP.pdf Published 2009. Retrieved January 2, 2012.

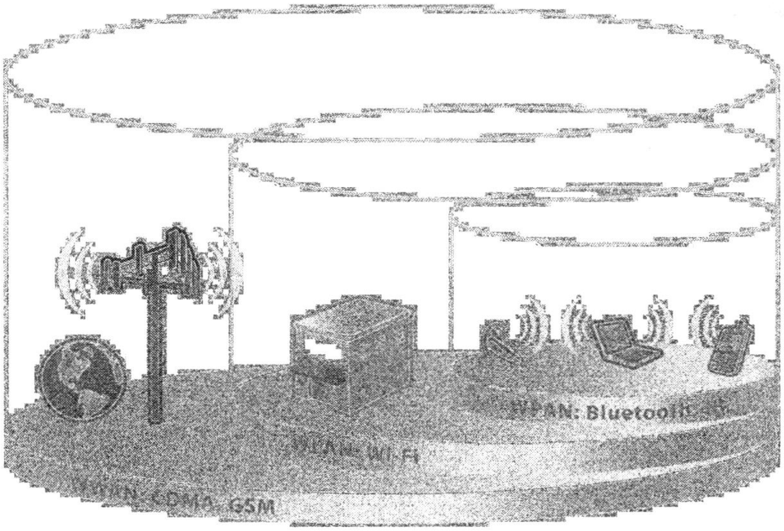

**Figure 20. Wireless protocols and coverage.**

### *Bluetooth*

Bluetooth connects devices that are in very close proximity to each other, such as a cell phone transmitter and a wireless earphone, or a laptop computer and its printer. The IEEE approved the 802.15 Bluetooth standard in 2001, but it wasn't until 2012 that the protocol was automatically built into smartphones and other devices.

Prior to Bluetooth, for a computer to connect with a printer wirelessly, they both needed infrared transceivers and required line-of-sight. Such an arrangement was often unreliable. Bluetooth is used in wireless personal area networks (WPAN), and its range is generally limited to a few dozen feet with a speed of between 1 and 3 Megabits per second. Compared to Ethernet, this is a very slow connection. This is why, for example, it is difficult to use Bluetooth to connect to a monitor or television; the bandwidth needs are higher than the maximum throughput for Bluetooth.

In addition to being a very short-distance protocol, Bluetooth devices have many limitations. Most can only connect to one device of the same profile at one time. For example, a laptop computer can connect to a Bluetooth keyboard, mouse, and speaker. But cannot connect to a second speaker or a second keyboard. A phone can only connect to one headset or earphones at one time.

### *Wi-Fi*

Wi-Fi is the Ethernet protocol for wireless LANs (IEEE standard 802.11), and has already been covered in more detail in the Ethernet section. Most laptop computers come with a Wi-Fi NIC already installed and ready to automatically sense nearby Wi-Fi networks. Businesses can install a Wi-Fi network, generally known as a *hot spot*, and allow customers to use it.

Wi-fi is vulnerable to hacking. While a wired network requires a special device called a *traffic analyzer* connected to the physical wire for someone to read the data transferred via the network, Wi-fi traffic can be read by any nearby computer with wireless access. Data sent across Wi-fi can be encrypted using WEP (Wired Equivalent Privacy), and WPA (Wi-Fi Protected Access). Encryption prevents unauthorized user of the wireless network from easily reading the contents of the traffic. In order to use a WPA protected network, the user must know the key code*.

* When Wi-fi first became available, few homeowners bothered to set their WEP or WPA key. People could just drive around until they found an open network and use it to access the Internet. Nowadays, the Wi-fi hub requires the use of a key, so homeowners don't have a choice.

## Cellular Networks

For the past twenty years, cellular networking has been moving toward a single all-devices-use-it standard. This year, in 2021, we may finally realize the goal of a single cellular standard; LTE, or Long Term Evolution.

### *Long Term Evolution (LTE)*

Starting around 2014, cell phones started using LTE (which stands for Long Term Evolution). This revolution in cellular technology was marketed as **4G** LTE, which stands for the fourth generation of cellular technology. Unlike the 2G (2$^{nd}$ generation) and 3G (3$^{rd}$ generation) technologies before it, LTE enables any phone device to work with any cellular network. Previously, there were two competing technologies, Global System for Mobile Communication (GSM) or Code Division Multiple Access (CDMA), and if you had one, you couldn't use the other.

### *Global System for Mobile Communication (GSM)*

Just a few years ago, more than ninety percent of cellular phones in the world were using GSM (which originally stood for the European group that developed it, Groupe Spécial Mobile[*]). GSM was the first protocol to utilize a SIM (Subscriber Identity Module) card imbedded within the phone. If you take a SIM card out of one phone and put it into another phone, the cellular identity (along phone number lists and voice mail connections) will be transferred to the new phone.

GSM was also the first to enable Short Message Service (SMS) which allows users to send text messages to other users. Originally, GSM used a time-sharing method of multiplexing (sharing the airwaves) which was too slow for smart phone communication, but in 1997 General Packet Radio Service (GPRS) was integrated into GSM, which enabled devices at that time such as iPhones and Blackberries to send and receive email messages. In 1999 Enhanced Data Rates for GMS Evolution (EDGE) was also integrated to enable Internet surfing or connections to corporate resources such as servers or databases. Most non-U.S. cell phone connect with GSM. In the US, only AT&T and T-Mobile utilized GSM.

As popular as it has been, now that most devices utilize LTE, GSM is going by the wayside. Both AT&T and T-Mobile will be shutting down their GSM networks in early 2022.

### *Code Division Multiple Access (CDMA)*

Used mostly in the United States by Verizon and Sprint, CDMA[†] has slightly higher bandwidth than GSM. After being developed by Allied forces to prevent jamming of cellular signals by the enemy, CDMA was further developed and supported mainly by Verizon. One of the disadvantages of CDMA was that the phones set up for CDMA cannot be used with any other network; since a SIM card is not utilized by CDMA, the phone itself had the identification technology that connected it to the network. Nowadays Verizon phones have SIM cards in order to connect to LTE, so Verizon shut down its CDMA network at the end of 2020.

### *5G*

Unlike the other cellular technologies listed here, 5G (which stands for fifth generation) doesn't appear to have a name, nor an acronym. And that may be because it is just beginning to be deployed (as of this writing) despite the fact that it's already being advertised on television as if it is already here. It also may be because 5G is not really a technology, but rather a vision for several different layers, so it involves several different standards. Presently the biggest obstacle to 5G is

[*] European Telecommunications Standards Institute. 2011. Cellular History. etsi.org. http://www.webcitation.org/5yRQXw2sv. Retrieved January 2, 2012.

[†] Do not confuse CDMA with CSMA/CD. They are completely unrelated. CDMA is a cellular phone technology and CSMA/CD is the communication protocol of Ethernet.

that in order to use it, you need to obtain new equipment (phone, tablet, computer) that is compatible. 5G is not backward compatible with earlier technologies.

One thing to realize about 2G, 3G, 4G and 5G is that these marketing terms generally have no direct relationship with the underlying technology. The cellular phone companies can use 4G and 5G to mean whatever they want it to mean. During the competitive development of LTE, some companies would tout parts of it as 3G, and others would tout parts of it as 4G. It is only after the fact, when some standardizing body has reviewed and approved a name and a number to a technology that it can be truly discussed with any definitive accuracy.

## *Middle of the Link Technologies*

Whether wired (Ethernet) or wireless (Wi-fi or LTE), these technologies only get you to the nearest router, which gets you on the Internet. The phone and cable companies act as the ISP (Internet Service Provider). They connect buildings and homes to the core routers in the middle of the Internet. Usually these connections to the phone company and/or the cable company is between the end device on both sides. For example, if you use your laptop computer to purchase something on Amazon, your laptop is connected to the Internet router through Wi-fi, and Amazon may be connect to the Internet router through a wired Ethernet cable, but those two routers are connected to each other in the middle with a long-distance technology.

Table 6 summarizes some of the more widely implemented long-distance links used by phone companies or cable companies acting as Internet Service Providers (ISP) or providing point to point data lines for corporations who need them to connect their computers outside of the Internet. The majority of these require special hardware, often supplied, leased, or bought from the long-distance carrier or the ISP supplying the service. Most people will recognize either DSL or cable modem because they are the two most common connections to the Internet that people use from their homes, but the other types of connections are utilized by business organizations more often.

**Table 6. Types of Long Distance Links (WAN)**

| Consumer or Business | Type | Device | Notes |
|---|---|---|---|
| Consumer Internet Connections | Dialup | Modem | Uses POTS (Plain Old Telephone S POTS was developed for analog voi but the line usually gets converted t signal by sampling it 8,000 times pe 8-bit resolution (64,000 bits per sec transmit data, the line must be modu (converted from digital to analog) a demodulated (converted from analo Very few people use POTS for data connect (Indeed, the number of POTS itself is decre people are giving up their landlines for cell wireless.) |
| | DSL (Digital Subscriber Line) | DSL switch (based upon the ADSL protocols) | A broadband service, which may be different download and upload spee sold by phone companies) |
| | Cable Modem | Cable Modem (uses Ethernet protocol) | A broadband service, which may be different download and upload spee sold by cable companies). May be b television services. Speed advertise theoretical maximum; real speed is much slower due to line sharing. |

* Mbit/s is Megabits per second and kbit/s is kilobits per second. Keep in mind that advances in technology will likely continue, result

| Consumer or Business | Type | Device | Notes |
|---|---|---|---|
| | FiOS (Fiber Optic Service) | Optical Network Terminal (ONT) | A high speed broadband service which may be sold with different download and upload speeds. FiOS is branded and sold exclusively by Verizon. FiOS uses a fiber optic connection to the home or business, and may not yet be available in every area. May be bundled with television services. |
| Business Services for Connectivity to Internal Computers | T1 (also known as Leased Lines) | Requires PBX (Private Branch Exchange) and/or special hardware provided by phone company called a NID (Network Interface Device) | Point-to-point connection at T1 connection speeds. The term "T1" is both a speed and a service. (While 1.544 Mbit/s may seem slow, this service is used by business computers for text transmission where bandwidth needs are not as high as for video or audio.) |
| | Fractional T1 | Requires PBX and/or an NID | Blocks of 64 Kbit/s bandwidth. Same as T1, but partial bandwidth service, not a full 1.544 Mbit/s. Multiple blocks can be multiplexed, possibly on demand. |
| | T3 | Requires PBX and/or an NID | 28 T1 lines multiplexed together |
| | Frame Relay | Requires PBX and/or an NID | Point-to-point but allows virtual connection speeds that change dynamically. |
| | ISDN (Integrated Services Digital Network) | Network Terminating Unit (NTU) | ISDN was touted for many years as a faster alternative to dialup lines, but was pushed out of the market by DSL and cable modem. Still available and used for special purposes, such as dedicated video conferencing links. |

| Consumer or Business | Type | Device | Notes |
|---|---|---|---|
| | ATM* (Asynchronous Transfer Mode) | ATM Switch | Intermediate long-haul access methc specialized services. Utilized heavil decades, but declining in usage toda |
| | SONET (Synchronous Optical Network) and SDH (Synchronous Digital Hierarchy). | Add-drop multiplexers (ADM) | Flexible, fiber optic backbone long-hau offered by phone companies. Also kno Carrier (e.g., OC-1, OC-3, OC-12, and OC-192, and OC-768). SONET technol the U.S. is known in other countries as legacy network still exists in 90% of in but is fast being replaced by distributed |

* The ATM access method should not be confused with an Automatic Teller Machine (also called an ATM). They are not related in an

## Speed and Size of Connections

You will note that the speed of a connection is rated in kilo, mega or giga bits per second. First, notice the difference between bits and bytes. There are 8 bits in a byte, so a megabit is 1/8th the size of a megabyte. Typically, we use kilobytes, megabytes, and gigabytes when talking about memory and storage (discussed in more detail in the *Bits, Bytes, K, Megs, and Gigs* section). We use bits per second when talking about data speeds, because we are trying to assess how much data (i.e. how many packets) can be transmitted over the connection in a given amount of time.

Speed of a connection is related to the size of the connection like the volume of water that can flow through a pipe (indeed, technical people often refer to the "size of the pipe" when discussing data connections). The larger the pipe, the more water can flow – and the more data can be transmitted.

Note that, like a water pipe, the volume (bits per second) can be estimated as a whole assuming the entire pipe can be filled to capacity. This would be the *data rate*. However; that is not the speed that data actually transmits. Each individual transmission on the line travels **at the speed of light**, but it is the number of transmissions the line can handle in sequence that determines how fast the data travels from one device to another device. Since packet switching is a shared line technology, the number of other devices on the connection determines how much room is left for the data transmissions of any individual device. As discussed earlier in the chapter, transmission problems can cause contention; or too many devices trying to transmit, resulting in a traffic jam that slows down all the traffic as each individual device resends, over and over, the same packets that can't get through. If the contention lasts long enough, eventually the network connection will stop trying, and the connection is lost.

Technologists try to prevent this from happening by *overprovisioning*, which is when they make the pipe so large (i.e., the line so fast) that many devices can share it without contention. But even so, the speed of each device is dependent upon how many devices share the same line. This is why a cable modem, which typically has hundreds of houses on the same Ethernet segment, can be rated at 100 Megabits per second, while only 6 Megabits per second is transmitted while watching a televised movie because there are several other people who are also watching movies on the same segment.

On the other hand, a DSL switch is generally dedicated to only one household per segment, which allows a rating closer to the advertised Megabits per second.

Also note that ISPs generally allow a download speed far greater than upload speed for home (but not business) networks. This is for a number of reasons, both technical and marketing. Generally homes don't need a fast upload speed, so there was no motivation for ISPs to develop or install technologies to increase upload speeds. Because of increasing television channels and streaming, however, the technologies to increase download speeds already existed. It is typical to have a ratio anywhere from 40:1 to 100:1 download speed to upload speed.

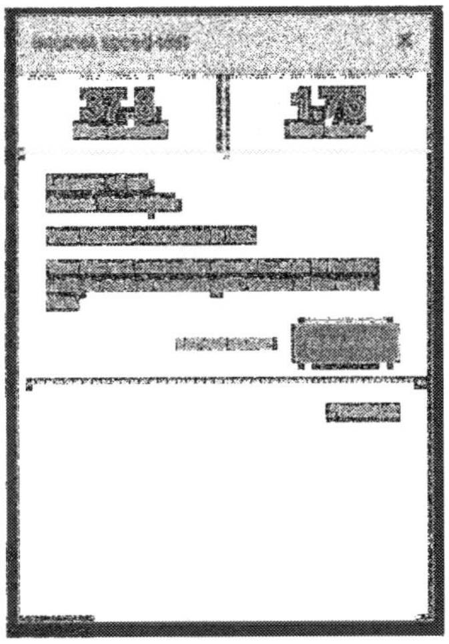

It is difficult to compare the speed ratings of different services because of the combination of different technologies, different media, different upload/download, and how many devices would be sharing the pipe. Even if only one device was on the line, there are other restrictions (e.g., the time it takes the processor to prepare the data to be transmitted and the time it takes for the packet to be evaluated and resent through each switch, hub, and router). When data is transmitted far across the world (say, from New York to Tokyo), then *latency* must be taken into account. Latency is the delay that occurs as the data travels long distances.

In any case, the only way to know what speed is actually achieved is to install the service and run a speed test. There are several available for free online; just type in network speed test and run it. Keep in mind that the speed test only tests the speed at that moment in time. Each time you run it the speed will be different (though it shouldn't vary drastically unless there is a problem) It will provide results such as the one in Figure 21.

**Figure 21. Internet Speed Test**

## 2-1. Interactive

1. **Classroom Activity**: In the computer lab, look at the back of the computer and notice the types of connections. Which wire connects the computer to the network? What kind of wire is it? What network protocol is it using? What device is your computer connected to at the other end, and where is it? If you are connected internally to a server, where is it physically located? How fast (big) is the connection to the server? How fast (big) is your physical connection to the Internet?

19. **Independent Activity:** If you have a computer at home, look at the back of the computer and notice the types of connections. How are you connected to the Internet? How fast (big) is your physical connection to the Internet? Who is your Internet Service Provider (ISP)? What type of device connects you to your ISP? What networking protocol is it using?

20. **Independent Activity:** Do you have a home network? If so, which wire connects the computer to the network, or is it wireless? If wired, what kind of wire is it? What network protocol is it using? If you are connected internally to another computer or device in your house, where is it physically located? How fast (big) is the connection to the server or device?

21. **Classroom Activity:** Take a long string (about three feet long, although the length doesn't make that much difference), and place it on your desk so that it makes a circle. Can you rearrange it so that it looked like a star? Discuss the issue of logical bus topology of Ethernet with physical star topology of computers on an Ethernet network.

22. **Discussion:** If you have a cell phone, does it use GSM, CMDA, or LTE most often? Can it also use Wi-fi? Why would you want to use Wi-fi instead of LTE? Can it also use Bluetooth? What would you want to use Bluetooth for?

23. **Discussion:** Why can't you use a BNC connector on a UTP cable? Why are UTP cables more common than coaxial cable? If fiber optic cable is so fast, why don't we all use fiber optic cable instead of copper cable?

## 2-2. Self Study Test Questions:

1. Is a network hardware, or software, or both?
2. What is a NIC?
3. What is Ethernet?
4. Describe the difference between packet switching and circuits. Which one is Ethernet based upon?
5. What do we mean when we say that an Ethernet packet is encapsulated?
6. What is contention, and what could an IT manager do about it?
7. Where does a MAC address come from, and what is it used for?
8. What does CSMA/CD stand for?
9. What are the three types of cable used most often in home networks? Which is most expensive? Which is least expensive?
10. Describe the differences between the UTP cables used for telephones and the UTP cables used for Ethernet.
11. Name at least four types of common transmission problems.
12. In Upper New York State, is February a good time to install a fiber optic line? Why or why not?
13. What is structured wiring? What is an MDF and an IDF?
14. What are the components of a Wireless network on the physical layer?
15. What is a topology? Give at least two examples.
16. What is the physical topology of Ethernet? What is the logical topology of Ethernet?
17. Why can't a cell phone use Bluetooth for calls to another state? What does it use?
18. What is Wi-Fi?
19. What is LTE, and why would you want it?
20. Which is faster; Wi-fi or LTE?
21. Which is faster; a T1 line or DSL?
22. Under normal circumstances, if a DSL line is 50 Megabits per second, and a cable modem line is 100 Megabits per second, which is one faster? Under what circumstances would the reverse be true? Why?
23. Name at least three business data services sold (usually) by phone companies?

3

# Chapter 3: Network: Internet Software

Overview: In this chapter, readers will learn the following topic:

- Internet Protocol
- IP Addressing
- How Routing Works
- Translating between binary and dotted decimal notation
- Dynamic Host Communication Protocols

**Table 7. Network Software Protocols of Technology Map.**

| MAJOR | MINOR | CATEGORIES |
|---|---|---|
| Network | Network Software Protocols | TCP/IP (Internet level)<br>Netbios (or Netbui) (Windows network level) |

The Internet is a world-wide network of connected computers. What makes the Internet the Internet is that every one of those computers is using the same Transmission Control Protocol on top of the same Internet Protocol (TCP/IP) . Just like Ethernet, TCP/IP is founded on the concept of packet switching, but it works at a higher level. The Internet runs *on top of* Ethernet. Each message sent between computers on a network is chopped up into packets and then transmitted to the receiving device, which reforms the message by combining the packets. Each device is connected to a router that handles the transmission. While Ethernet is at the physical layer, TCP/IP is at the *logical* layer, above the physical layer.

There are several characteristics of Internet technologies that combine to form what we know as the Internet. They are: mesh topology, IP addressing, routing, subnetting, Domain Name Servers (DNS), and Dynamic Host Configuration Protocol (DHCP).

## Mesh Topology

The Internet is based upon a mesh topology at the logical level. The mesh topology utilizes many-to-many links (with each "node" connected to all or some other nodes). A many-to-many links is when many different devices (in this case routers) connect to many other devices. In other words, everything links. There is no structure - it's just that every single node links to every single other node, everywhere. The basic concept is illustrated in Figure 22). Of course, there are more than five computers connected to the Internet, so rather than the simple clear topology shown in Figure 22, it actually looks more similar to the topology shown in Figure 23. This topology was developed so that even if a whole section of the internetwork were damaged, the internetwork would still be connected.

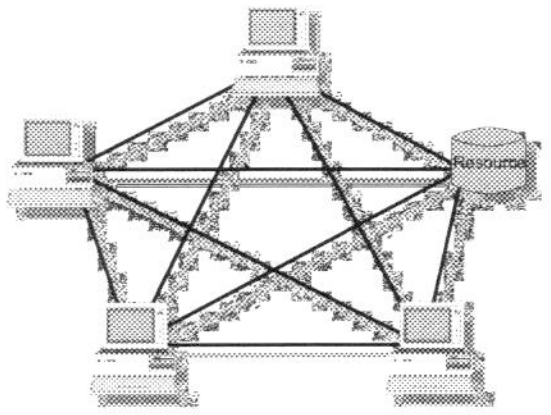

**Figure 22. Simple mesh topology.**

***The Story of the Internet*** [*]

During the Cold War, the United States military became concerned with ensuring success in the event of a nuclear war. At that time, it was thought that the main deterrent against a Soviet nuclear attack was the threat of total nuclear retaliation by the Americans. During the late 1950s and early 1960s, in the midst of the Korean and Vietnam wars, a think tank project (paid for by the US Department of Defense and run by the RAND Corporation) led the way in strategic thinking about nuclear attacks. Paul Baran (who joined RAND Corporation in 1959) was particularly focused on finding a way to enhance the reliability of the U.S. military commanders communication links, which were, in his mind, dangerously weak and fragile. At the time, the only type of data communication was point-to-point, which meant that an actual wire connected from one computer to another computer. If the single connection was broken (as it would be, undoubtedly, if a bomb were dropped on it), then no communication could take place. The military was afraid that at a critical moment for nuclear retaliatory forces, point-to-point communication links would fail. So these military commanders were very interested in the work that Paul Baran was doing.

For years the think tank at RAND had been trying to invent a robust communications system - but with no success until Paul Baran joined the team. He came up with the three concepts that would give a communications system the robustness necessary to stay connected even in the face of a cataclysmic event:

1. a distributed network instead of a centralized one.
2. massive amounts of redundancy built into the network.
3. breaking the message up into separate "packets," each of which would contain some of the file.

The thinking was that a distributed network could keep working, even after a lot of damage. If a message sender did not immediately receive an acknowledgement of the message sent, the sender would simply send it again and again until acknowledged or timed out. Additionally, a packet would also contain both the source and final address, and a sequence number to reassemble the packets in the right order once they all reached their destination.

It was these concepts that were used by Bob Kahn (at Bolt Beranek and Newman) and Vinton Cerf (at Stanford University) when they were hired by the Department of Defense to develop a communication method to share research. They operationalized Paul Baran's ideas by developing Transmission Control Protocol/Internet Protocol (TCP/IP ) that is now the basis for all Internet connections. (It was also the concept Bob Metcalf and Dave Boggs used when they developed the original Ethernet protocol.)

In the early 1990s, the United States government handed over management of TCP/IP to a private company and opened it up to commercial ventures, mainly due to the influence of Al Gore. (Many people fault Al Gore, thinking that he boasted imaginary credit for the public Internet, but the fact is that Al Gore helped create the public Internet by opening up the management so that it could be used by everyone, not just the government.)[†]

---

[*] ***Internet Genesis*** (2003) by Karl Kruszelnicki at http://www.abc.net.au/science/k2/moments/s42478.htm, published on August 9, 1999 and accessed on Sept 11, 2020. **The 'myth' of the invention of the Internet** by Nick Schulz written on July 23, 2012 https://www.aei.org/technology-and-innovation/the-myth-of-the-invention-of-the-internet/ accessed on September 11, 2020.

[†] ***Did Al Gore Invent the Internet?*** by Matt Novak published on October 18, 2013. https://paleofuture.gizmodo.com/did-al-gore-invent-the-internet-1447761524 Accessed Sept 11, 2020. Includes info on Cerf, V., and B. Kahnand Al Gore's support of the Internet.

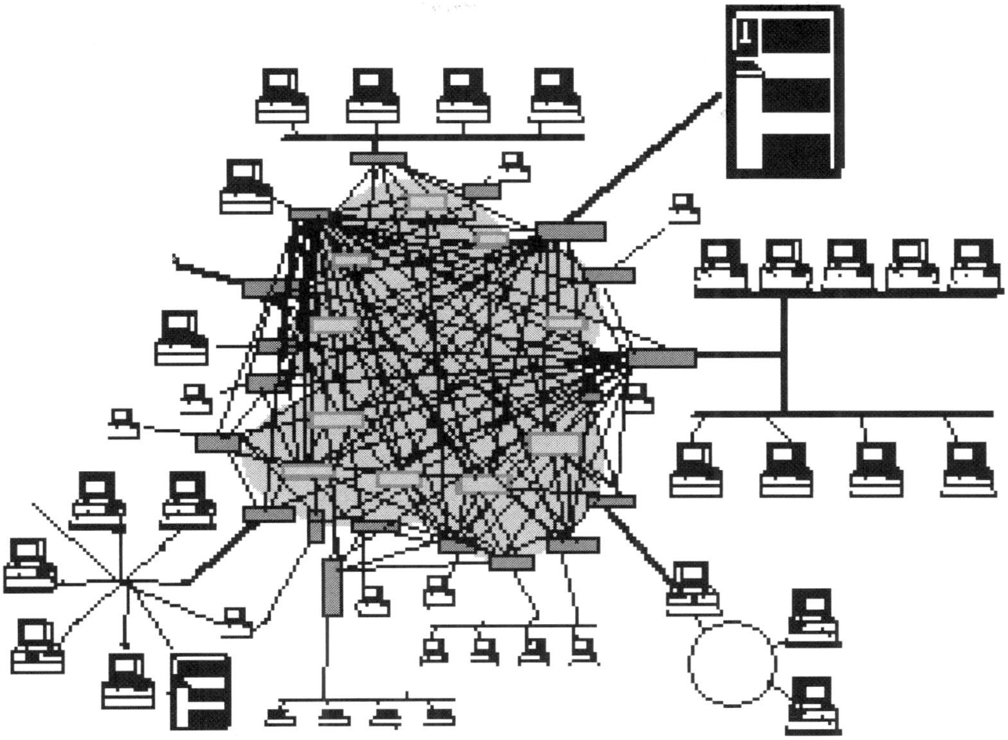

**Figure 23. The Internet mesh.**

## IP Address

In an internet each device is identified through a unique number called an IP address (Internet Protocol address). In the Internet, all of the numbers are centrally controlled and managed. Until 2012, all of the addresses on the Internet were 32-bit numbers (32 digits in binary notation) known as IPv4 (the fourth version of the IP standard). In 2011, however, the IT world passed a critical point; all of the IPv4 addresses, (all 4.3 billion of them) had been assigned. The next version of the protocol (IPv6) has been around since 1994, but changing over to it means purchasing new routers and changing IP assignments, which is a lot of work and is very expensive. However, when we ran out of IPv4 addresses, IT vendors had no choice but to switch. Many large ISPs now use IPv6 exclusively, though it is more common to use both an IPv4 address and an IPv6 address.

### *IPv4 and IPv6*

An IPv4 address is a 32 bit number. 192.46.35.67 is an example IPv4 address (using decimal notation, discussed in further detail in the next paragraph). 192.46.35.67.53.23.45.87.102.208.234.23.45.86.133.24 is an example IPv6 address. IPv6 is a 128-bit number (128 digits in binary notation). Note, however, that binary is not used for IPv6 typically. IPv6 is expressed in hexadecimal notation, which is based on 16 rather than 10 (using both letters and numbers)*.

IP is represented in a special format known as *dotted decimal notation.* Dotted decimal notation breaks the 32 (or the 128) digits into four (or 16) groups of eight. Each group is referred to as an *octet.* To translate from each group of eight-bit binary digits to the decimal number, each bit is given a value that corresponds to its location in the string. For example, if the binary code was

* Don't worry. We won't be doing hexadecimal translations in this book. We will, however, go over decimal translations in detail which will give the reader an idea of how hexadecimal might be done.

11111111, then the first value would be 128, the second value would be 64, etc. You can see two binary numbers translated in Figure 24.

| | | | | | | | |
|---|---|---|---|---|---|---|---|
| **1** | **1** | **1** | **1** | **1** | **1** | **1** | **1** |
| **128** | **64** | **32** | **16** | **8** | **4** | **2** | **1 = 255** |

| | | | | | | | |
|---|---|---|---|---|---|---|---|
| **0** | **1** | **0** | **1** | **0** | **1** | **0** | **1** |
| **0** | **64** | **0** | **16** | **0** | **4** | **0** | **1 = 85** |

**Figure 24. Translating IP address octet from binary to decimal.**

You add the values together (128+64+32+16+8+4+2+1) and you get 255, the largest possible value for an octet within an IP address. In Figure 25, the binary number 10000111 translates to 135, the number 00001111 translates to 15, the number 00000010 translates to 2, and 00011110 translates to 30. Each group of binary digits (bits) is then converted into decimal notation independently of the other places, and separated with dots (decimal points), as shown in Figure 25.

| | |
|---|---|
| **32-bit Binary -** | **10000111000011110000001000011110** = 2265907952 (decimal) |
| **4 groups of 8-bit Binary -** | **10000111 00001111 00000010 00011110** |
| **4 Decimal numbers -** | **135 15 2 30** |
| **4 group Decimal (padded) -** | **135 015 002 030** |
| **Dotted Decimal Notation (padded) -** | **135.015.002.030** |
| **Dotted Decimal Notation (no padding) -** | **135.15.2.30** |

**Figure 25. Dotted Decimal Notation**

## *More Than an ID Number*

The IP Address is unique - if the computer is on the Internet, the IP address of the computer will be different than any other computer in the world*. Because every computer on the Internet has a different IP number, one might think that it is an identification number. It is, but it is not **just** an identification number. The number itself holds the location of the specific device which is identified. The IP address actually provides the routing information for a packet. That's why people can be found if their IP address is known. (Hence why criminals use IP address "spoofing" to pretend to be an IP address they are not.)

A computer needs to know three items in order to connect to the Internet 1) the IP address, 2) the Default Gateway, and 3) the Subnet mask. They work together to get messages from one computer to another computer. Getting messages through is called routing.

* If two different computers on the Internet are accidentally given the same IP address, the network traffic will randomly go to one or the other, meaning that neither one will transmit or receive properly. Back in the day when IP addresses were manually assigned to computers this was a common problem because people didn't know what they were doing and didn't realize they couldn't use the same number twice. Now IP addresses are universally assigned by a DHCP (Dynamic Host Configuration Protocol) server, so it is less of a problem.

# Routing

The Default Gateway is another name for the *router*. A router is the device that connects a computer to the Internet. A router is like a door to the Internet. Every computer on the Internet must have a connection to a router. Many computers on the same network can share a single router, but the router is the device connecting to the Internet. There are many ways to see the IP address and router address, but simply typing IPCONFIG from a command prompt is usually the easiest way. (See the inset *Looking at your own IP address*)

A router has a specific task; when a router receives a packet, it reads the destination IP address and sends the packet in the direction of the destination (i.e. routes the packet). A router does not necessarily send packets on the best or quickest path. It will just pick a path from among the list of other routers that will get the packet closer. Any packet may go through dozens of routers before it finally reaches its final destination. Each time a packet gets passed along by a router on the Internet, it is called a *hop*. The packet keeps hopping to the next router until it finally reaches a router that is directly attached to the computer with the destination IP address.

***Looking at your own IP address***

If you get into the command line on your computer (on a Windows machine you can right-mouse on the **Windows Start** and choose the **Run** and type **CMD**.) you can type in **IPCONFIG**. The system will display three items: the IP address, the Subnet mask, and the Default Gateway.

Keep in mind that your IP address may change each time that you reboot your computer. Your system may be set up to be assigned an IP address as soon as it attempts to connect to the Internet.

Also keep in mind that each device on your home network may have an *internal* IP address (typically starting with 192 which is reserved for private networks) and an *external* IP address. To get the external IP address you go to any browser and type in **"My IP"** and your IP address will be displayed. The external IP address will also change, but is under the control of your ISP.

When we look at the list of routers that a packet has gone through, it is called a *trace route*. Figure 26 shows the trace route showing the hops from a computer in Ashburn, Virginia to ebay.com. Notice that the packet travelled from VA to DC to AZ back to DC and then to Denver, Colorado (where eBay's servers must be connected to the Internet). Notice also that the packet went across five different networks on its way. Ten hops is pretty good; more than 30 would have been typical of a longer distance or a congested Internet. You might also notice that the packet got "hung up" at the sixth hop (which took 41 milliseconds) and ninth hop (which took 53 milliseconds). Either the routers were too busy to deal with the packet right away, or they were not working properly, or they were extremely security conscious and not willing to respond to a tracert command. Nonetheless, the delay was still so fast (53 milliseconds is 53/1000 of a second) that we would barely notice it.*

Each time a packet is sent (and remember that millions of them might be needed to send a full message), the sending application waits for an *acknowledgement* packet. The destination device sends an acknowledgement packet back to the sending device after it receives it to let the sending device know that the message packet was received safe and sound. If either the packet going out or the acknowledgement packet coming back have a problem (such as interference or contention on the network), the message does not go through and must be resent.

* The utility that will provide the tracert can be downloaded from http://www.visualroute.com/lite.html, although you can also type in *tracert domain.nam* from any command line or DOS prompt. (You cannot tracert to ebay.com anymore. Since this trace route exercise, Ebay has changed its configuration as a security measure and that prevents the trace route from being shown.)

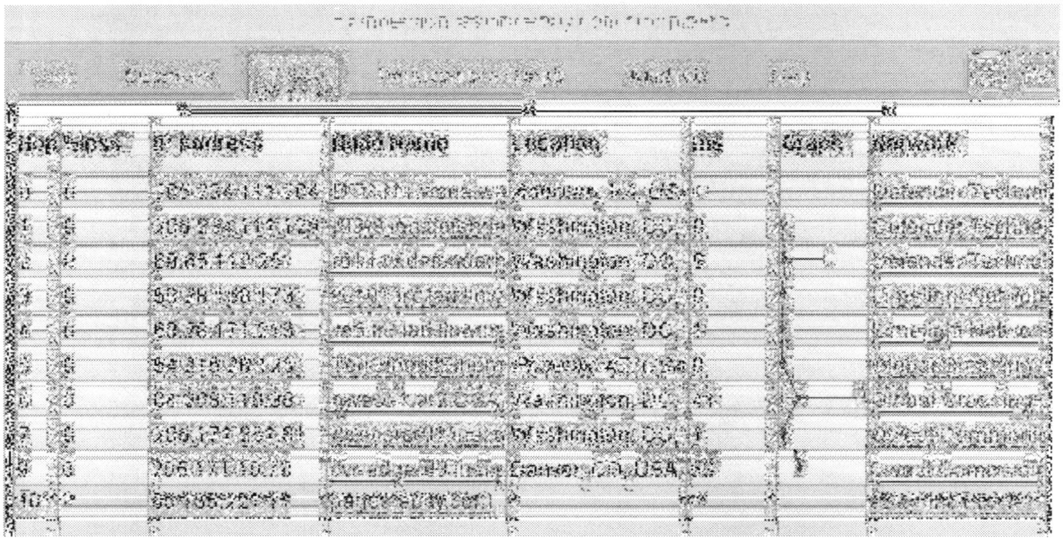

**Figure 26. Trace route showing hops to Ebay.com.**

There are other reasons why the packet doesn't get where it is going. Each packet on the Internet has a *TTL* (Time To Live) *counter*. When the TTL counter runs out, the packet stops looking for its destination (i.e. the routers stop forwarding the packet). The sending application never receives the expected acknowledgement packet, so it simply sends the data out again.

How does a router know where to send the packet so that it gets closer to its destination? The key to routing is the IP address itself. A router knows which devices are on its own network. In other words, it has a list of them. That list is called a *routing table*. The routing table for the router 156.33.22.1, for example (near the bottom of the figure) would be two devices: 156.33.22.2 and 156.33.22.50.

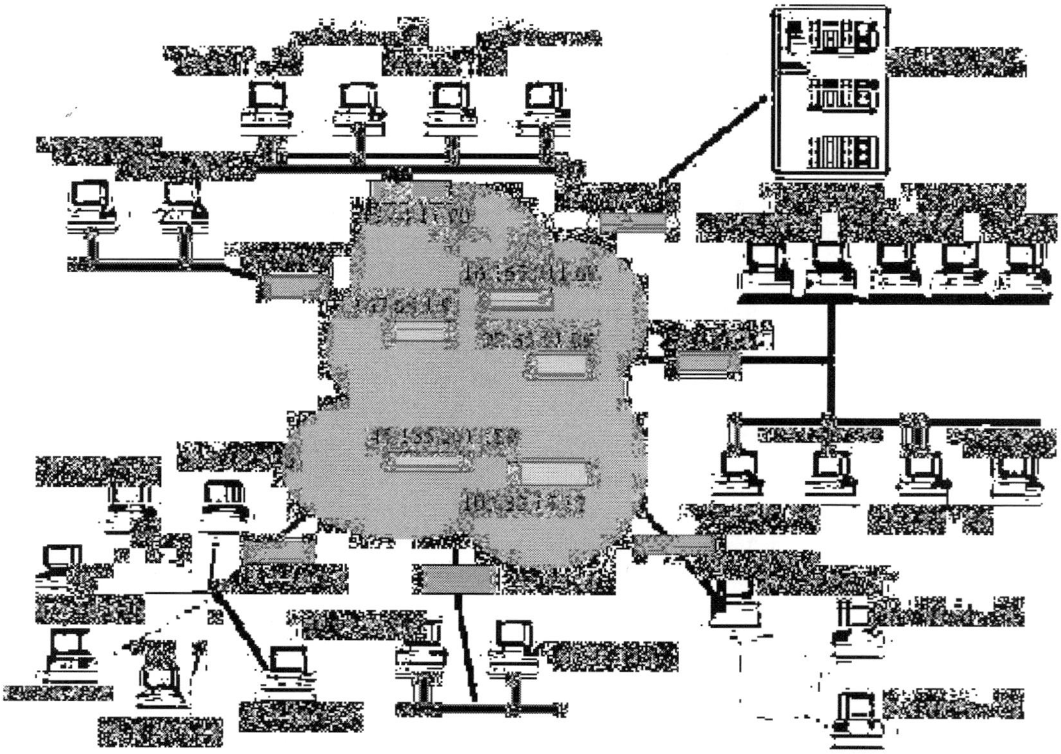

**Figure 27. IP addresses on the Internet.**

Notice in Figure 26 that when the packet went from Phoenix to Washington, DC, it was still in the same network (which you can tell from the network number 64 that starts the IP address and the network name "Global Crossing"). Notice in Figure 27 that the IP addresses of the routers (the small gray boxes) begin the same as all the devices connected to the router. In order for the router to figure out exactly which part of the IP address determines the network, and which part determines the actual device **on** the network, the router needs to know the subnet mask.

## Sub-net Mask

The subnet mask was the middle number in the IPCONFIG information and needs to be identified when the computer first gets on the Internet. Some part of the IP address identifies a specific device on a specific network. Some part of the IP address identifies which network that device is on. The subnet mask tells the router which part is which. In the case of the previous example, router 156.33.22.1, the subnet mask must be 255.255.255.0. That means that the first three octets (156.33.22) is the network portion of the IP address, and the last octet (in this case, .2 and .50) is the device portion of the IP address.

There are several levels within a network. The first level is the Internet Service Provider (ISP). The ISP was assigned a block of numbers from Internet Assigned Numbers Authority. The network administrators at the ISP decide for themselves what subnet mask to use based upon whether they need more networks or more devices. But they also might assign a block of their numbers to a corporation. Then it would be up to local administration of the business to assign specific addresses within the block again, based upon whether they need more networks or more devices. The way that the subnet addresses are assigned can be seen in Table 8.

**Table 8. Assigning Subnet Masks**

| Type | Dotted binary notation | Dotted Decimal Notation |
|---|---|---|
| Original address assigned | 10000000.11111001.00000000.00000000 | 128.249.000.000 |
| Network mask | 11111111.11111111.00000000.00000000 | 255.255.0.0 |
| Subnet mask | 11111111.11111111.11100000.00000000 | 255.255.224.0 |
| Subnet 1 addresses | 11111111.11111001.00100000.00000000 to 11111111.11111001.00111111.11111111 | 128.249.000.001 to 128.249.063.255 |
| Subnet 2 addresses | 11111111.11111001.01000000.00000000 to 11111111.11111001.01011111.11111111 | 128.249.064.000 to 128.249.95.255 |
| Subnet 3 addresses | 11111111.11111001.10000000.00000000 to 11111111.11111001.10011111.11111111 | 128.249.128.000 to 128.249.159.255 |

Basically, network administrators can set up the number of networks and the number of devices any way they want. A subnet mask of 255.0.0.0 will give them 125 networks and 16,777,214 devices. A subnet mask of 255.255.0.0 will give them 16,382 networks and 65,534 devices. A subnet mask of 255.255.255.0 (the most common in corporations) gives them 254 devices but allows for 2,097,152 networks. From the perspective of the user, it does not matter which subnet mask the network administrator has set up if all the machines on the same subnet are using the same subnet mask. If you change any of the numbers shown in the IPConfig, you will no longer be able to connect to the Internet (or to any other computer on the subnet).

To recap: every computer on the Internet has a unique number, and the subnet mask tells each computer and each router where it exists in the hierarchy of networks and devices. When a computer wishes to send information to another computer, it breaks up the information into smaller pieces and encapsulates them into a packet. It includes within the packet both the source IP address and the destination IP address. Then it sends the packet to the default gateway (i.e., the nearest router). The router reads the destination IP address and sends the packet towards the network that contains that IP address*.

But wait a minute: How does the computer know what IP address to put into the packet in the first place? We, as human beings, don't go around memorizing the IP addresses of every web site we wish to visit. Nor do we have at the tip of our tongues the IP addresses of the people to whom we wish to send emails. Somehow the computer must have a way of figuring out how to translate JoeSchmoe@kutztown.edu into 156.12.1.115 and https://amazon.com into 176.32.103.205.

## Domain Name Servers (DNS)

A computer can look up the IP address of any domain name by sending a message to the nearest Domain Name Server (DNS) and asking where it can find the domain. The Domain Name Server responds with the IP address. Any web address you type into the browser, any hyperlink you click on, any email message you try to send - they all require that the words be translated into IP addresses.

***Finding out your DNS***

> If you go back to the command prompt on your computer and type in IPCONFIG /ALL (the same command as before but this time asking for all of the information instead of just some of the information), you will be able to see your DNS, since it will be listed. Like your IP address, your DNS may change depending upon your ISP.

There are thousands of DNSs on the Internet. The DNSs each have a list of every IP address and its corresponding domain name. Each DNS also gets updated from the other DNSs frequently - every few minutes or so. That way, even as domains are being registered or expired or changed by their hosting companies, all the DNSs in the world will know about the change within a few hours.

Most of the time your computer will list two or three different DNSs in case the first one is busy or out of commission. If your DNS address is not configured correctly, then you will be able to connect to the Internet, but you won't be able to find any web pages or click on any hyperlinks.

You may be wondering how your computer gets the IP address and the DNS address. In the old days, you would have been directed by your ISP or corporation to type them into the TCP/IP options on your PC. But most of the time, we get the IP address, the subnet mask, and the DNS from our ISP through a process called DHCP (Dynamic Host Configuration Protocol).

## Dynamic Host Configuration Protocol (DHCP)

DHCP is the process that provides your computer with needed information to connect to the Internet. When your computer is turned on, it usually doesn't have an IP address. (If it did, that would be known as a static IP address. Your ISP or IT technician would have directed you to type in the IP address, the subnet mask, and the DNS address.) Most computers are assigned this information when they initially try to connect to the Internet.

This is how DHCP works. The TCP/IP utility included in your device's operating system sends out a message (to any other device that is listening, since at this point the device doesn't know where the router is). The message is the electronic equivalent of "I am here." Somewhere on the network (usually the router, but not always) there will be a DHCP program listening for that message. When it hears it, DHCP responds with the electronic equivalent of "Hi! How ya doin'?

* For a mind-bender: this encapsulation, because it is happening on the logical level, is *inside* the encapsulation that is happening at the physical (Ethernet level on a LAN or a X.25 on a WAN) layer. In case you were wondering, when a packet travels between a LAN and a WAN the Ethernet frame around the packet is discarded and replaced with the X.25 frame (or vice versa) because they are both on the physical layer. The contents, however (i.e. the IP encapsulated data) is kept intact in both.

Here's your IP address, your subnet mask, your default gateway. Oh - and by the way, here is your DNS, too. Your welcome."

The TCP/IP utility of your OS remembers the information and uses it the next time it wants to communicate with any other computer on the Internet. Once assigned, this information stays the same until the computer is turned off (or it expires, at which time it automatically renews, but may end up being a IP Address). If your computer is using DHCP, the properties of your TCP/IP settings would look like the ones in Figure 28. If your computer has a static IP address, then the options would be filled in rather than being set to **Obtain IP address automatically** and **Obtain DNS server address automatically**.

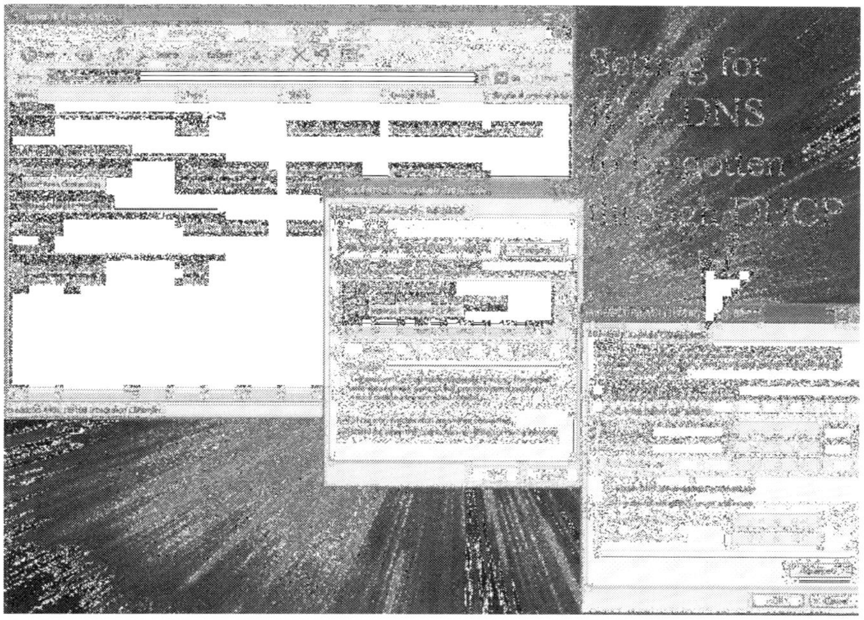

**Figure 28. DHCP configuration.**

Now that we have fully covered the Internetworking concepts, let's move up one more layer, to the networking software.

# 3-1. Interactive

1. **Class Activity**: On a Windows 10 machine, click on ⊞ and type *Run*. In the command line, type in **cmd** and press [Enter]. Using the command line interface, enter the following commands. (IPConfig shows the TCP/IP configuration of your computer. Ping lets you know if you can reach a domain. Tracert traces the routers encountered by a packet on its way to the domain.) After each one, discuss what the command did and what it means.

```
IPConfig
IPConfig /All
ping whitehouse.gov
tracert whitehouse.gov
ping hplconsortium.com
tracert hplconsortium.com
```

2. **Class Discussion:** What would happen if two different computers on the Internet have the same IP address? What happened when we ran out of IPv4 addresses?
3. **Class Activity:** Convert the eight digit binary number 10101010 into its decimal equivalent.
4. What IP address in dotted decimal notation does this number represent? 11001101001100100101010100000001
5. Can you convert the following number back into its binary equivalent? 254.35.67.102

# 3-2. Self Study Test Questions:

1. What is an IP address?
2. What type of logical topology does the Internet use?
3. Describe the process to translate between an 8 digit binary number to a decimal number.
4. Describe the process to translate between a decimal number to an 8 digit binary number.
5. What three items of information does a computer need to connect to the Internet?
6. What additional item of information does a computer need to find a web page or the host computer for an email address?
7. Who assigns IP addresses?
8. What critical Internet event happened in 2011 regarding IP addresses? What is the impact of the event?
9. How can you find out what IP address has been assigned to a computer?
10. What would happen if you changed the IP address (or the Default Gateway or the subnet mask)?
11. Describe the process of routing a packet on the Internet.
12. What is TTL?
13. What is a trace route? What is a hop?
14. What is DHCP? Why does a computer need it?
15. What is a static IP address?

4

# Chapter 4: Computer Hardware

Overview: In this chapter, readers will learn the following topics:

- ➪ Why You Need to Know Hardware
- ➪ The Basic Hardware Components
- ➪ Types of Computers
- ➪ Power: Processor, Memory and Storage
- ➪ Hardware Peripherals

Just above the networking hardware layer is computer hardware. This chapter will start by explaining why it is important to understand the different types of computer hardware and describe the different components within that hardware. Then we will discuss hardware types and the components. The technology map places computer hardware into three categories Types of hardware, main components, and peripherals, as shown in Table 9.

**Table 9. Technology Map Hardware Categories**

| MAJOR | MINOR | CATEGORIES |
|---|---|---|
| Hardware | Types | Smartphone, Phablet, Tablet, Personal Computer, Server, Mainframe, Supercomputer |
| | Main Components | CPU, HD, SSD, DASD, RAM, ROM, BIOS, Motherboard, Ports, Heat Sink, Fan, Power Supply, Video Interface |
| | Peripherals | Mouse, Keyboard, Monitor, Scanner, Camera, Printer, Flash Drive, DVD, Smartphone, Touchpad, Game Controller, Video Camera, Speaker, etc. |

## Why the User Needs To Know Hardware

Imagine that you towed your car to a mechanic and explained that you couldn't turn the steering wheel. The mechanic spends three hours working on your steering and then says it can't be fixed; the whole car must be replaced. When you come to get your car and you notice that your tires are flat. Who would be the "dumb" one - you for telling the mechanic that something was wrong with your steering wheel, or the mechanic for not noticing that the tires were all flat?

In the world of automobiles, this would never happen. Even though you are just a driver, not a mechanic, you know how to recognize a flat tire. You understand what part the tires play in allowing you to turn the steering wheel so that you can drive around, and you wouldn't confuse the cause (the flat tire) with the effect (steering wheel won't turn).

Furthermore, it is pretty easy to tell when your tire is flat, so any mechanic would see it. But what would happen if different mechanics only looked at different parts of the car? If you took the car to a mechanic who would only look at steering wheels and would have no responsibility whatsoever for the tires, how would that mechanic know that the flat tire caused the steering wheel problem? If that were the case, you yourself would have to know the parts. You would have to know how to figure out which type of mechanic to go to - the one who fixes steering wheels or the one who replaces tires.

In the world of IT, there are multiple possibilities for hardware and software combinations for each type of computer system. IT technicians are specialists, and none has enough expertise to diagnose problems for all of the different layers of technology. The complexity of IT systems makes it especially difficult to determine the cause of a malfunction, which could be the result of a problem in any one of its many connected parts. Therefore, an end user must be able to recognize the different components of a system and what type of technical support is available for each. Basic knowledge of how the components work together is necessary in order to communicate with the technology professionals troubleshooting the system.

The technology map introduced in the first chapter is very helpful here. The first question to figure out is whether or not the problem is with the network, the hardware, or the software. It would do no good to take your computer to a computer repair shop if the problem is that your network connection doesn't work. Complaining to the publisher of a new program you purchased to publish DVDs will do no good if the problem is that the DVD drive on your computer has malfunctioned. It is the user's responsibility to identify in which general area the problem has occurred, in order to identify which repair specialist may be able to fix it.

## Types of Computers

There are several types of computers.

- Mainframes and supercomputers.
- Servers (also called minicomputers or midrange computers).
- Personal computers (or microcomputers including laptops, game consoles, and desktops). Also workstations (a more powerful device than a personal computer, generally used for system development or intensive database or graphic design work).
- Mobile computing (smartphones, tablets, and phablets).

### *Mainframes*

Mainframe computers were the first computers and are typically much larger than any other type of computer. Even though the first mainframe computers were the size of a large room, they weren't as powerful as today's personal computers (or even as powerful as today's smartphones!), but as micros and servers have evolved, so have mainframes. Now mainframe computers are smaller, but more powerful, than ever before.

Mainframes have a few brands, models, and types. IBM started the field in the 1960s and is still one of the top mainframe manufacturers. They started with the 700 series, increased their dominance with the 360 series, and retain the top spot with the zSeries, which was developed in partnership with Hitachi, a Japanese manufacturer.

Other than IBM, Unisys is the only other manufacturer left in the U.S., and just a few years ago it launched a new brand of mainframe called ClearPath Libra. Hitachi, Fujistu, Oki and NEC are all in Japan, while Siemens and Telefunken hail from Germany. While most mainframe computers are designed to work with PCs connected to them, they also still work with terminals. Mainframe performance is often referred to as MIPS (millions of instructions per second), and the different systems are compared to each other based upon that measurement. Additionally, software for mainframes is often priced based on the MIPS of the machine upon which it has been installed.

***The Mainframe Industry Story***

In the early 1970s, the mainframe market was very competitive, although IBM always had the lion's share. At the time, there were often references to *IBM and the Seven Dwarfs*, referring to Burroughs, UNIVAC, NCR, Control Data, Honeywell, General Electric and RCA. Later, as the market began to shrink, the phrase changed to *IBM and the BUNCH*, referring to Burroughs, UNIVAC, NCR, Control Data, and Honeywell. Today, the only survivor as a mainframe manufacturer is IBM. The remnants of General Electric, RCA, UNIVAC and Burroughs can be found in Unisys. Honeywell was bought out by Groupe Bull, and NCR was bought out by AT&T before being spun off again in 1996. At that point, NCR refocused on its core business, point-of-sale devices (cash registers, now combined with bar code scanners).

## Supercomputer

There is a sub-class of mainframe computers called the supercomputer. The most well-known supercomputer manufacturer is Cray (though its founder, Seymour Cray, never used the term supercomputer). A supercomputer differs from a mainframe computer in that it is generally a one-of-a-kind system devoted to a single program, such as forecasting the weather, or doing intense calculations for research or graphics animation. Additionally, supercomputers performance is measured in FLOPS (Floating Point Operations Per Second)*.

## Servers

A term that has fallen by the wayside is minicomputer. In the early days of computers, the minicomputer category referred to those systems not large enough to be mainframes, but too big to be considered personal computers. With the exception of IBM, all the vendors of minicomputers have been acquired or closed: Digital Equipment Corporation (DEC), Data General, Prime Computer and Wang. IBM now uses the phrase "midrange systems" to refer to their iSeries (formerly known as the AS/400, the epitome of a minicomputer). HP and Sun both refer to their systems as servers, and so what was previously called a minicomputer is now called a server. (See inset for a discussion of the term server and the multiple ways the term is used).

In business, the majority of the computers that the information technology department deals with directly are usually servers.

***Server as hardware or software***

The term *server* can be very confusing because it is used in two completely different contexts. When talking about hardware, server means a more powerful computer than a PC, one designed to have many different PCs connected to it. It is a physical box designed to serve something (usually files or applications) connected to a network. A server and can also be called a midrange system or minicomputer.

However, when talking about software, the term server is also used. But in this case, server is referring to an application running on the system. These server applications are not limited to physical servers, nor are they limited to one per system. A physical server might be running an application server, a web server, a fax server, a print server, and a mail server all at the same time. But so can a mainframe computer, or even a desktop computer. If I have three computers in my home, but only one of them is connected to a printer, when I print to that printer *through* the connected PC, it is acting as a print server.

It is generally this software server that we refer to when we discuss *client server* architecture. Client server is concept that a piece of software (client) running on one machine (usually a desktop) works in tandem with a piece of software (called a server) running on another machine (also called a server). For example, when you look at a web page, your client software (Chrome or Firefox) interacts with the web server (Apache or IIS).

* More details on the difference between supercomputers and mainframes can be found on https://www.quora.com/What-is-the-difference-between-mainframe-computers-and-supercomputers which was a FAQ answered by Sushil Kumar on July 14, 2016. Accessed September 9, 2020.

## Personal Computer

The term microcomputer is still in use, but not used as often as a variety of other terms that may refer to the same thing; personal computer (PC), desktop, tower, laptop, notebook, or just computer. The term PC usually refers to all personal computers, including those manufactured by Apple computer which uses the Macintosh operating system. But sometimes PC is used to differentiate computers with the Windows operating system (which are manufactured by many different companies) rather than the Apple systems with the Macintosh operating system. In other words, sometimes PC includes Macs, and sometimes it only means Windows-based machines.

The term *desktop* is used to refer to a personal computer because often the computer sits on top of a desk. Sometimes this term means all personal computers, and other times it is used to differentiate the personal computer from computers that are in data centers, such as servers. It can also be used to distinguish desktop computers from those that are mobile, as in laptop or notebook computers. (Although notebook computers were originally smaller than laptop computers, the difference has largely disappeared as laptop computers have become extremely small. Whether the newer smaller tablets (like a Chromebook or a Surface) are considered laptop computers or notebooks or tablets is still a question. They don't have the full capability of installing and running software that a laptop computer has, but they have more than a smartphone or iPad tablet.*)

The term *tower* came into use because sometimes computers sit wide and flat (as in desktop), and sometimes they sit narrow and tall (as in tower). However, some people use the term tower to refer to the CPU (central processing unit), the main box of the personal computer, as opposed to the peripherals.

In recent years, the personal computer market has been dominated by Apple, Dell, Lenovo (an IBM spinoff), Acer, Toshiba, and Asus. Also, recent innovations in technology has meant that laptop computers can now be as powerful as desktop computers (whereas previously, heat and power requirements for the most powerful CPUs required the larger box of a desktop computer).

## Smartphones, Tablets, and Phablets

Smartphones started out as devices called PDAs (Personal Digital Assistants). The first commercially produced PDA was the Apple Newton in 1992. The Apple Newton was not a commercial success, but introduced important innovations such as the touch screen, stroke recognition, synchronizing with a calendar and desktop applications such as word processing and spreadsheets – all available on a device that could fit in your pocket.

Smartphones sales have eclipsed computer sales. There are now competitors to the iPhone that run the Android operating system (from Google), produced by Samsung, T-Mobile, HTC, Motorola and Nexis.

Apple also introduced the iPad in 2011, and created a whole new category of devices known as *tablets*. A tablet is not a phone, but neither is it a full-fledged computer. It can allow the user to independently surf the web, send and receive emails, post on social media sites, read books, take and edit photos and videos, videoconference with other users, communicate with phones and computers, and perform many other useful

***Smart Phones and their Ease of Use***

It is obvious that telephones have become smarter because they are no longer easy to use. Even technical people have difficulty. Bjarne Stroustrup, designer of the C++ computer programming language, famously said; "*I have always wished for my computer to be as easy to use as my telephone; my wish has come true because I can no longer figure out how to use my telephone.*"

---

* Ackerman, D. *Should the iPad be Considered a Computer?* https://www.cnet.com/news/should-the-apple-ipad-be-considered-a-computer/, posted January 27, 2010, accessed October 28, 2020.)

tasks. Newer tablets, include Amazon Kindle Fire, Acer Aspire, Barnes and Noble Nook, and Samsung Galaxy Tab. Microsoft, previously a software-only company, started manufacturing its own tablet, the Surface (although some classify the Surface as a laptop).

There is even an in-between sized personal device, larger than a phone but smaller than a tablet, known as a *phablet*. Popular brands include Samsung's Galaxy Note, Apple's iPhone 6plus, and Nokia's Lumia. In 2014, phablets outsold both laptop computers and desktop computers, and is almost half of all phones sold. The screen is large enough to allow the user to surf the web, do word processing, work on spreadsheets, and edit photos and videos, yet it is still small enough to fit in a large pocket. At the same time, a phablet has phone capabilities, a feature lacking in most tablets.

## System Configuration Types

Mainframe computers of the seventies and eighties were less powerful than smartphones of today - but at the time they cost millions of dollars and were considered very powerful. They could, more quickly and more easily, do the work that had been previously done by hundreds of people. Because of their importance to the business they were *centralized.*

The main system itself was stored in a secured, temperature-controlled room, often with a raised floor in order to easily access the wiring between the different devices. This room was known as a *data center*. Only technical personnel were allowed access to the computer itself, though usually others could see into the room through large glass windows (leading to the nickname *glass house).* Programmers and specially trained data entry operators used devices called *terminals* in order to enter or review the data. Unlike computers of today, terminals did not have any processing capability or storage. They were later referred to as *dumb terminals* to differentiate them from personal computers used as terminals, which had become the norm. Dumb terminals were like remote keyboards attached to a single computer. The terminals could be anywhere - in the same building or in a different city - but all the equipment and processing was centrally controlled from the central computer in the data center.

As computers got smaller, however, they became more *distributed*, which means that instead of all the computers being found in a central location in one room, the organization might have computers distributed in various locations throughout different facilities (but usually on the same *network*). At the same time that the server hardware was becoming more distributed, the processing function within the software itself became more distributed. Instead of just a central processing unit (the main computer) and dumb terminals, part of the program would run on the main computer (the *server*) and part of the program would run on the smaller personal computer (the *client)* accessing the larger computer. This configuration is known as a *client/server system,* and was the most common configuration for processing during the eighties and nineties.

When the main computer does the majority of processing, the system is said to have a *thin client.* If the majority of processing is done by the personal computer attached, the system has a *fat client.*

Then, around the turn of the century, the capabilities of the Internet were made available to commercial enterprises that changed the scope of the entire system. Instead of a few hundred computers on a wholly-owned-and-controlled network, computerized function needed to be available to thousands and millions of computers anywhere in the world across an uncontrolled network of networks (the Internet). The system still had a server, and still had a client, but because this configuration became the standard, we don't generally call them client/server systems anymore. Very few systems still use terminals, so there is no other kind of configuration; the only question is whether the client is thin or fat.

As the number of clients increased, the requirements for centralized processing power of the main system increased. A single server (even a large mainframe) could no longer respond quickly enough, so the centralized computer would be attached to another centralized computer and they would share the processing requests. Instead of one large machine, smaller connected computers called *server farms* began to proliferate (as shown in Figure 29. A server farm is often housed in a data center. Each computer in the series connected together serves various parts of a transaction process called *tiers*. A tier is just one component of a transaction process. For example, if you purchase a book online, you would first deal with the web server to process the http: request. Then you would deal with the database server to get a look at the inventory for the item. Once you decide to buy, you would deal with a secure authentication server for the credit card transaction, and then with the application server to record the order and payment. Together, they form a complete transaction, though you had to go through four different tiers to do it.

**Figure 29. Server farm.**

## The Basic Hardware Components

All computers (mainframes, servers, and personal computers) contain the same basic components although they are often called different things. Storage in a personal computer is called a hard drive, or a solid state drive, whereas storage in a mainframe is called DASD (Direct Access Storage Device, pronounced *dazdee*). Because most people only come into contact with the personal computer, we will focus on that type of hardware in this chapter. Keep in mind, however, that though the terminology will be different (i.e., you *boot* a personal computer, but you *IPL* [Initial Program Load] a mainframe), the components are the same.

## Computer Guts

Inside a computer will be:

- A motherboard - the main integrated circuit board that connects everything together, usually made of green plastic with many tiny components attached, either with solder (melted metal) or plugged in.
- A power supply - provides electric power.
- A cooling fan - computer chips need to be cooled because they stop working reliably when they get hot, as they sometimes do. There may also be a *heat sink*, designed to absorb heat and keep it away from the chips.
- ROM (read only memory) chip – usually containing the beginning instructions to the computer that it reads when it is turned on. ROM chips are actually firmware, a combination of software and hardware.

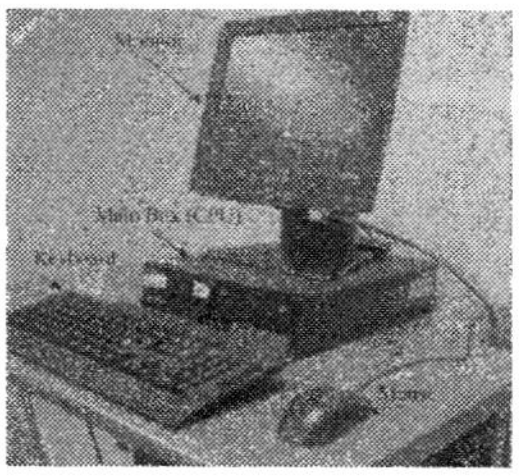

**Figure 30. Personal Computer**

- RAM chip (Random Access Memory chip) - empty storage for information that is only available when the computer is on. RAM can usually be added to the motherboard in order to improve performance.
- CPU (Central Processing Unit chip) - the thinking part of the computer (i.e., the brain). The speed of a CPU is measured in gigahertz.
- A hard drive (HD – also known as the C: drive*) - to store all the software and all the data. Also called a hard disk. These days, some computers have internal flash drives called Solid State Drives (SSD) instead of mechanically spinning hard drives. The hard drives of mainframe computers are called DASD.
- Video Interface - a connector for the monitor. Sometimes this connector is on the video card instead of directly on the motherboard.
- NIC (Network Interface Card) - a card that allows your computer to connect to other computers or devices. Most NICs these days are Ethernet 100 (Mb/s) or Gigabit Ethernet (1000 Mb/s).
- USB Interface - Universal Serial Bus - a type of connector that is more universal than either of the older parallel or serial interfaces. Most printers, external hard drives, cameras, flash memory and other peripheral devices today use USB connections.
- HDMI Interface (High-Definition Multimedia Interface) – a type of connector used with monitors, external media systems and projectors.

As computers have gotten smaller, the number of drives and other devices installed have gotten less common. A computer may have these devices inside the main box (thereby becoming a main component) or connected to the main box (thereby becoming peripherals). A desktop computer will often have them connected, while a laptop computer will more often not include them.

- A DVD drive. (Digital Video Disc) - a type of media that can store between 4 and 7 gigabytes of data. DVD is not just the media; it refers to the format that enables the DVD to present menus and play video on a stand-alone DVD player outside of a computer. Twenty years ago, most computers would come with a CD-ROM or a CD-RW drive (and DVD

* All computer hard drives are called C: because in the early days of the computer, the floppy drives were called A: and B:, and the hard drive was called C: even if there was only one floppy drive. Software was programmed to be installed on the C: drive, so for backwards compatibility, all computers must continue to use C: even if there are no other drives.

drives were rare). CD-ROM stands for Compact Disc, Read Only Media and CD-RW stands for Compact Disc, Read-Write, but since a DVD drive can also read and write CDs, it no longer makes sense to have both. A CD can hold between 650 and 850 megabytes of data.

- An external speaker – so you can hear sounds and music.
- Video Card – a card that makes your video monitor look better or work faster (especially if you want to play high-quality video games).

Even though there is a specific chip called the CPU, sometimes people call the entire box the "CPU". Anything that is inside the main box (also referred to as a desktop, or a tower, or a workstation) is considered internal to the computer. Anything outside the computer is considered a peripheral (discussed later in the chapter). Older computers may still have *parallel* or *serial* interface port for peripherals to connect, but these days almost all the peripherals connect through a USB or HDMI port.

Laptops generally have other peripherals that are not normally included in a desktop computer:

- A webcam so that you can easily videoconference.
- An integrated microphone/speaker port so that you can connect a headphone or earbuds.
- An internal speaker.
- A touchscreen.

A typical personal computer is shown in Figure 30. Personal computers have been getting smaller and smaller. Some manufacturers combine the monitor and the CPU into a single device, such as the iMac and the VAIO VCG-V620G all-in-one PC.

It is important to be able to identify what is inside of a computer. A picture of the inside of a desktop computer is in Figure 31. (While laptops are more prevalent these days, the inside is so compact it would be difficult to see the components, so the picture shows a larger, older, desktop). Usually, laptops, tablets, and smartphones are not designed to be opened up and be updated, whereas desktops were expected to have upgrades for things like video cards, speakers, NICS, and RAM, so they were designed to be opened and components added and removed.

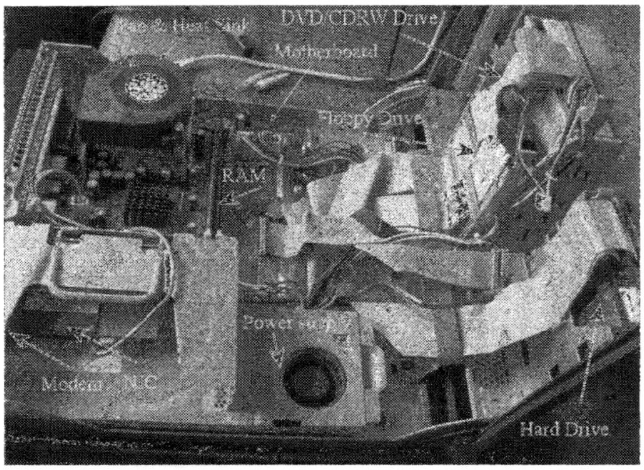


**Figure 31. Inside a microcomputer.**

A close-up of some of the inside components are shown next in Figure 32 and the connections typical on the back is shown in Figure 33.

**Figure 32. Close-ups of RAM, CPU, internal battery, hard drive.**

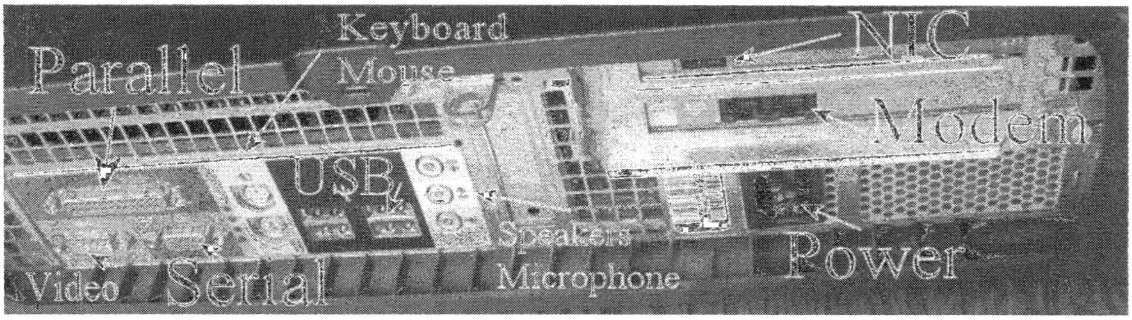


**Figure 33. Connections and ports of desktop computer (previous generation of PCs).**

## *USB Cables*

As can be seen from the pictures, the previous generation of PCs had many different types of peripheral cables. Parallel ports were used for printers, serial ports were used by modems, video ports were used for monitors, and keyboard, mice, speakers, and microphones each had their own round port.

Nowadays, the most common peripheral cable is USB (Universal Serial Bus), which is generally used for everything. But USB cables have several different kinds of endings to enable various needs.

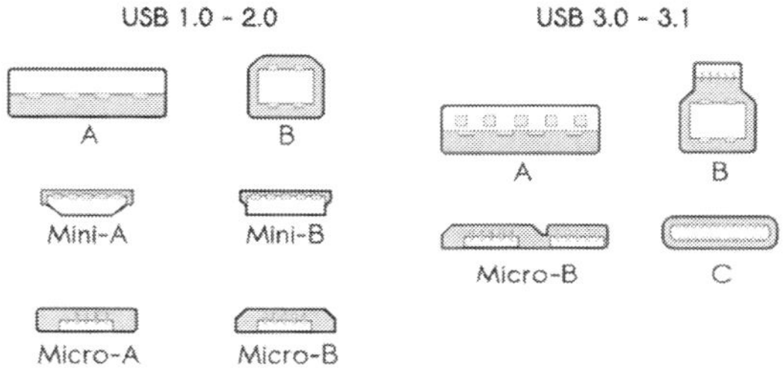


**Figure 34. Different types of USB cable connectors***

# Power: Processor, Memory and Storage

Regardless of the physical size of the phone/computer, one of the most important things for a person to know is: *How powerful is it?* A computer has three attributes that identify its power. The first is the speed and volume capability in the central processing unit (CPU). The second is the amount of space in the random access memory (RAM). The third is the amount of space in storage (either hard drive or solid state drive) to store applications and data. The next section will talk about the three indicators of computing power: memory, storage, and CPU.

When you buy a computer (or a cell phone these days), or talk to others about a computer, you should have a basic idea of how powerful the computer is since the power is the most influential factor in price. You should know how much storage space you have, how fast your processor is, and how much memory you have. In order to understand all the different measurements, however, we need to discuss bits, bytes, k, megs, and gigs.

## *Bits, Bytes, K, Megs, and Gigs*

There is nothing more confusing than the terminology and usage for measuring technology concepts such as memory, hard drive space, and data transfer speeds. The confusion comes from the fact that sometimes these items are measured in binary measurements (such as k byte), and sometimes they are measured in decimal measurements (such as kilobyte). The difference

* https://commons.wikimedia.org/wiki/File:USB_2.0_and_3.0_connectors.png, Original graphic by Milos634, Augus 2015. Last accessed September 27, 2020.

between a k byte and a kilobyte are often ignored (until you try to figure out why the numbers never add up).

Furthermore, common usage of terms is different depending upon the industry. The hard drive industry measures hard drive space in decimal gigabytes or terabytes, which translates into a smaller number of binary gigabytes. For example, a 500 decimal gigabyte hard drive, (500GB HD), only has 465 binary gigabytes. A 32 decimal gigabyte flash drive has only 29.8 binary gigabytes.

A bit (binary digit) is always a bit, and a byte is always a byte. Both are the same whether being measured in decimal or binary notation, and eight bits always equals one byte. In decimal notation, however, there are 1000 bits in a kilobit, 1000 kilobits in a megabit, 1000 megabits in a gigabit, 1000 gigabits in a terabit. Additionally, there are 1000 bytes in a kilobyte, 1000 kilobytes in a megabyte, and 1000 megabytes in a gigabyte.

In binary notation, on the other hand, there are 1024 bytes in a k byte, and 1024 k bytes in a megabyte, 1024 megabytes in a gigabyte, and 1024 gigabytes in a terabyte, 1024 terabytes in a petabyte and 1024 petabytes in a exabyte. (There are also 1024 bits in a k bit, 1024 k bits in a megabit, and 1024 megabits in a gigabit) [*].

Unfortunately, many people refer to these measurements with a shortened name: k, meg, and gig - without specifying whether they are talking bits or bytes, or binary or decimal. For technical folk, that's not a problem because they use the context to understand which one was meant. If you are dealing with memory or files sizes, meg refers to megabytes using the k byte conversion (binary). If you are dealing with hard drive space, meg refers to megabytes and gig refers to gigabytes using the kilobyte conversion (decimal). If you were dealing with data speed, meg refers to megabits per second (decimal) and gig refers to gigabits per second. Most people are not technical enough to understand the nuanced differences, however, so most just accept whatever nomenclature they are given by the vendors and don't bother trying to convert between the different measurements.

In the earliest days of the computers it was easy to tell the difference between memory and storage because storage (the hard drive) was usually large (in megs) and memory (RAM) was usually small (in k). Both were very expensive. In 1985 the IBM PC cost around $4000 and would traditionally come with 128 k bytes of memory and a 10 megabyte hard drive. These days, that's not even enough memory for a dumb flip phone![†] In 2020 a typical personal computer would cost around $1500 and come with a 500-gigabyte hard drive and 16 gigabytes of RAM. Now that both memory and storage are cheap, they are both measured in gigabytes[‡].

---

[*] In an effort to decrease the confusion, a new nomenclature had been proposed using kibibyte instead of kilobyte and mebibyte instead of megabyte. However, the verbal similarity between the b and the g made it difficult to determine which word is being spoken, and the industry overwhelmingly rejected the idea.

[†] For an interesting discussion of the way that storage sizes have changed, read http://www.makeuseof.com/tag/memory-sizes-gigabytes-terabytes-petabytes/ originally written by Joel Lee on August 14, 2012 but updated by Ben Stegner on May 8, 2020. Last accessed September 9, 2020.

[‡] In 2020 I purchased a Dell computer with a 2 Terabyte internal hard drive, 64 Gig of RAM, and a 10 Terabyte external hard drive for less than $3000. I contrast that with the VLDB (very large database) project I ran at Bank One which cost 30 million dollars for a 10 Terabyte system.

**Table 10. Binary Measurements of Memory and Storage**

| Full technical name | Commonly called | Numeric Multiplier | # of binary Bytes | # of decimal Bytes |
|---|---|---|---|---|
| k byte | K | $2^{10}$ | 1,024 | 1000 |
| megabyte | Meg | $2^{20}$ | 1,048,576 | 1000000 |
| gigabyte | Gig | $2^{30}$ | 1,073,741,824 | 1000000000 |
| terabyte | TB | $2^{40}$ | 1,099,511,627,776 | 1000000000000 |
| petabyte | | $2^{50}$ | 1,125,899,906,842,620 | 1000000000000000 |
| exabyte | | $2^{60}$ | 1,152,921,504,606,850,000 | 1000000000000000000 |

The abbreviations for the references are shown in Table 11.

**Table 11. Terms and Their Abbreviations for Measurements**

| Term | Abbreviation | Reference measurement |
|---|---|---|
| bit | b | 0 or 1 |
| byte | B | 8 bits |
| kilobyte | kB | 1024 bytes (binary)<br>1000 bytes (decimal) |
| megabyte (decimal) | MB | 1024 kilobytes (binary)<br>1000 kilobytes (decimal) |
| gigabyte (decimal) | GB | 1024 megabytes (binary)<br>1000 megabytes (decimal) |
| terabyte (decimal) | TB | 1024 gigabytes (binary)<br>1000 gigabytes (decimal) |
| petabyte (decimal) | PB | 1024 terabytes (binary)<br>1000 terabytes (decimal) |
| exabyte (decimal) | EB | 1024 petabytes (binary)<br>1000 petabytes (decimal) |

Now that we have discussed the differences between bits, bytes, megs, and gigs, we can move on to understanding the three different aspects of computing power; the processor, the memory, and the hard drive space.

## The Processor

The CPU (central processing unit) has a *clock speed,* which is considered a measurement of how fast the CPU could process instructions. A low clock speed might be a limiting factor because if your application is ready to roll, but it must wait for the next "internal pulse," it could be slowed down a little. But in general there are many other factors that determine how "fast" a computer could process an instruction that are much more influential than clock speed, so unless you are crunching billions of numbers, or trying to render video, clock speed is immaterial. Many CPU chip manufacturers (Intel, Motorola, AMD, etc.) have waged an educational campaign to eliminate *The Megahertz Myth.*[*]

Most of the time when a computer user experiences a delay it is because the application is waiting for the hard drive, the monitor, or the network, all of which are perceptibly slower than the CPU. The memory, hard drive, monitor, internal temperature, and/or network will affect your computer's speed much more than clock speed. (Remember that the next time your neighbor brags about a 3.6 Gigahertz clock speed – it really doesn't matter!)

Additionally, the newer CPUs are composed of multiple chips (Duo, or Quad CPUs). The processors share the load (but also need to be coordinated, which is called "overhead," so you actually get less than two times the performance). In order to maintain higher performance, however, the combined CPUs must run at a lower clock speed.

The next big leap is 64-bit architecture, which means that the CPU has addressing registers (internal "boxes" for storing digits) that are 64-bits wide instead of the more common 32-bits. Until recently, 64-bit architecture was only found on supercomputers and powerful business machines. Normally, applications written for 32-bit computing don't work on 64-bit machines, and vice-versa, so an end user using a 64-bit CPU might have a little difficulty because not all applications will run on the system. However, 64-bit computing is becoming more popular and easier to find. Although there are still some difficulties in running applications on a 64-bit computer, for the most part, major applications will run without a problem (even if they aren't able to take advantage of the additional speed). As more people adopt 64-bit computers, more applications will be re-written to take advantage of the additional speed and power, and there will be fewer limitations on applications that require a lot of processing (such as databases and video editing).

## Memory

Memory is the "working space" your computer uses for immediate processing, also known as RAM (random access memory). In traditional computer textbooks, you will see RAM referred to as primary memory and hard drive space referred to as secondary memory, but it is more common (and less confusing) to refer to RAM as memory and hard drive space as storage.

Memory is *volatile,* which means that when you turn your computer off, everything in memory disappears. If you are writing a letter in a word processor and haven't yet saved it to disk, for example, then it is only in memory. If someone were to trip over the power cord of your computer, you would lose the letter you are working on and would have to retype it.

Too little memory is often the cause of slow computers. More memory means that the processor doesn't have to dip into the hard drive as often to pick up information. Hard drives are much slower (even at 300 revolutions per second) than integrated circuit chips (which go at the speed of conduction – effectively as fast as the speed of light)[†].

The amount of memory you should install is based upon the needs of the OS and the types of applications you would be running. The minimum amount of memory needed by various selected OSs is shown in Table 12. Computers that will be doing word processing would not need any more memory, but a computer that is intended to run databases, graphics, or video would need quite a bit more.

---

[*] Tony Smith (February 28 2002). "Megahertz myth: Technology". *The Guardian*. http://www.guardian.co.uk/technology/2002/feb/28/onlinesupplement3. Originally retrieved December 24, 2010. Last accessed September 9, 2020.

[†] This distinction may disappear with SSD because the solid state drives are much faster than hard drives that have to spin to access the file on the media.

**Table 12. Operating Systems with RAM Requirements**

| Operating System | Recommended minimum RAM (using Decimal bytes) | Really need |
|---|---|---|
| Microsoft Windows 10 | 1 Gig (or 2 for 64-bit architecture) | 4 Gig |
| Apple MacOS Catalina* | 4 Gigs | 8 Gig |
| Linux (open source) | 1 Gig | 16 Gig (if used as a server, the most common use) |
| Google Android | 512 Meg | 1 Gig |

## Hard Drive

The other determiner of your computer's power is the amount of space on your hard drive. These days, a typical PC will come with 250 to 500 Gigabytes (decimal) of hard drive space. In 1995, the typical hard drive was 20 or 40 Gigabytes. (And remember, in 1985, a typical hard drive was 10 megabytes or non-existent as many computers only had floppy drive storage.) One reason for the dramatic increase in hard drive space is the space requirements of the OS. In 1985 the operating system was DOS, and it took up less than 2 megabytes of space. In 1998 a typical hard drive was 100 megabytes, and the Windows 98 OS took up about 20 MB. Of the 200 gigabyte hard drive of today, the Windows 10 OS will take up between 9 and 16 GB, depending upon the options. Furthermore, Windows uses a portion of your hard drive dynamically for virtual memory, and that can use up space as well. (If you are ever looking at the root of your hard drive and notice a huge file called hiberfil.sys and/or pagefile.sys that you cannot delete, that is where Windows is storing the virtual memory).

Another reason for the larger hard drives is that people nowadays need space for graphics, music, and video. Plain text files are tiny. You can fit the whole Library of Congress on a couple of DVDs if you leave out the pictures. At the same time, a single photo uses up over one megabyte. One minute of video uses up 10 megs of your hard drive. A full-length 90-minute movie uses a couple of gig unless it is HD, in which case it can be much more. You can pretty quickly use up your hard drive if you download photos and videos frequently.

The practical aspect of understanding the general size of different types of files is in knowing how much space you need to store information. In Table 13 you can see how much space typical files might take up on a hard drive or other storage medium. Notice that the chart gives you both the accurate assessment of the names and amounts and relationships for the terms. It also includes "about how big" references that can help gauge how big or small something is. The third column compares the size of the item to a typical 32 gig flash drive that might hold the item.

* Prior to 2013, Apple named its OS versions after big cats (Snow Leopard, Lion, etc.). After that it named them after landmarks (Yosemite, Maverick, etc.)

**Table 13. Amount of Space Needed for Different File Formats**

| What? | About How big?* | How many on flash drive... |
|---|---|---|
| Single page of text only | 2 KB | You can fit 16,000,000 of them on a 32-gig flash drive. |
| Single eight-color graphic picture | 500 KB | You can fit 64,000 of them on a 32-gig flash drive. |
| Single Word document (ten pages with graphics) | 1 MB | You can fit 32,000 of them on a 32-gig flash drive. |
| Full color photo | 4 MB | You can fit 8,000 of them on a 32-gig flash drive. |
| A three-minute 2X3 inch video | 10 MB | You can fit 3,200 of them on a 32-gig flash drive. |
| A five-minute full screen video in standard | 150 MB | You can fit 213 of them on a 32-gig flash drive. |
| A database of thousands of records | 200 MB | You can fit 160 of them on a 32-gig flash drive. |
| A PowerPoint presentation (with graphics, photos, and animations) | 500 MB | You can fit 64 of them on a 32-gig flash drive. |
| A database with millions of records | 2 GB | You can fit 16 of them on a 32-gig flash drive. |
| A high-density full screen full length video (movie) | 7 GB | You can fit 3.5 of them on a 32-gig flash drive. |
| An accounting system for a Fortune 1000 company (for a couple of years) | 250 GB | Can't fit on a flash drive. Need external hard drive. |
| The amount of storage needed for Facebook posts each day. | 600 TB | Server farm. 150,000 DVDs. |
| Amount of storage on Google servers | 15 EB | Server farm. Amount the equivalent of 30 million computer hard drives. |

Years ago, to transfer information from one computer to another, data would have been transferred on a CD (which holds about 750 megabytes) or a DVD (which holds about 4 gigabytes). But these days, people utilize flash drives when it is too time consuming or impossible to transfer the data across the network. A flash drive can hold anywhere from 8 GB to 128 GB of data.

## Hardware Peripherals

Peripherals are devices attached to the main computer. As noted earlier, many devices can be found inside the main box, or outside of it – attached through a cable and port. Peripherals are wide and varied, and may include:

* Keep in mind that none of these sizes is meant to be exact as the actual size of files varies wildly. But the sizes will give you an understanding of the *typical relative* size for each type of file.

Printer
Keyboard
Mouse
Monitor
Speakers
Digital Camera
Digital Video Camera
Cell Phone
Scanner
Smartphone
Game Controller
Cable Modem
Router
DSL Switch (Digital Subscriber Line)
Hard Drive (External)
Projector
Rollerball
CD-WR Drive
DVD Drive
Flash Drive (also known as a key drive, thumb drive, or memory stick)

Peripherals can be categorized as either input devices or output devices. An *input* device enables people or other devices to enter information into a computer. An *output* device enables a person or other device to receive information from a computer. A mouse, keyboard, camera, and scanner are input devices. A monitor, printer, and projector are all output devices. Some devices can perform as either input or output, including flash drives, hard drives, or networking devices such as routers or DSL switches.

## 4-1. Interactive

1. **Class Discussion**: How many different components inside the main system box of a computer can you name? How many different peripherals? For each peripheral you name, state whether it is an input peripheral or out output peripheral.
2. **Class Discussion:** What types of media have you used to copy, move, install, or use files? Which were the most convenient? Which were the least costly?
3. **Class Discussion:** How have input devices changed over the past ten years? How have output devices changed? What do you think input and output devices of the future will look like?
4. **Class Activity**: Find an old computer, or using a computer provided by your instructor, open up the case. Identify all the major components.
5. **Independent Activity**: Interview a business person and ask them what kinds of computer systems they use. Categorize them as mainframes, servers, or personal computers. About how many of each does the business use? Does the IT department of the business manage smart phones?
6. **Independent Activity:** If you had to estimate the space needed for taking 12 hours a day, 365 days of video of your life, how much room would it take? Add all the movies you will ever watch, and every book you've ever read, and assume you live to the age of 100. How much data would that generate?

## 4-2. Self Study Test Questions:

1. What is a mainframe computer?
2. What is a server? How is that term related to minicomputer?
3. Give at least 4 names for personal computers in use today.
4. Name at least 5 basic components inside the main box of a computer.
5. What is a CPU? What is a motherboard?
6. What is the difference between RAM and ROM?
7. Why does a computer need a cooling fan or heat sink?
8. Why is it easier to open up a desktop or tower computer than a laptop?
9. Name at least 5 devices that are peripherals, and identify whether they are input, output, or both.
10. How many bits in a byte?
11. What are the two methods for translating between bytes and kilobytes, megabytes and gigabytes?
12. Why does a hard drive sold as a 300 Gig hard drive have only 279 Gigabytes?
13. What are the three factors that determine how powerful a computer is?
14. Why don't we use CDs and DVDs much anymore? What took its place?
15. What is the difference between memory and storage?
16. What is the most convenient method for transferring less than 64 Gigs of data between one computer and another?
17. What is a data center? How has the equipment in a data center changed over the years?
18. Describe what is meant by client/server configuration.
19. What is the difference between a thin client and a fat client?
20. What is a tier in a server farm?

5

# Chapter 5: System Software

Overview: In this chapter, readers will learn the following topics:

- ⇨ Operating Systems (Windows, MacOS, Linux, MVS)
- ⇨ Cloud Drives (Dropbox, iCloud, Google Drive, and OneDrive)
- ⇨ Utilities (such as anti-virus and backup software)
- ⇨ Development Software (languages like Java, C++, Visual Basic, and workbenches such as Visual Studio, PHPStorm, and Java Enterprise)

Now that hardware has been covered, let's move on to software. Software can be categorized as System Software or Application software. In the technology map, System software is <u>below</u> Application software. Application software will be covered in the next few chapters, while this chapter is about System Software which has four sub-categories:

Development

Utilities

Cloud Drives

Operating System

The lowest layer of software is the code that runs your device – Operating System (OS). On top of that are Cloud Drives (a new categorization), and above that are Utilities and Development. Subcategories as well as examples of each category can be found in Table 14.

**Table 14. Technology Map System Software Categories**

| MAJOR | MINOR | CATEGORIES |
|---|---|---|
| System Software | Development | Languages (Java, C/C+/C++/C#, JavaScript, Objective-C, PHP, Visual Basic, Ruby, Python, Rust, Elixir, ASP, Javascript, BASIC, RPB, Informix, Fortran, Cobol, Progress, Access, FoxPro, etc.) |
| | | Web Development Tools: Visual Studio, Netbeans, Eclipse, Javascript Libraries (JQuery, BackboneJS); Front End Frameworks (Bootstrap, Foundation, Semantic UI, uikit); Application Framework (Ruby, AngularJS, Express, Meteor, Django, ASP.net, Zend Framework, CakePHP, CodeIgniter) Code Editors (Atom, Sublime, PHPStorm) |
| | Utilities | Backup (Genie, Acronis, NTI, GRBackpro, Paragon, Veritas, NovaBackup); Firewall (SMS, Check Point, PICS); Antivirus (Windows Defender, Bitdefender, Kaspersky, Panda, F-Secure, AVG, Avast, Trend Micro, McAfee, Webroot Secure Anywhere, CA); Text Editors (Vi, Notepad+++, TextEditor); Enterprise Network Management (LogicMonitor, ManageEngine OpManager, Spiceworks, Idera Uptime Infrastructure Monitor, Ipswitch WhatsUp, HP BTO, Tivoli Netview) |
| | Cloud Drives | Dropbox, iCloud, Google Drive, and OneDrive |
| | Operating System | Single User (Windows 10, Windows 7, MacOS) Mobile (Android, iOS) |
| | | Multi User (Unix and all its flavors including Linux, Multics, HPUX), MPE, Windows Server, z/OS, z/VM, z/VSE, and z/TPF |

# Operating System Software

The base software of any computer is the Operating System (OS). Without an OS, the computer makes a really good doorstop, but it can't do much else. The Operating System tells the computer what the keyboard is, what storage is, what memory is, how to display text and graphics on the monitor, how to get files, how to store them, etc. OSs can be categorized as multi-user or single user.

## *Multi-user Operating Systems*

Mainframe computers* and servers were designed for hundreds of people to use at the same time. Mainframes and servers usually come with a proprietary OS that does not work on any other hardware. For example, International Business Machines (IBM) sells an OS called i for its AS/400 lines†. Other OSs like z/VSE & z/VM are for IBM's mainframes. Hewlett Packard (HP) uses the MPE OS to run its HP3000 servers.

Unix is a multi-user OS that enabled many different devices to connect and share resources such as file storage and printing. Unix originated in Bell Labs at AT&T (see the story of Unix), but had been morphed into dozens of proprietary OSs by hardware manufacturers. For example, AIX was sold by IBM for its RS/6000 models, Sun sold Solaris to run all its systems, and HPUX runs on HP workstations. And then, of course, there were dozens of other commercial versions of Unix made by various vendors.

***The Story of Unix‡***

Back in 1969, a couple of computer researchers, Ken Thompson and Dennis Ritchie, were working for AT&T (called Bell Labs at the time) on a computing project called MULTICS, (Multiplexed Information Computing System) with MIT (Massachusetts Institute of Technology) and Honeywell (part of General Electric at the time). Being very bright and relatively young, Ken and Dennis liked to play games. Ken utilized his abilities to write a multi-user computer game called Space Travel on the GE/Honeywell mainframe computer. The game was a simulation, allowing players to guide a ship to land on various planets and moons of the major bodies of the Solar System. Unfortunately, the cost of the timeshare for the mainframe was about $75 to play a single game. So Ken found a castoff minicomputers (a DEC PDP-7), and Dennis helped him rewrite the game so that it would work on the PDP-7.

Things went well until AT&T decided to drop out of the joint project. Ken and Dennis were devastated. They did not want to give up their Space Travel game. Thinking quickly, with the help of manager J.F. Ossanna, they convinced AT&T to purchase a PDP-11 (ostensibly to write a word processing program). Ken and Dennis then wrote an entire operating system (as well as a C compiler) so that they could port their game to the new computer.

Eventually they had to name the operating system they wrote. A friend, Brian Kernighan, suggested the name. He suggested they called it "Unics" for Uniplexed Information Computing System (which was actually a pun on an emasculated Multics). Eventually the name was changed to Unix, and the first multi-user portable operating system was born.

---

* More details on mainframe and other types of computers will be found in *Chapter 4: Computer Hardware* .

† Naming an OS with a single letter appears to me to be either arrogant or stupid. Consider trying to find out information on the OS by googling the OS name i. Not expedient. But then again, IBM did not ask me before naming their new OS.

‡ Most of this story was documented on ***History and Timeline***, by the Open Group at http://www.unix.org/what_is_unix/history_timeline.html (Last updated on January 29, 2003, last accessed on June 30, 2016) and on ***The Evolution of the Unix Time-sharing System*** by Dennis Ritchie at http://cm.bell-labs.com/cm/cs/who/dmr/hist.html (last published *AT&T Bell Laboratories Technical Journal* ***63*** *No. 6 Part 2, October 1984, pp. 1577-93.* ) and last accessed on Sept 20, 2020).

***More About Unix***

Ken and Ritchie, of course, wrote Unix as employees of AT&T, so AT&T owned it, not them. But U.S. trade restrictions prevented AT&T from (at that time) selling computer operating systems. AT&T decided, instead, to make the Unix source code available to universities -- cheap. Universities had to pay the grand sum of $1 a year to lease Unix from AT&T.

Students who learned Unix while at the University took their knowledge with them into the business community after they graduated. Since that was what the students knew, businesses had no choice but to use Unix as well.

The commercial versions of Unix were unable to compete with the open source version of Unix called Linux that was developed and released (for free) by Linus Torvalds,[*] thereby starting the Open Source Movement (see Figure 35 and inset box on open source). Internet protocols were built into Linux OS, so it was an obvious choices for webservers.[†] Most webservers on the World Wide Web running on the Internet today are Linux-based servers, though it is difficult to say exactly how many because statistics are gathered for operating systems sold, and Linux is not sold but simply downloaded and installed by anyone. A system sold with Windows Server, for example, may have been deleted and reinstalled with Linux – but none of the marketing services would know to count it.

**Figure 35. Linus Torvalds, originator of the Linux operating system.**

***Open Source Movement***

The open source movement started in April 1998 during a meeting in which several influential developers argued for a new business model for software. Instead of a company paying people to develop a software product and then charging users to install it, programmers donate their time to develop and install the software so that others can use it free of charge. The "source code" is openly and freely available to anyone

## Single-user Computer and Mobile OSs

A single user OS is intended for one computer, to be used by one person at a time. Early OSs could only have one profile, but OSs such as Windows 10 allowed multiple profiles that store configuration information for each user, although only one user can log in at a time.

The most recent change in interfaces is the use of the multitouch screen. Using multitouch gestures such as press, long press, pinch, swipes, and two or three finger press, the touch screen multiplies the number of options over the simple mouse clicks and drags (even with left, right, and scroll). Touch screens were originally only found on mobile systems such as tablets and phones (using Android by Google or iOS by Apple), but now can be purchased on the majority of laptop and desktop systems (though not yet standard).

## Command Line Interface vs. Graphical User Interface

Initially, single user OSs were all command line driven, just like multiuser OSs such as Unix. *Command line driven* meant that the user typed commands onto a blank line (a command line interface) at the prompt on the text-based computer screen. But then Xerox researchers invented a

[*] This picture came from December 2002 issue of Linux Magazine, distributed under the GNU Free Documentation license, (GFDL). Permission of Martin Streicher, Editor-in-Chief, LINUXMAG.com. Oddly enough, though most people pronounce Linux with a short *i*, Linus is from Finland, and his name is pronounced as if it was spelled with a long *e* (i.e. *Leenus*). He, personally, pronounces the OS *Leenux,* similar to his own name.

[†] O'Reilly, Tim. ***Linux/UNIX Server Market Share Beats NT/Win2K*** written on August 9, 2000 http://www.oreillynet.com/manila/tim/stories/storyReader$56. Accessed on May 18, 2008, but no longer publicly available.

different kind of system; one that used a mouse. This class of OSs were called graphical user interfaces (GUI, pronounced *gooey*). The Apple Lisa was the first commercially sold personal computer with a graphical user interface, and it used menus and pictures called icons that were clicked on with a mouse. (See the story of the GUI).

The advantage of a command line interface is power, speed, and flexibility. If you already know the commands, it is much faster to type them into the computer than to use a mouse or your finger to choose among the limited choices on the menu. Additionally, (again, if you already know the commands) there are hundreds of thousands of commands and their options that will allow you to do exactly what you need to do. For example, in Figure 36, the DOS commands are written to create a new folder, open the folder, and list everything that is in the folder. (You can get to the DOS prompt by either choosing DOS Prompt from the Accessory menu or choosing Run from the context menu of the Start button and typing `cmd` into the dialog box.)

As noted, if you already know the commands, a command line interface is faster, more flexible, and more powerful. That, however, is the rub. Very few people already know the commands. And since every command has dozens (and sometimes hundreds) of options, and those commands and options have to be typed in with a very particular *syntax* (the way the command and options must be formatted), unless you are typing in those commands on a frequent basis, it is difficult to remember them all.

***The Story of the GUI***

Many people believe that Steve Jobs from Apple computer designed the interface that we now know as Graphical User Interface (GUI), but he did not. Instead, Jobs was convinced by some of his employees to visit Xerox labs in 1979, where he saw the Alto computer. The Alto was different than all the other computers of the day; it used a mouse and menus or icons (little pictures) instead of a keyboard to enter most commands. It also displayed windows on the screen and had a trash can in which users could "drop" files they wanted to delete.

Steve Jobs liked what he saw, and he made a deal with Xerox so that the researchers could work with Apple to develop the Lisa computer (named after Steve Jobs' daughter, although ostensibly it was an acronym for Local Integrated Software Architecture). The Apple Lisa did not sell well because it was expensive, slow, and beset with unreliable hardware, but it formed the foundation for both the Macintosh computer and the Windows GUI interface.

In 1988 Apple sued Microsoft for infringing on the "look and feel" of the GUI but lost in court at the end of a four-year battle because they had earlier licensed the technology to Microsoft for Windows 1.0. In the meantime, Xerox sued Apple since Xerox was actually the originator of the look and feel of the GUI. Their suit was never pressed, however, because the statute of limitations had run out.

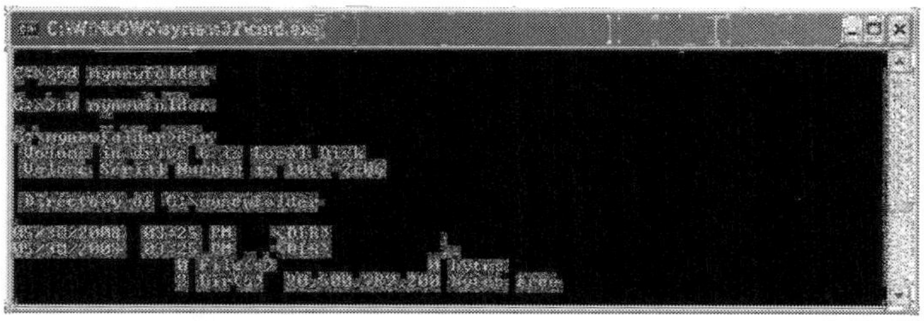

**Figure 36. Command line interface (DOS).**

The main advantage of the GUI (both mouse-driven and touchscreen-driven) is that you don't have to memorize the commands. It is much easier to use, and therefore accessible to many more people. People don't have to know esoteric and difficult-to-remember computer commands or syntax.

Additionally, GUIs standardized the way programs did simple actions like opening and saving files. All GUI programs open a file by going to the File menu and clicking the Open choice. All GUI programs save a file by going to the File menu and clicking the Save choice. Before GUIs, each software designer had to choose which keys would open or save a file, and they all tended to choose different keys (making it very difficult for the users to learn how to use their programs)*. The disadvantage of GUIs is that, compared to command line driven OSs, a GUI is slow, inflexible, and does not have very much power. The user is limited to the few hundred menus and icons instead of thousands of commands and options. These disadvantages persist whether the GUIs utilizes a mouse or a touch screen. GUIs may appear fast to the typical user, but for true power and flexibility, having all the commands and their syntax memorized cannot be beat. (This is one of the reasons hackers work almost exclusively with command line interfaces. Unlike menus and icons which are typically limited to "normal" activities, hackers need to go "below" the allowed-activities and perform actions that would be prevented by the GUI.)

## OS Basics: Folders, Files, and Windows

While most students have picked up the basics of using the operating system to store and retrieve files, it is important that all students understand file structures and application windows. We will take a brief foray into some of the concepts and techniques regarding folders, files, and windows including:

- windows location and size;
- switching applications;
- help;
- keyboard shortcuts for copy, cut, and paste;
- selection (including contiguous and non-contiguous);
- applications versus data;
- folders and files;
- network storage versus local storage;
- search and replace;
- styles and templates.

### Windows Location and Size

Graphical user interfaces are composed of many screen elements. Unfortunately, the names for these elements are not consistent, so be aware that one application may call a screen element one name while another might call it something else. One of the basic screen elements is a window†.

A window can be maximized to fill the screen, minimized to appear as a small picture at the bottom of the screen, or somewhere in the middle. When it is somewhere in the middle, the window can be moved or resized by using the mouse. Every operating system has controls on the window that allow you to do these actions. As can be seen in Figure 37, in Microsoft Windows Notepad the controls are in the top right corner of the screen. In the Macintosh operating system they are in the top left corner of the screen.

* For example, WordPerfect (one of the first wordprocessors), saved the file using the F10 key. F3 was the help key. Lotus 123 (one of the first spreadsheets) saved the file by pressing the slash key, then f, then s. F1 was the help key.

† Of course, there are windows on all GUI operating systems, including MacOS, Ubunti, Solaris, etc. Windows is not the only OS with windows.

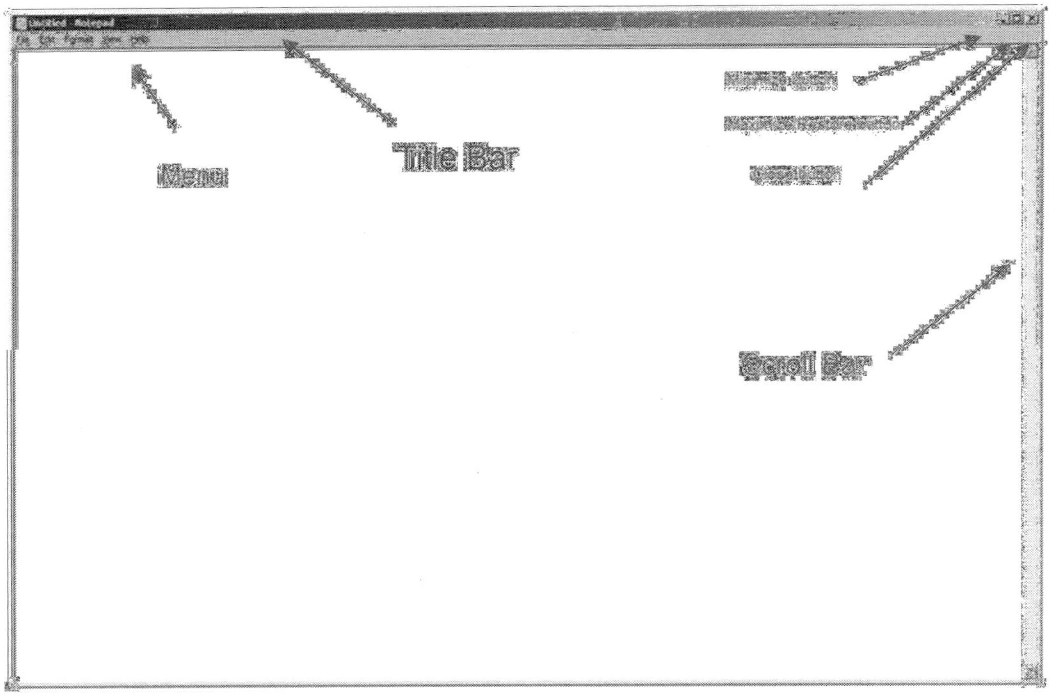


**Figure 37. Window controls and menu.**

Windows are helpful because they allow you to see more than one program at a time or more than one file at a time; you can arrange the windows so that you can see both applications at the same time.

### *Switching Applications*

When you have multiple programs running at the same time, they can either appear in windows on the screen, in small pictures (called *icons)* at the bottom of the screen (in the Microsoft Windows task bar, as seen in Figure 38, or the Apple Mac Dock bar in Figure 39 ). When more than one program is running at the same time, the easiest way to switch between programs is to use the [Alt -Tab] or [⌘- Tab] key combination. Press and hold the [Alt] or [⌘] key, and a small application choice box will appear in the middle of the screen. Tap the [Tab] key as the selection box cycles through all the application icons. Whichever app the box has stopped on when you release the [Alt] key will be the application that opens up next on your screen, becoming the *active* window. While you may have as many applications open as your computer resources can handle, you can only have one application active at any one time. You can tell when an application is active because the window is a different color, usually brighter or more solid than the other inactive applications.

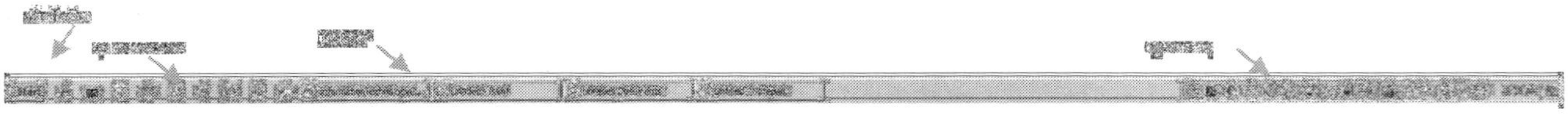

**Figure 38. Windows Task bar.**

**Figure 39. Macintosh Dock bar.**

### *Help*

There are many ways to get help with any program. Keep in mind, however, that *Help* is often less helpful than we would like. Usually there is a help button in the upper right corner of the window  or under the hamburger icon  which will lead you to a panel where you can type in a help search term. (Sometimes you must choose the *index* tab before you can get help, and other times you must choose whether you want help from the local help program or from the vendor's wider help website on the World Wide Web).

Additionally, you can usually get context-sensitive help. If you *hover* your mouse or finger over something (let the pointer sit for a second or two over a screen element without moving or clicking), you may get a help flag which describes the object over which you are hovering. Additionally, you can usually press the [F1] key (or the Control key on a Mac) when something is selected or when you are in a dialog box. This will either bring up either help specific to that object or the general help panel.

Furthermore, many objects have a context sensitive menu option, which appears when you press the right mouse button (or, on a Mac the Control or Ctrl key) on an object.

### *Keyboard Shortcuts*

There are hundreds of keyboard shortcuts for both the Windows and the Macintosh operating system. Keyboard shortcuts are often faster than using the mouse because you don't have to lift your hands off the keyboard and move the mouse, which takes time. While it is very difficult to memorize all the keyboard shortcuts, there are a few which are useful in many different programs - some which enable you to perform a task for which there is no menu. For example, you can copy with [Ctrl-C] (or ⌘-C on a Mac) and paste with [Ctrl-V] (or ⌘-V on a Mac) even when the copy and paste menu choices don't appear. (In general, the ⌘ key on the Mac* is the equivalent of the Ctrl or Alt key on Windows). Additionally, you can cut with [Ctrl-X] and paste with [Ctrl-V]. In most programs you can boldface with [Ctrl-B], italicize with [Ctrl-I], indent with [Ctrl-M] and underline with [Ctrl-U]. You can start a new file with [Ctrl-N], save with [Ctrl-S] undo with [Ctrl-Z] and print with [Ctrl-P]. [Ctrl-Home] will take you to the beginning of a file and [Ctrl-End] will take you to the end. Memorizing the keyboard shortcuts of your most frequently used tasks on the computer can save you boatloads of time over the years.

In addition to Ctrl shortcuts, many menu choices are available through an [Alt] keyboard shortcut. When you press the [Alt] key (and hold it for a second) you will see little letter flags appear on the menu. When you press that letter, that button, or tab or menu or ribbon choice gets selected, and the next set of letters will appear. For example, you can save a file by pressing the following key

---

* Whereas the Control or Ctrl key on a Mac is the equivalent of a Right Mouse button on a Windows. The Mac Ctrl is not equivalent to the Windows Ctrl. The Option key on a Mac is the equivalent of the Alt key on a Windows, except for Alt-Tab, in which case the Command ⌘ key is the equivalent. *(Could they make it more confusing?)*

combination: [Alt-F-S]. To describe the same combination with words, we would have to say: *Use the mouse to choose the Office button. Click on the Save menu choice.*

One of the most important things to remember about keyboard shortcuts is that they exist; which means that you might be blithely typing along and suddenly your whole screen changes and you find yourself somewhere you didn't want to go - because you accidentally pressed the [Ctrl] or the [Alt] key instead of the [Shift] key without realizing it. A common accidental keypress is [Ctrl-N] which opens up a new file. If you don't realize what happened, you may think that the file you *had* been working on disappeared and was replaced with a blank screen. You may *think* that you had lost your work when all you have to do is close the new file you accidentally opened, and your original file will reappear.

### Selection

Understanding *where you are* and *what is selected* will avoid countless problems and help you figure out what the program is doing. Normally *where you are* is demonstrated by a blinking cursor or an outline border of some sort. You may be in a datafile, in a dialog box, on a menu, in a cell, in a data entry box, on a tab, etc.

Additionally, you may have nothing *selected*, or you may have something *selected.* Again, you will be able to tell what is selected because it appears in a different background or color or has a border of some sort. You can select just one object at a time, or you can select multiple objects at a time. Generally, when you select a second object, the first object gets unselected, so you have to press either the [Ctrl] key or the [Shift] key (depending on the program) to select more than one object at a time.

When you are in some programs, such as ***File Explorer*** (previously known as *Windows Explorer* or *My Computer* or *File Manager*[*]) you can select *contiguous* objects by clicking on the first one, scrolling to the last one, holding down the [Shift] key, and clicking on the last one. Contiguous select will select all the objects between the first click and the last click.

You can also select *non-contiguous* objects by using the [Ctrl] key. To select non-contiguously, Click on the first object. Then hold down the [Ctrl] key and click on the next object, and the next, and the next. The objects don't have to be next to each other. In this way, for example, you can select the third, seventh, and tenth file in a list of files to copy. Finally, you can also select all the objects in a file by using the keyboard shortcut [Ctrl-A].

*Where you are* and *What is selected* determines what you can do. Some menu items are only available when you have selected the proper type of object. Other menu items are only available when nothing is selected. Some menu items are only available when you have multiple objects selected. Eventually you get used to which types of actions you can do with which types of objects, but the first step is recognizing that all programs are controlled by the concepts of *where you are* and *what is selected.* Experienced technology users pick these concepts up intuitively, but inexperienced users have a hard time until they are pointed out.

## Cloud Drives

The concept of cloud computing is relatively new. While it was introduced about 15 years ago, making it technically feasible took more than 10 years, so it's only been recently that people were able to use cloud computing with any sense of reliability. Here's the concept: rather than worrying about purchasing software and installing it on a specific local device such as a desktop computer, people would, instead, access the information they want from any device from anywhere in the world because it is stored *virtually*, which means that the physical location of the information is immaterial. A cloud drive is easily understood (and made very attractive) when demonstrated on

[*] The fact that technology vendors keep changing what they call things is one of the reasons technology is so hard to learn, covered in a later chapter.

television commercials by showing a picture being taken and edited on an iPhone that magically appears on the large monitor desktop computer at the same time as appearing on the laptop. And sometimes, for simple needs, it works that way – easily. Unfortunately, cloud drives are not quite that simple for most people, and come with issues as well as benefits. Just ask anyone who lost all their pictures because they thought they were backed up on a cloud drive (when they were actually *synced* and there wasn't enough room for all of them).

## Common Cloud Drive Services

Currently (2021) there are four main cloud drive services: Dropbox, Google Drive, iCloud, and OneDrive. The last three are each offered by a major hardware/software player in the technology world. Google Drive is, of course, offered by Google. iCloud is offered by Apple, and OneDrive is offered by Microsoft. Dropbox is the only one that is independent, which gives it an edge for people who use all three: Microsoft, Google, and Apple products and services.

## Origination of the Term "Cloud"

For years a picture of a cloud was used to represent the physical equipment and data lines leased from telephone companies but used by businesses. (You can see the illustration in Figure 19 in Chapter 2: Networking Hardware). The original phrase was used in a business plan in 1996, but Cloud Computing in the modern context wasn't applied until Google CEO Eric Schmitt used the term in a speech at the *Search Engine Strategies* conference in 2006.[*] What Eric Schmitt was alluding to wasn't exactly what we mean by it today, but a more encompassing *virtualization* of computing.[†]

## Issues with Cloud Computing

Cloud Computing is used daily by millions (many of whom aren't even aware that they are using it). There are still issues.

The cloud services that come with hardware is often intrusive and unrelenting in its aggressive configuration. Often computers and mobile computers purchased today have the cloud set as the default storage location and won't let the user change over to store things locally on their hard drive without a lot of technical knowledge and expertise. While this is fine for users who only take a few pictures and surf the web on their devices, business users with heavy-duty applications such as video editing, database, analytical spreadsheets, or book writing find themselves in a real bind. These resource-intensive applications do not work well when the source documents are not local on the same hard drive as the application. For example, an Access database placed on a cloud folder will continually create multiple "conflicted" copies of the file because the file is left open for long periods of time with constant "saves" while the individual records and components are edited. A video-editing program will think the file is "updated" because locally it was, but the cloud version is still in the middle of updating, so the video program will overwrite the updated version with a previous version without realizing it.

Furthermore, each of these cloud services handles the "local versus non-local" aspect of the files differently, causing disconnects when trying to apply the rules of one to another. OneDrive does not even have a local version of the file; all the files are stored on the cloud unless specifically told to download it, and even then there are problems; the file often does not stay in sync (especially if

[*] Regalado, Antonio (2011) *Who Coined 'Cloud Computing'?* Published October 31, 2011, http://www.technologyreview.com/business/38987/, Retrieved December 28, 2011.

[†] Reuven Cohen (2009) *Google Trends - Cloud Computing Surpasses Virtualization in Popularity.* Published April 2009. http://www.elasticvapor.com/2009/04/google-trends-cloud-computing-surpasses.html. Retrieved December 28, 2011.

it is a larger file). If the person installs the desktop version of the software, the other three (iCloud, GoogleDrive, Dropbox) have two copies of each file – one on the local drive and one on the cloud server. They do a relatively good job of letting you know when the file has been synced (i.e. changes uploaded onto the cloud server or downloaded to the local server), but the user has to be savvy enough to look at the status of a file before it is known whether or not it has been updated. Multiple devices complicates the issue as the files can be updated on any of the devices and then synced with the cloud server, which then downloads the changes on each device as it logs into the service.

Generally, cloud services work best when the local space is large enough to hold the entire cloud drive. If the local drive is smaller however (such as on a Chromebook or newer laptops), then the user must choose which files to sync, and which files will remain only on the cloud. Again – often users are unaware of where their files are, so they often lose file simply because they can't find them. The proliferation of "aliases" such as Documents and Pictures (folder names that actually point to a different folder which may or may not have the same name but which may be in a very different place) make it even more difficult for non-tech-savvy users to understand where their files are located.

Additionally, understanding how to clean up or delete files is an issue. If someone deletes a file from the cloud to save cloud space, the file will automatically delete locally from all files. Or if they delete it locally from their hard drive, it will automatically delete from the cloud. This is not a problem as long as the person realizes it, but if they delete a file thinking that there is a copy on the cloud, they will run into trouble. Or, a person wants a file to delete, and they delete it, but it keeps popping back on because it exists on another device, so the file "resyncs" when that device connects.

I've personally experienced a major problem when connecting a device that has been off for a long time. When a device attaches after a long period of disconnection, as the cloud service tries to sync the older files to the newer, it will delete a bunch of files (because they don't exist on the local device) and replace newer files with older versions (thinking they are newer).

Finally, there are still many legal issues of responsibility of liability and ownership of data which are still unresolved in the courts. When you store your data on a remote server, who is responsible for it - you or the owner of the server? What happens if someone steals the information? Or what if it gets lost or corrupted? Certainly, those things can happen to your data sitting on your own hard drive, but there the responsibility for it is clear. The information is clearly owned and controlled by you. You are responsible for backing up your own data. If the hard drive fails, you cannot sue the manufacturer of the hard drive for the loss of your data. But if your information is stored elsewhere on someone else's computer, it is more difficult to back up and control. If the company were to go bankrupt and close its doors, you would have no way of getting your data back. You cannot be considered 100% responsible for it since technically anyone with access to the server has access to your data, and whether you alone own the data is not yet fully established.

## Benefits of Cloud Computing

These technical, political, and legal issues, however, have not prevented the growth of the cloud computing industry. The ease with which multiple devices can all access the same data has proven to be such a convenience that issues of understanding, ownership, responsibility, and security have paled in comparison. Furthermore, storing files on a cloud drive makes those files much easier to share with others. Groups and teams have come to rely upon the ease with which they can now share a single file – often editing the file live during a meeting – rather than deal with the difficulties of attaching copies of the files to email and sending them back and forth. In the final analysis, cloud computing appears to be here to stay.

## Utilities

In order to keep software and hardware working at peak efficiency and safety, a computer might need additional software that keep the computer or data safe and running. Sometimes that utility is a separate program, and sometimes it is included as part of the operating system. Common utilities include file manager, backup, antivirus, antispyware, firewall, or conversion utilities.

### *Evolve and De-Evolve*

The Utility category changes more often than other categories because useful utilities tend to be absorbed into the operating system over time. For example, in the early days of computing (when DOS was the current operating system) many people purchased third-party programs such as Norton's Utilities. These programs provided an easy-to-use file manager, a program to do backups, and a program to undelete deleted files. When Windows 3.1 was released, however, it included a file manager and a backup utility. When Windows XP was released, it included a trash bin that could undelete files as well as a very powerful search utility. Of course, it is also possible for utilities to be separated out from the operating system again (see inset paragraph *De-evolvement*).

***De-evolvement***

While most of the time utilities have gradually been absorbed into the OS, Windows has de-evolved. Microsoft took utilities that had been useful in Windows XP, and removed them from the next version of the OS (Windows 7) and hasn't returned them. For example, Windows XP had highly useful and effective *search* and *backup* programs. The search and backup features included with Windows 10 lost many of the advanced features upon which users had come to rely. For example, the new Windows backup could not back up more than 250 gigs, and could not do an incremental backup at all, forcing users to invest in a separate backup utility*.

### *Virus Prevention*

Unfortunately, there are some people who enjoy causing havoc just for fun, and they use our reliance on computers to do so. When someone writes code that destroys or alters data, it is known as *malware*. Malware is any software designed with malicious intent.

One type of malware, a *virus,* is directly designed to do damage. Viruses can be whole programs or just little bits of code, but they usually self-replicate and spread. A common type of virus hijacks your browser so that it will always take you to a website (usually where you can buy things) and prevents you from going to where you actually want to go. Another type of virus is called ransomware. The virus takes your computer data "hostage" and won't let you access it until you pay up. Sometimes this starts with a fake report that you have a virus, which then gives you a link, a website, or a phone number to call to "eliminate" the virus (which is actually when they plant the virus, which then pops up as ransomware a few weeks later).

The most common way for a virus to spread is through email via an attachment or a link to an infected website. (This is why it is important not to open up any attachments that you are not expecting, or click on any links that you don't recognize - even from friends, who may be unknowingly infected.) Hackers have become very sophisticated in their approach. It is difficult to ignore an email that says "Here is the important information you requested. Please review and get back to me right away" as if a co-worker is sharing info that you need. Or "your package has been delayed – please read these instructions for receiving it." Or "your account has been frozen – click here to reopen it." These emails sound so legitimate, and the websites look so legitimate,

* And don't get me started on the extremely poor search capabilities of Windows 10 compared to the excellent search utility on Windows XP. I now use a separate utility called Search Anywhere which is far superior to Windows 10 Search.

that even technology veterans can be fooled. It is important that you learn how to tell the real emails from the malware emails.

Another kind of malware is *spyware* - a program designed to watch what you do on your computer, including watching which websites you go to. Ostensibly designed to sell the information to marketing brokers, spyware often does damage as well by slowing down your computer to the point where it is unusable.

In the past people were forced to purchase Antivirus utilities such as BitDefender, AVG, Norton's Antivirus or McAfee's Antivirus, but Microsoft now provides a free utility called *Windows Security* that is more than adequate for protecting Windows computers from both viruses and malware. Windows Security had previously been called *Defender Antiviris* or *Security Essentials,* and (despite the claims from the competitors who lost marketshare pretty fast), Windows Security turned out to be just as good as third party software. Because new viruses and spyware programs are constantly being developed, an antivirus program must also be constantly updated with information about the new viruses so that they can be prevented from infecting the system.

Businesses must ensure that all of their user systems are not infected with malware. An IT department might place an unprotected computer on the network purposefully to see how long it takes to get infected with malware. An unprotected computer is called a *honeypot.* A typical honeypot gets attacked within 2 minutes, and gets infected with a virus within twenty minutes.

Another common utility (now included in Windows Security) is a firewall. The Internet was developed without any built-in security. Instead, security was provided through the use of a firewall, a program that would only allow traffic through if it came from certain trusted addresses. Keep in mind that a firewall does not do the same thing as an antivirus program. A firewall will prevent anyone using a system that doesn't have an authorized IP address from connecting to your system. An antivirus will prevent a malicious virus from running on your system. These are two different things, and both are needed to keep your system safe from harm.

Of course, there are many people who still purchase utilities such as file managers, backup programs, anti-virus, and firewalls because they do not like (or trust) the utilities that come with the OS, or their needs go beyond the capabilities of the version of the programs included for free in the OS. For example, an IT manager backing up dozens of servers would need a dedicated backup system due to the complexity of the task. Someone who has to convert data frequently from one system to another might purchase a conversion utility.

### *Fraudulent and Unnecessary Utilities*

It is a sad fact that utilities is one of those categories of software that leads to a great deal of confusion. Software that promises to clean up the system, speed up the system, keep information private or help manage unwanted programs is usually unnecessary. Operating systems are not designed to work well with third-party applications changing registry options or moving programs or deleting drivers. Most of these types of utilities tend to cause damage, and may even be fraudulent. A common ploy is to popup on the screen with a "warning" about viruses or other problems when, if fact, no such problems exist. Clicking on the popup will likely result in the purchase of a product or subscription that is completely unnecessary. The best advice is to be absolutely sure before installing any utility on your system, and never respond to unsolicited warnings about viruses or compromised systems.

## Development Software

Of all the categories of software, development software has evolved the most over the course of the past thirty years. Whereas application software like wordprocessors are still basically the same capabilities as thirty years ago, development software of today would not even be recognizable to a programmer from thirty years ago. Development software includes programming languages and development tools that enable programmers to write code so that other people can use the software that results.

## Languages

Programming languages constitute the "code" that developers write so that computers can understand the instructions. Languages can be categorized in "generations" and are often abbreviated as 1GL, 2GL, 3GL, 4GL, or 5GL with GL standing for generation language.

A computer can, at its very core, only understand two things; on or off (or 1 or 0) also known as binary code. The first generation (1GL) of computer coding was machine language or binary code. In general, people don't program in binary or machine language.*

The second generation (2GL) was composed of short commands that represented a series of binary codes. This was called assembler or assembly language, and only the very earliest programmers coded in it. Assembler language is specific to only one specific configuration of hardware and cannot be *ported* (copied to run on other computers with a different configuration).

The third generation (3GL) was the first "real" programming language. Third generation programs could be written with any text editor, but then were compiled by a program that would take the text and convert it into machine code. Examples of *legacy* languages (languages that are still used in some places but are no longer installed in new systems) are Fortran, Cobol, Pascal, and BASIC. All legacy languages are *linear* in that they ran their instructions in order, one command after another. Examples of linear programming languages that are in use today include PL/I, C, C+, C# (pronounced C-sharp).

More often, programming languages of today are object oriented rather than linear. Object oriented programming came about as a natural evolution from the text-based user interface of DOS to the menu and icon driven user interface of Windows and Macintosh computers. Object oriented programming languages use basic components called *objects* which have *attributes* and *classes* which can be inherited at different levels. Object oriented programming languages are also event based; which means they wait for something to be clicked, hovered over, or touched before launching into a command. Examples include JAVA, PHP, Visual BASIC, Tuple, Python, C++, Objective-C, Rational Rose, Ruby, Elixir, and many others.

It takes considerable knowledge and dedication to learn a programming language and is not something non-programmers can do easily. So the fourth generation (4GL) was designed to be easier to use, opening up development to the power computer user rather than just dedicated programmers. Often 4GLs directly managed a relational database and contained English-like commands. Examples include dBase, RBase, XBase, RPG, Informix, SQL, SQLSybase, Focus, Mapper, Mark-IV, Progress, FoxPro, Ramis, Powerbuilder, Dataflex, Forte, and Access. Additionally, fourth generation languages were often interpreted instead of compiled. They required a program on the user's machine that would read and convert the code instantaneously instead of requiring the programmer to convert the code to machine language. This was a big deal because programs written with compiled languages could only be run on the type of computer for which it was compiled, whereas interpreted languages could run the same program on many different computers, relying upon the interpreter to feed the right machine language to the processor. In other words, they were easy to port.

When they were first introduced, fifth generation languages (5GL) were touted as the next level of artificial intelligence - with the ability to understand natural language and learn through the use of a sophisticated set of conditional commands called a neural network. Examples included Logo, OOPS, LISP, and R2, Smalltalk, scikit-learn and NLTK. When it became obvious that computers

* At least not anymore. My father became a computer programmer in 1963 and worked on some of the very first mainframe computers. Back then, there were no programming languages, and he programmed in pure binary code, editing thousands of pages of zeros and ones.

were still decades away from understanding natural language or learning independently, fifth generation languages sort of fell by the wayside. Very few people program in them today, although some useful aspects of artificial intelligence can be found in many applications such as natural language processing (SIRI, Cortana, Google Talk, etc.)

## Development Tools

In and around the programming languages themselves are many other "tools" that are necessary to turn the code into usable programs. The earliest of these simply partnered a text editor with a compiler so that developers didn't need to exit out of the text editor in order to compile and run the program. As software development became more sophisticated, more capabilities were added into the development environment. Additionally, as most development moved to the web, traditional LAN-based or single machine programming decreased. Most development these days is web-based.

Many of these development tools require the use of a powerful workstation and a specialized development network environment. The first of these development systems was called a CASE workstation. CASE stands for Computer-Aided Software Engineering. Now these environments are called Integrated Development Environments (IDE) or Application Lifecycle Management, or (ALM) )systems. Typically these systems included a text editor, a compiler, and a front-end system that helped the developer to envision the application. The developer would use the front end to design the underlying database, plan the screens and menus, and write the reports. Then, instead of typing thousands of lines of code, the developer would have the workstation generate the program code automatically. Examples include Netbeans, Eclipse, IntelliJ, and Visual Studio.

***Multiple Uses of Program***

> Another common point of confusion is the term *program* which can be used as a noun (as in "I want to purchase that software program") or as a verb (as in "Please program the system to ask this question"). Programmers program programs is a grammatically correct sentence.
>
> Furthermore, *to program* generally means *to write code in a computer language.* In recent years, due to the proliferation of sophisticated workbenches and integrated development environments, programmers are often called *developers* because they don't actually write code anymore; they develop programs instead. The actual coding is done by the workbench software.

Sometimes IDEs would include middleware. *Middleware* was the name given to the "missing piece" between the different software components in a distributed environment. For example, a relational database running on a dedicated database server (such as mySQL, Oracle, PostgreSQL, or DB2) may need to connect to the application running on the application server written in a programming language (such as Visual Basic, Cobol, Java, or C). Since programming languages had no built-in commands to work with relational databases, middleware was needed to enable them to work together. Examples of middleware include PHP, ASP, and Javascript. (There are some who argue that these systems are programming languages themselves, that they have evolved beyond middleware, but it is difficult to draw the lines. For purposes of this textbook, we will consider middleware to be a special type of programming language[*])

Over time, however, the term middleware was more often used for the standardized architecture used within an enterprise such as Simple Object Access Protocol (SOAP), Web services, Representational State Transfer (REST), JavaScript Object Notation (JSON) and Enterprise Application Integration (EAI). The nuances of the different types of development software and

[*] More details can be found at https://www.keycdn.com/blog/web-development-tools/ written by Brian Jackson and last updated on June 14, 2016. Last accessed on September 20, 2020. Additionally, you may be surprised not to find HTML or CSS in this list. As we will explain in the next chapter, HTML and CSS are markup languages, not programming languages.

middleware is beyond the scope of this book, but managers of technical staff will need to be very familiar with the webtools being used by the technologists.

In addition to the software used directly by the developers for writing code, IDE systems might include programs to help with change management, project management, requirements development, documentation development, testing, and quality assurance/defect tracking. Many of today's systems were designed to integrate with Github, or Apache Subversion, which are third-party version control websites.

## 5-1. Interactive

**1. Class Activity**

On a Windows 10 machine, click on Cortana (O) or Search and type ***Run***. (On older Windows machines, press on Start and choose *Run*. You can also right-mouse on the Windows icon and choose *Run* from the context menu).

In the command line, type in `cmd` and press [Enter]. What appears is a command line interface that is similar to the DOS prompt which was used before Windows became so common. The symbols at the beginning before the blinking cursor is called the *prompt,* and it shows the current default *directory* (viewed through Windows as a *folder*). You will notice the prompt change as you move around the disk. Type some of the following commands to get a feel for how it works, looking at the screen and trying to figure out what it is showing you for each command.

Here's what the commands do: `C:` moves you to the C: drive (if you are not already there. You may be on a network drive, which would be called by another letter. C: is reserved for the main hard drive on any personal computer*. The command `cd \` changes the directory to the root (the top level of the directory hierarchy). The command `dir` lists all of the files and directories in the current directory. The command `cd Windows` changes the directory to the subdirectory Windows, which contains all of the system files. Adding the `/w` option to the dir will give you a wide view of just the file names. Adding the `/p` option stops the scrolling after each page and waits for you to press a key to continue. The command `cd ..` takes you "up" one level, to the parent directory of whatever directory you are currently in.

```
c:
cd \
dir
cd Windows
dir
dir /w
dir /p
cd logs
dir
cd ..
dir
cd ..
dir
```

Navigate around and look at a few other directories than the ones in the list. Can you see the files in the directories? Close the command box by typing `exit`.

**2. Independent Activity.**

Look around at the files on your computer until you find several files that have only text in them (word processing files) or have only pictures in them (graphics files).

(You may have to turn on your file extensions in order to see which kind of files they are. To turn on the file extensions, go to **This PC --View** and check the box for **File name Extensions** in the Show/Hide group. You can also Right Mouse on the Windows Button in the lower left corner and choose **File Explorer** then choose **File—Change Folder and Search Options—View** and uncheck the option **Hide Extensions for Known File Types**).

Compare the sizes of the graphics files to the text files. What do you find?

---

* Why is C: the hard drive and not A:? Because in the early days of computers, they usually had 3 drives, an A floppy drive, a B floppy drive, and a C hard drive. The floppy drives have disappeared, but the C: drive has stuck.

3. **Independent Activity:** Mapping Network Drives

A drive mapping is associating a drive letter with a folder on the network, and is the way that you can access your folders whenever you are on a network. To see the drive mappings, open up **This PC** (on other versions of Windows, it's called **My Computer)** and click on the **Folders** button so that you can see all the drive letters. You may have to "close up" the folder detail in order to see the mapped drives by clicking on the **V** just to the left of the folder name. The **>** indicates that the folder is already collapsed – if you click on the **>** the folder detail will expand.

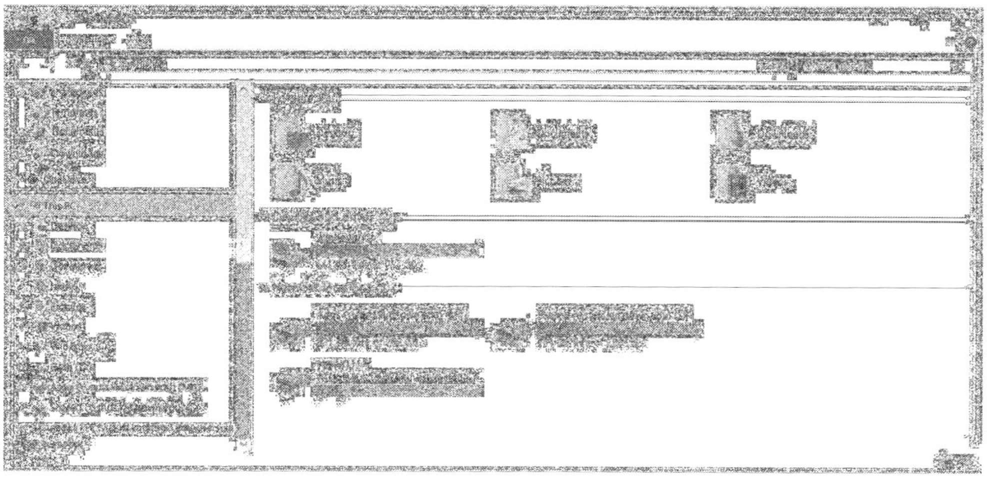

**Figure 40. Drive mappings under this PC.**

Take a look at the drive letters in the list. You can see that the S: drive is mapped to the **Shared** folder. If you are in one of the computer labs, the drives will automatically map. If you are on the KU network, but in one of the residential labs or in the library, the drives are not mapped automatically (i.e., do not appear when you look at the **This PC** or **My Computer** list), then you can map the drive in order to view it.

In order to map a drive, select This PC. Choose **Computer--Network—Map Network Drive**

A dialog box will appear that will look something like the dialog box below. Click on the button with the **V** to the right of the word **Drive:** and choose an unused drive letter. The drive letters with something after them are already used for something and should not be changed.

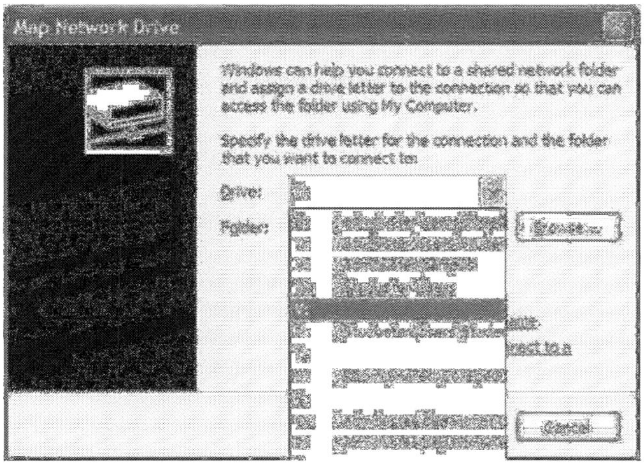


**Figure 41. Mapping choices.**

Once you've chosen the drive letter, type the path into the **Folder** box. For example, on the Kutztown University network, the path for a student's personal network drive is:

`\\stufs.kutztown.edu\users\students\username`

If my username were aaaaa111, I would choose to map the U: drive, and what I would type in would be:

`\\stufs.kutztown.edu\users\students\aaaaa111`

For example, if I wanted to get to the folders with the team projects which were on the Apps volume of the same network server, I would choose to map the M: drive, and what I would type in would be:

`\\stufs.kutztown.edu\Apps`

If I wanted to get to the shared folders on the same network, I would choose to map the S: drive, and what I would type in would be:

`\\ stufs.students.kutztown.edu\shared`

At this point, a dialog box may come up and ask you for your username and password. Enter your normal username and password as you would from any computer on the network. There may be additional security (such as authenticating through a wireless network or connecting to the network through a VPN [virtual private network]), but the process for mapping is similar for all networks. (This is just an example, Kutztown University is slowly moving away from mapped network drives, so by the time you read this they may no longer be active.)

4. **Independent Activity**: Do you use any cloud computing services? On how many devices? Do you know where the folder is on each device that contains the cloud data? Make a list of all your data, the locations, and what type of data it is (i.e. how important it is).

## 5-2. Self Study Test Questions:

1. What are the major categories of system software?
2. Name 2 or 3 examples of system software and which subcategory they would fall under.
3. What are the different generations of computer languages?
4. Name at least two examples of each generation of computer language.
5. Explain the meaning of the sentence *Programmers program programs.*
6. What is ALM, and how does it relate to development tools.
7. What is meant by utilities evolving or de-evolving?
8. Give several examples of utilities.
9. What is malware? How can malware be avoided?
10. What is a firewall? What does a firewall usually protect?
11. What is an operating system? Give several examples.
12. What are the advantages and disadvantages of GUI versus command line driven OS?
13. What are the issues and benefits of cloud computing?

6

# Chapter 6: Collaboration: World Wide Web and Other Applications

Overview: In this chapter, readers will learn the following topics:

- ⇨ Difference between the Internet and the World Wide Web
- ⇨ How Businesses Collaborate Using World Wide Web
- ⇨ Other Collaboration Software (groupware, community webs, social media, search)
- ⇨ How Email Works
- ⇨ Email Etiquette and Setup
- ⇨ About Web Pages
- ⇨ How Web Pages Work

## Technology Map Placement of Collaboration

Now that we have covered networking, hardware, and system software, we can begin to talk about some of the applications and devices related to how businesses and people connect to each other using the World Wide Web. The World Wide Web is collaboration software*. Like system software, application software is considered an instruction for the computer (though at a higher level than operating systems, cloud drives, utilities, or development software). Like all software, you can't see it or touch it like you can hardware. It is generally in digital format, except when someone prints out the code. Even then, what you are seeing is a representation of the software, not the software itself.

Understanding the difference between the Internet and the World Wide Web is an important precursor to understanding Collaboration. Another important concept is how businesses utilize the collaborative capabilities of these technologies. Then, finally, we can talk more about the collaboration software itself.

Collaboration is a strange category because it is not actually software, per se, but rather websites and services where people can share information. For our purposes, the collaboration category is any software and hardware combination, which results in people being able to connect with other people. The category includes the World Wide Web, Email, Groupware, Community Webs, Social Media, and Search.

---

* Previous versions of this book called the World Wide Web networking software, but because it has become so ubiquitous in all use of software, it has been promoted to collaboration application software. Collaboration software is changing so rapidly that categories are difficult to establish and may change drastically in a matter of months. Many pioneering collaboration systems (such as Novell's Groupwise) have been rendered obsolete and others have been swallowed by other collaboration systems. For example, IBM's Lotus Notes became a component of Domino/Websphere and Microsoft's Exchange and Groove became components of SharePoint Services.

**Table 15. Technology Map Collaboration Software Categories**

| MAJOR | MINOR | CATEGORIES |
|---|---|---|
| Application Software | Collaboration Applications | Browser (Microsoft Internet Explorer or Microsoft Edge, Mozilla Firefox, Google Chrome, Apple Safari, Opera) |
| | | Email (Web-based email such as Yahoo, MSN, AOL, Gmail, Hotmail, Comcast, etc, or Client-based email such as Outlook, Windows Mail, Pine, Thunderbird) |
| | | Groupware (IBM Domino, Microsoft SharePoint, Novell GroupWise) |
| | | Community Webs (Amazon, Ebay, Digg, StumbleUpon, Kayak, Youtube, TikTok, Blogspot, Wikipedia, Craigslist, Taobao, Pandora, etc.) |
| | | Social Media (Facebook, Linkedin, Twitter, Instagram, Snapchat, TicToc, MySpace, Friendster, Diaspora) |
| | | Search (Google, Bing!, Yahoo, Baidu, Ask) |

## The Difference between WWW and Internet

The public Internet is a worldwide internetwork of computers and computer networks. It was developed through the efforts of hundreds of people, but primarily Bob Kahn at Bolt Beranek and Newman (a consulting firm) and Vint Cerf at Stanford University. (See the Story of the Internet inset in Chapter 3: Network: Internet Software.)

Shortly after the Internet came about, Timothy Berners-Lee and Robert Cailliau developed an easy-to-use interface (a system of hyperlinks) that sat on top of the public Internet. The system of hyperlinking is called the World Wide Web. A hyperlink is an object (a picture or word) that is programmed so that if you click on it with a mouse or press it on a touchscreen, you will jump to a new location to view a new page. Suddenly, someone with no technical training at all could connect to any other computer on the public Internet network and could retrieve information from any of those computers. (To be accurate, it felt like suddenly, but it took many years. The Internet* started functioning in 1982, and the World Wide Web was invented in 1990, though it didn't come into widespread use until after 1995 when Windows 95 was release with TCP/IP built into the operating system.)

Many people use the terms World Wide Web and Internet interchangeably, but the truth is they are two completely different things. The World Wide Web is an application (a giant shared application, but an application nonetheless) that runs on top of the Internet, no different than email or search engines. Because of the World Wide Web, all we do to connect to any computer anywhere in the world is click a mouse or touch a screen.

In this chapter, we will focus on the applications that run on top of the Internet and how those collaborative applications are utilized by businesses today.

## Business Use of World Wide Web

Businesses have been around for thousands of years and the World Wide Web has only been around for twenty-five years. Nonetheless, businesses have reacted speedily to the new capabilities that never existed before. The early years were very rocky, but eventually most businesses realize

* If you count ARPANET, the precursor of the Internet that was controlled by the military, then you would need to say it started functioning in the seventies, but universities started using it publicly in the early eighties.

that they needed to utilize the Internet in some form or other and could easily take advantage of the expanded market the World Wide Web could deliver.

There are several different ways that businesses can utilize the World Wide Web and all its technologies:

- Corporate presence (brochureware site)
- Product/service information
- E-commerce
- Development of Online Marketing Community, Intranets & Extranets
- API (Application Programming Interface)

## Corporate Presence

A corporate presence web site is often known as the business card web site or a brochureware site. Some companies spend millions designing just the right look and feel of their sites; others get a high school kid to put up a free WIX site with their company logo on a welcome page and a list of their products and services below. But in both cases the Web site contains static information about a company. It can be considered the equivalent of a Yellow Pages advertisement[*]. These days, a brochureware web site is even more essential. Most people (97%) go to the world wide web to find products, services, and businesses[†]. While some businesses try to get away with just a Facebook or Instagram page, that would be a mistake. Social media is a temporary time-sensitive sort of contact that requires followers or members. A website, on the other hand, is completely open to the public and provides the same information all the time.

## Product Information

Once businesses master the brochureware site, depending upon the businesses, they may be ready to start providing detailed product information on their web sites. This is more complicated than you may think. Anytime there is information that must change relatively frequently, you either must have someone who knows how to code in HTML (the HyperText Markup Language of the World Wide Web), or paying someone to make simple text changes to the website, or setting up the website for *content management*. The most common content management application is Wordpress[‡]. A website set up with a content manager is easy to modify and doesn't require that the end user be able to program in HTML code.

Some businesses place the product information online and then provide a toll-free number for people to call when ordering. This enables customers to purchase "online" (kind-of) without going to the added expense of installing a full ecommerce system. Ecommerce, of course, would be the next step.

---

[*] If you are *very* young, you might not even recognize the term *yellow pages*. Before the World Wide Web, businesses had a business phone number, for which they received a listing in a directory the phone company called the *Yellow Pages Directory* (as opposed to the *white pages*, which had everyone's residential phone number listed). Businesses could pay extra money for a larger advertisement in the yellow pages. When people wanted to find a product or service, they would look up that product or service in the phone book, so it was a major source of revenue for the businesses advertising in it.

[†] **Top 10 Reasons Why Your Business Needs a Website in 2020**, by Mukesh K. Singhmar. Posted on December 08, 2019 https://www.seeresponse.com/blog/top-reasons-business-needs-website/. Last accessed on September 23, 2020.

[‡] There were, for many years, many different types of content management systems including Joomla, Drupal, Expression Engine, TYPO3, MediaWIKI, and dozens more. By 2016, generally most of them have fallen by the wayside, as WordPress seems to have been adopted by most websites that require a content manager. Even professional web developers utilize WordPress as it makes putting up a website for others very easy.

## E-commerce

E-commerce occurs when people can order products and services directly from a Web site. This typically involves putting a catalog of items online and providing screens for ordering and paying for the items.

### Setting up E-commerce

Setting up E-commerce is complex. One solution to limiting the complexity is to simply place products and services on someone else's Web site like Amazon.com or eBay.com—both of which are like distributors for online vendors. For Amazon, the costs range from 6 percent to 15 percent plus per transaction charges (a few more bucks). For eBay, the costs range from 6 percent plus an insertion fee of up to $5. These charges are in addition to the normal 3 percent to 6 percent charges for a credit card or PayPal sale.

Paypal* is a payment processing company that is very popular for businesses doing e-commerce. Customers can pay with a credit card using Paypal as a merchant account or directly with a Paypal account (an online alternative to traditional banking). Paypal is popular as an alternative to paying for a merchant account (the ability to take credit card transactions) because Paypal is much less expensive. Unless a business generated hundreds of thousands of dollars every year in online transactions, monthly merchant account fees are atrociously high (proportionately). A Paypal account allows a business to transfer the funds paid by the customers to their bank for just the tiny transaction fee to Paypal without a monthly account fee. Paypal has come full circle and now offers a physical credit card to its business account holders.

As a merchant account, Paypal is still only for online transactions. In order to take credit cards in a storefront, a business needs to get a regular merchant account and rent or buy the equipment for the credit cards to be "swiped" or inserted. There are mobile versions that connect to tablets or laptops (such as Square) or installed versions. Having an offline solution for credit cards is generally inexpensive (comparatively speaking) because card-present fees are lower than card-not-present fees.

However, fraudulent transactions are much more prevalent with online transactions. Since the vendor is responsible for fraudulent transactions, companies often purchase various fee-based security authentication schemes, which add to the cost of the credit card services but saved money by avoiding more fraudulent transactions.

The next consideration, of course, is the financial software being used. Most of the time, financial software has some kind of Point Of Sale (POS) module that provides a link between the accounting software and the merchant account and payment gateway. Generally, financial software also connects directly to banks as well, so that deposits and payments are done online too. Many financial packages, such as Quickbooks, are moving from off-line desktop versions to on-line versions in order to more easily integrate with banks and credit card authentication systems.

As the systems become more and more integrated, the expectations of customers have risen. In addition to the back-end considerations for how people can pay for transactions they have conducted on the world wide web, there is also consideration for how easily the online shopping cart works and premium features such as links to the websites of the shipping companies to track the packages. In order to put on the web real-time dynamic information about products, a company must think about databases, middleware, and links to inventory systems. This requires custom programming and a full-service web development provider. It is not easy to do well.

* Paypal started out as Confinity, an online payment system as an alternative to traditional banking, in 1998. It had an initial public offering in 2002, was purchased by eBay right afterwards, and then spun off as a separate company again in 2015.

In the early days, in order to avoid the high cost of integrating their Web inventory with physical or in-store inventory, many businesses simply set up their Web sites as completely different storefronts with separate inventory. Customers complained bitterly. They couldn't return something to the store that they had purchased online. They might buy something on the web only to find out later that the item was out of stock (unknown at the time of the order because the web only gets updated in the evening). They might want something that is out of stock on the web, without knowing that they could have easily gotten it in-stock at a local store.

These days, it is not uncommon for a business to reflect on a website their real-time inventory at all their brick and mortar stores as well as their websites online inventory. But synchronizing inventory is not an easy task (see the inset *The Challenge of Integrating Web Inventory with Live Inventory*.) None of these problems are insurmountable, but problems like this become troublesome barriers to e-commerce for some businesses.

***The Challenge of Integrating Web Inventory with Live Inventory***

Any operational manager responsible for inventory control can describe the myriad of difficulties in keeping track of inventory. Let's discuss just a smattering of common problems, mostly surrounding the difficulty of granularity.

Granularity refers to the level at which a unique item exists. Imagine that we own a mom-and-pop knickknack store. We sell decorative candles. We can order a skid of candles from any one of three different vendors. We take the skid and break it apart into three levels of inventory: cases, boxes, and individual candles. The candles come in either small or large, and either red, blue, green or white. For a small knickknack store none of this is any problem. When we sell the items, we know that the individual small candles are $2 (regardless of color) and the individual large candles are $4. A box of 12 small candles is $20, and a box of large candles is $40, a case of small is $240, and so on. When we want to reorder, we just take a look at the inventory that is left. If all the red candles are gone, we put them on the sheet to reorder.

This process is easy for a human, but for a blind and dumb computer system it is very difficult. It is very difficult to properly figure out what items are in stock, how much to sell each candle for, which vendor to purchase from, and which items need to be reordered. Most inventory systems in use today don't work at that level of granularity, choosing to ignore color, for example, when identifying the SKU (stock keeping unit), resulting in only 18 unique identifying numbers instead of 72. These systems rely upon a human being who can easily recognize which color needs to be ordered. Unfortunately, that wouldn't work in an online system because people (oddly enough) don't want to be surprised by what color candle they ordered. Some online systems get around this by adding an "attribute" field and expecting the buyer to type in the color or choose the color from the list. This, then, gets into a huge programming issue because each product would need a different attribute list. This is just one example of the complexity involved when trying to provide an online purchasing experience.

Ease of ecommerce is not just integration with inventory. The look and feel of the screens are important. Even more important is the usability of the site. How easy is it to order? How quickly does the screen download? Does the site remember what customers ordered before, or do they have to enter their name and address every time they get on? Can they configure complex items when they order them? How easily can they find what they want? How secure is the credit card information? How secure is the connection? Can others comment on the quality of each of the products or services? Companies usually spend a great deal of time and effort in improving the usability of their web sites.

## HTTP versus HTTPS

Website security brings up another issue. Let us reiterate that the Internet itself was built without any sort of security. In order to get "security" (i.e. the ability to hide the information being transmitted from prying eyes), security had to be added on. One way to do that is to utilize *Secure HyperText Transfer Protocol* (HTTPS) instead of simply *HyperText Transfer Protocol* (HTTP) to transfer data back and forth between the client browser and the server website. Although HTTPS does not encrypt the data with a private key the way a VPN (Virtual Private Network) does, HTTPS does at least encrypt the data sent with TLS (Transport Layer Security, formerly SSL [Secure Socket Layer]). This level of encryption is actually only necessary when sending credit cards or passwords or other data vulnerable to theft, but a few years ago the browser companies (Google, Microsoft, and Mozilla) decided that all websites should utilize HTTPS whether they needed it or not. As a result, perfectly safe websites are now "off-limits" because the browser incorrectly reports the website as unsafe because it does not exclusively utilize HTTPS. This has caused an extra level of stress for small business owners who either must learn how to implement HTTPS on their websites or pay someone else to do so.

***HTTP versus HTTPS for Non-Profit organizations***

I have a lot of personal experience with the frustrations of the decision by Google, Microsoft, and Mozilla to require all websites to utilize HTTPS. I run a non-profit organization called Health, Prosperity, and Leadership Institute that helps small groups and causes with health, prosperity, and leadership programs. One of the services we offer is free websites to our groups. Shortly after Chrome, Firefox, Internet Explorer and Edge began preventing browsers from opening up pages on these non-profit sites (despite the fact that they didn't have any ecommerce or need for security), we started getting calls from our organizations complaining that no one could get to their websites. Huge dialog boxes with dire warnings would appear on the browsers of their constituents and donors telling people that the site they were about to go onto was unsafe and could have viruses or malware. The first year, because we didn't know enough, we paid hundreds of dollars to our web hosting company to install SSL to avoid the problem. The following year, when the web hosting company increased the price for SSL precipitously (17 times, because now they wanted to charge for EACH domain instead of just the top level domain), we had to find a solution or give up providing the free service.

And there was a solution – we just had to learn it. There is a company called ***Let's Encrypt*** which provides a free, automated, and open certificate authority (CA), run by the Internet Security Research Group (ISRG). It did take us over a year to figure out how to install the certificates (because our web hoster was not supportive [since they were losing thousands of dollars they wanted to charge] and most certificate installers required the domain to have superuser administration rights [something our web hoster did not allow us to have]). Eventually we figured out how to use a third party program called **acme-challenge** to install the certificates and then renew them every three months automatically. Special thanks goes out to the wonderful people at **Let's Encrypt** who were patient and kind during the learning curve of trying to get the SSL certificates installed so that all of our websites could be viewed with HTTPS instead of HTTP.

## *Online Marketing Community*

Having an online community is almost a business mandate at this point. Every business seems to have a Facebook page, or a place for people to post reviews of their products in addition to a webpage that provides basic information on the company, its products and services. Even if the vendor themselves don't create a community, there will be another aggregate website out there allowing people to post their review of the products and services. Ninety percent of consumers find businesses on the World Wide Web, and 82% read online reviews of businesses on a variety of review sites before purchasing.*

Online marketing can be very successful. Some products have grown entirely due to the virtual buzz surrounding them created by vendor-supported communities. One of the earliest online community marketing success stories was the Big Green Egg (Figure 42), which is a high-priced

* *Local Consumer Review Survey* by Rosie Murphy posted on December 11, 2019. https://www.brightlocal.com/research/local-consumer-review-survey/ Last accessed on September 27, 2020.

ceramic grill. Devoted fans post frequently on www.biggreenegg.com and that attracts others who purchase it.

**Figure 42. Product with devoted online fans: The Big Green Egg.**

Online communities provide a prime opportunity for companies to get to know their customers more intimately and keep a finger on the pulse of their needs and behaviors. Online communities offer a chance for companies to break down geographical barriers by connecting people globally in many ways through online interaction. They allow for more detailed and sustained conversations and deepen customer relationships. They offer online interactive access for people, marketing collateral, information and products.

In the early days, setting up a community was an expensive process. Now that there is Facebook and several other websites that already attract people, setting up a page is easy. But is it effective? That depends entirely on the company, the product or service, and the target market. There are some online communities that have stood the test of time and have survived, bringing increased business and customer engagement. Thriving online communities seem to have the following characteristics that enable them to endure:

- Have a target customer base that is relatively tech-savvy.
- Don't cost the members any fees (i.e., are free) and don't have too much advertising.
- Are focused on hobbies and interests of people.
- Have long-term leaders (perhaps even volunteers) who post frequently and keep everyone interested and civil.
- Focus on relationships, not information technology or products.
- Have enough of a base membership to keep active posting.

While it seems obvious that online communities and social media sites are going to be a permanent part of the future, just exactly how businesses can best take advantage of them is still a question.

Another way to utilize the community aspect of the World Wide Web is to place usually-private resources at the hands of an employee through an intranet, or a vendor or customer through an extranet.

## Intranets and Extranets

An intranet is when a company uses Internet technologies (such as TCP/IP , and HTML*) to provide their employees with access to needed information. For example, if a company puts absence request forms on a web site for employees to download and fill out when they want a day off, they are implementing an intranet.

An extranet is using public Internet hardware, software, and protocols for making resources available to partners (such as vendors and representatives) and customers outside of the internal network. The actual hardware and software are the same as an intranet.

To some extent, an extranet is used every time an employee emails someone outside of the company. But an extranet can be even more useful. By providing passwords and a secure encryption protocol (provided by most web hosting services) and an encryption transfer method

* HTML stands for HyperText Markup Language, which is covered in more detail later in this chapter, and TCP/IP (Transmission Control Protocol, Internet Protocol) was covered in the previous chapter.

(often called a VPN or virtual private network), we can allow select people to use the public Internet to access private data that stays private.

The productive uses are infinite: allowing employees to work from home, providing literature to representatives, delivering promotional material to resellers, sharing project milestones with partners, sharing inventory needs with vendors, conducting online focus groups about products and services, and so on.

Both intranets and extranets are more complex than typical Web pages, so they require the use of a full-service hosting service or internal teams of IT support people or both. But the value can be tremendous.

### API (Application Programming Interface

These days, it is very common for a website providing web services to publish an API, or an application programming interface . When a company provides an API, they are basically allowing any other developer to reach the back-end data of their website through an API key (a unique combination of letters and numbers that the company provides to any developer who wishes to use the back-end data of the website). For example, if a company wishes to email all their customers with a coupon, they generally can't do it directly because most email systems limit the number of emails that can go out through their servers (in an effort to decrease the amount of spam). But they can use an API to transfer all of their customer info to Mailchimp, a website that provides email marketing services. Every time a new customer is added to their website, the customer is automatically (through the programming of the API) also added to the customer list on Mailchimp so that they would be included in the next coupon mailing.

APIs have mostly taken over the role of EDI which stands for Electronic Data Interchange. Before the Internet, EDI was the only way for two different companies to share electronic data such as purchase orders, invoices, and payments. Because EDI dominated for so many years, and was the result of (often) years-long negotiation between multiple companies, it stayed in wide use for many years after the Internet became available[*]). At first, just the connectivity between the companies became replaced with an Internet connection instead of the proprietary setup of leased lines and the EDI data protocol stayed the same. But eventually XML (Extensible Markup Language) became the protocol of choice and the EDI data protocol has mostly fallen by the wayside.

## Collaboration Software

The category of Collaboration Software includes the Email, Groupware, Social Media, Search, and the technology of World Wide Web itself.

### Email

In the early days of personal computers, a corporation would install an email system so that employees could send messages to each other. It wasn't until the Internet that employees of one company could send email messages to an employee of another company. Today, of course, email is ubiquitous. For many businesses, it is as hard to imagine doing business without email as it would be to run a business without telephones. But email is still a relatively recent process. In 2003, a survey of 268 businesses[†] revealed that 35% still did not have Internet email (although 27% of them had corporate email within the company). Almost eight percent did not have any email at all. The graph is shown in Figure 43.

---

[*] Based on a survey discussed on http://www.eft.com/edi-dead-we-asked-200-supply-chain-executives-their-thoughts by EFT Infographics on September 8, 2015. Last accessed October 2, 2020.

[†] Courtesy of ETM Associates, Inc in conjunction with Berks County Chamber of Commerce. Used by permission.

Even in companies that had email, not everyone got an email address. Only 50% of the companies gave most of their employees' email addresses, as can be seen in Figure 44. The same survey, done just four years later, showed that 98% of the companies had Internet email, and most gave email addresses to all their employees.

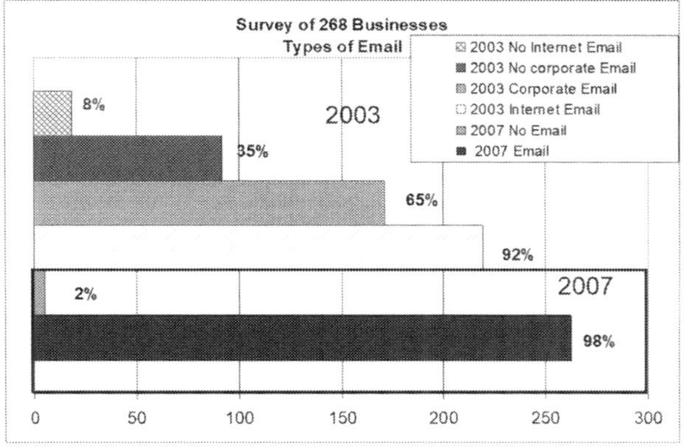


**Figure 43. Survey of types of email.**

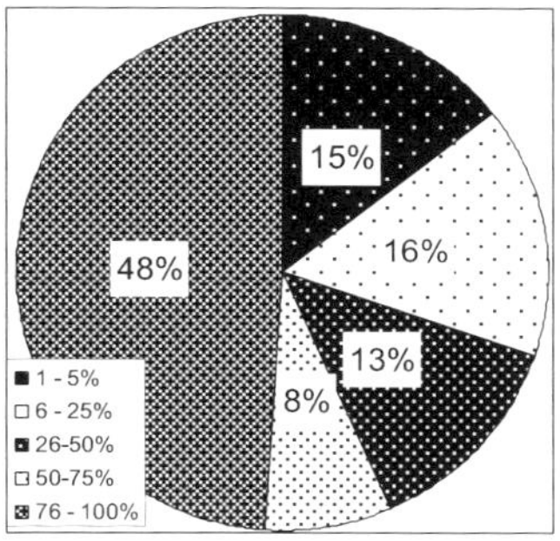


**Figure 44 Survey of employee percentages with email in 2003.**

Now, less than twenty years later, it wouldn't even make sense to do the survey; *of course* every employee has an email address! Email is unequivocally the "killer app" (a phrase that refers to the one most important application that "sells" a technology).

### *How Email Works*

We rarely think about the actual process of getting an email message from our computers to the computer of the person for whom the email is intended but knowing how email works is important for everyone.

An email system needs a sender and a receiver. At each end, the sender needs to have an application that is capable of composing and addressing the message.

An email address has three components: the address, the @ sign, and the domain name. The address and the domain are separated by the @ symbol, as in "Mary@Tuffet.com[*]". The user composes the message and addresses it using a program generically known as the email client. The email client stores the message until it gets the instruction to send the message (which may be immediate, or may be based on time, such as every 10 minutes, or when the user presses a button or makes the choice to check email). When instructed, the email client connects with the server email system (or email server) and sends the message to it. The email server stores the message until it gets instructed to send it elsewhere or delete it. The email client of the user connects periodically with the email server, notices that a new message is stored in the mailbox of the user and downloads the message so that the user can see that there is a new message. When users see their messages, they can click on them to open and read them.

[*] The SMTP protocol that sends emails is case insensitive – which means it does not matter what case you use when you *send* email. Mary@Tuffet.com, mary@tuffet.com. MARY@TUFFET.com, and Mary@Tuffet.COM are all the same. However, if you use your email address as your login name (as many systems do), then case matters. When someone tells me their email address and specifies the capital letters, I know that they don't understand this simple fact. They think because their login name is case sensitive that ALL email is case sensitive, and that is not the case.

Underneath all Internet email systems is SMTP (simple mail transfer protocol) which contains the instructions for the email clients and email servers to deal with email messages. The email system a person uses might come from a variety of sources. A company such as IBM, or Walmart or BankAmerica provides email services to their employees. Schools such as Kutztown University provides email services to their staff, faculty, and students. The email services might also be provided by a free website such as Google, or Yahoo, or MSN/Hotmail (Microsoft) or AOL*. An Internet Service Provider (ISP) such as Comcast, PenTeleData, Windstream, AT&T, or Verizon might also provide your Email. Or it might be provided by a company outsourcing it to a web hosting company or ISP.

***Story of the @ Symbol***

Ray Tomlinson was one of the programmers working on figuring out how to transfer files over ARPANET (Advanced Research Projects Agency Network, the precursor of today's Internet) when it occurred to him that he could modify the SNDMSG (send message) command to attach files (i.e. an email). At the time, (in 1971) SNDMSG could only send a message from one user to another on the same machine. To attach files outside the system, he needed something to distinguish a local user mailbox from a mailbox on another computer. He noted that it seemed natural to use the @ sign, which was commonly used in commerce as in "6 baskets @ $2 each". In English, the sign was called "at", as in 6 baskets at 2 dollars each. Sending a message to "George" would send a local message, but sending a message to "George@Anothercomputer" would send it over ARPANET to another computer.

Email can have several different configurations (listed here):

- **Direct connection** - for example, Exchange and Domino are both email servers that can connect directly with clients that were designed specifically for the purpose. This is also the description when you use the webpage for email set up by the email provider.
- **HTTP** - (Hypertext Transfer Protocol) connection through a web server onto a web page. Email messages stay on the web page and are viewed as if going to the website.
- **IMAP** - (Internet Message Access Protocol) connection through an Internet site into your email client. Email messages *synchronize* but do not *download.* They remain on the server until you delete them there.
- **POP3** - (Post Office Protocol) a connection between an email server and an email client. Email is downloaded, and only stays on the server if you have configured the settings to leave the messages there after downloading.

Sometimes the email client and the email server both reside on the same computer and work together simultaneously - the email client might actually just be a form on a web page that provides a window into the messages stored on the web server. This is *http connection.*

Sometimes the email client resides on your smart phone, tablet, or computer. Then the email client connects to the email server through the cell phone lines or Wi-Fi, or the Ethernet network, and downloads the email so that you can see it on your device. If a proprietary protocol is used to download that email to your device using a special client software written and distributed by the email provider, then that would be a Direct connection.

The most common setup is the IMAP protocol. The messages never leave the server; you can just synchronize different devices to the server to see your email. The disadvantage of IMAP is that most browser-based clients don't have advanced features like rules or subfolders, and there is no way to archive or offload the emails to a separate folder.

* And yes, there are still many people with an AOL email address. You can tell when someone has been on email for a very long time if they still use an AOL email address since the company was at the height of its popularity in 1989.

A more flexible method (but becoming less and less common) is the POP3 protocol. With POP3 the messages are removed from the server and moved to your device*. The disadvantage is that in order to view your email from multiple devices (such as both your computer and your phone) you must configure the account to leave the messages on the website. The advantage is that you have more control over your email. In addition to utilizing rules to organize your email, you can sort, offload to folders, and archive your email on a local hard drive.

### Email Etiquette

People who email frequently have developed protocols, or practices that have become customary regarding email. Most of the rules for email etiquette are unwritten. Like an unzipped fly or bad breath, violations of the unwritten rules will seldom spark email recipients to say anything. (But leaving a bad impression with others may not be a good thing).

One important factor; rules for business-oriented emails are different from rules for informal social emails. If you are emailing a friend, there's no problem with shortcuts, text codes, poor grammar or spelling. As long as your friend understands the message, you can pretty much write whatever you want.

However, when we use social email rules for business communication, it's like showing up at a business mixer in jeans and a T-shirt when everyone else is wearing a suit. No one will complain, but neither will they take us very seriously. Here are some of the unwritten rules for professional email†. See how many of them you already know.

- Use a meaningful topic in the subject line.
- Begin with a greeting.
- End with a signature.
- Use good grammar.
- Use the spell checker.
- Don't use ALL CAPS.
- Use a professional email address.
- Properly configure your email program with your real name.
- Share your email address; put it on your business card.
- Deal with spam professionally.
- Don't expect that your email will be read (it may end up in a spam folder).
- Don't write anything in an email you wouldn't want to see written on the first page of The New York Times.
- Avoid emotionless email escalation.
- Use CC or BC judiciously to copy others.
- Don't use email for time sensitive information.
- Don't use an email distribution list as a discussion board.
- Don't send jokes, virus warnings, or other extraneous information.

An email message is just like a business letter; impressions count. When you don't use a meaningful subject line, or you skip the greeting and signature, you are conveying the underlying message that the recipient of your email is not very important to you. If you use poor spelling or grammar, you are further portraying yourself as uneducated or unintelligent. Additionally, using all capital letters conveys the impression THAT YOU ARE YELLING AT THE TOP OF YOUR

* The advantages and disadvantages of IMAP and POP3 are discussed in the section on setting up multiple email addresses and/or devices.

† A personal note of advice; always use good grammar and full business-formal rules for email when you are communicating with one of your teachers. It reflects very poorly on you to email a teacher and not put forth that extra effort necessary to demonstrate respect.

LUNGS AT THE PERSON TO WHOM YOU ARE WRITING, which does not convey the calm, thoughtful persona you would probably prefer to convey.

While it may be fun to use ChickInHeat@Hotmail.com or PhillyFan@aol.com, your business communication will not be seen as professional. Also, if you don't fill in the "Name" field of your email program, you make a bad impression. Take the extra step and set up your real name. (Send yourself an email to see how it looks to others when you send them an email, and to see if your real name appears in the From line.)

One of the downsides of email is spam (unwanted bulk email) and scam. Even worse than the hundreds of messages that you get trying to sell you something you don't need is emails that are [gasp!] fraudulent and filled with lies. Phishing is the name given to an email that tries to elicit private information from you for nefarious purposes such as stealing your money or identity. Experienced emailers quickly learn to ignore anything that sounds too good to be true. (No, you did not really win the Irish Lottery, and you haven't really been offered a percentage of the $30 million that some expelled Nigerian king would like to transfer to your U.S. bank account.) You must also be careful with email requests from your friends for money because they are traveling and were robbed and need to pay for the hotel or get out of jail or get a ticket home. And the IRS or the Social Security Office will never email you and demand that you send them money. These are all well-known scams.

But spam and phishing can be dealt with professionally. First, don't be a whiner. Don't complain about spam. Spam is not going to go away. Railing against it is like railing against the humidity. Learn to use your spam filter and don't respond to spam or phishing. Just like junk mail, accept that every day you will have to quickly look through all your messages and delete most of them because they are unwanted bulk messages not addressed to you (even with your spam filter). And don't forget to check your spam filter for the real messages that got caught there! Learn to whitelist email addresses from which you wish to get email unencumbered by being waylaid into the spam bucket.

Why is spam filtering so hard? Like junk mail, spam is in the eye of the beholder. The most common definition for spam is "unsolicited email," but that definition doesn't work well. If I send a "how are you doing" message to my sister in California, it was unsolicited, but no one would call it spam. "Unsolicited bulk email" doesn't work either. If I send the same message to all my friends and family with email addresses, the number is well over 200 (I have a very big family). But again—no one would call it spam (though many anti-spam programs would refuse it because it is addressed to more than 50 names at one time).

In business, if you send a newsletter to 300 people that you know and have met personally even if they didn't ask for it, that is not spam. If you send the same newsletter to 30,000 people, most of whom you've never met, that is spam. If you send a "how are you doing" message to 300,000 people, it would definitely be spam no matter what the content. But if the IRS sends out 2 million emails to people to let them know some information they need to get their refunds, that would not be spam no matter how many people receive it.

Unwanted bulk email is, by definition, based on a human judgment (unwanted). While a human being can look at any message and pretty much figure out immediately if it is unwanted, a computer cannot. As a result, your email messages to others may get caught in the spam filter (also known as the junk folder). Because emails can get caught in the spam filter, it is a bad idea to assume that someone has actually received and read your email. When sending an important email, always follow up with a phone call so that the recipient can get the message out of the spam bucket if it landed there. And most importantly, don't send spam.

Also remember that email has the tendency to come across with a harsh and aggressive tone. In a face-to-face conversation, over 90 percent of a message's intent comes from facial expression and tone of voice, which doesn't exist in email. When receiving a negative-sounding email, reply by phone or a personal visit, not a reply email. If it is from someone you do not know, ignore it.

Be judicious in copying others on your email communications, and circumspect in the topics you choose to email. Filling in the CC: (carbon copy—and yes it is an anachronism) seems like a quick and efficient way to copy third parties on an email message. But copying can be a problem. You don't know how the recipient will react to your copying someone else on the message.

Blind copy (BC:) can also get you in trouble. Blind copy hides the email addresses of the people receiving the email. One legitimate reason to use blind copy is to avoid annoying the reader with a long list of email addresses through which they must scroll before they get to the message. Another legitimate reason to use blind copy is when you are sending to a confidential list of people.

The problem occurs when you use blind copy to hide the fact that someone was copied on the message. That's risky by itself, but it is even riskier if you don't communicate with the person who was copied. They may very well forward (or reply all) to the message—letting the original target person know that you've been duplicitous. It's better not to use blind copy to hide the fact that the message has been copied, but if you do, ensure that the recipient knows why you sent it as a blind copy. Alternatively, if you receive a message without your email address in the To: or CC: fields, treat the message as confidential, and do not forward or reply.

The fact remains that email (alone) is not appropriate for information of a time-sensitive nature. Since there is no guarantee that people will see your email in time (or even at all) it makes more sense to use the phone or send a text for time-sensitive issues, and expect others to do the same.

Forwarding jokes, virus warnings, or other items that urge you to send to everyone you know is a very bad idea. You will note that several people in the government of Pennsylvania lost their jobs or were forced to retire because they forwarded a joke. And you might be surprised by how few people actually appreciate it (and don't just grudgingly put up with it because they like you). In the end, don't joke. Just don't do it.

Finally, be careful what you write in email. Email is very different from postal mail. Email can be read in plain text by anyone with a "sniffer" network device as the email travels over the public Internet. Furthermore, email is stored on several servers as it passes on its journey from writer to recipient. Those servers may be backed up several times throughout the day, resulting in several permanent copies of any email message. Instead of thinking of email as a sealed letter, think of it like a postcard. There is literally no way to keep an email private. By the way—the same is true texting. So think before you text and send.

***Proliferation of Email Addresses***

It is common for a single individual to have dozens of different email addresses (even if they don't use them all.) Let's take an example; Joe Schmoe.

Joe Schmoe graduated from Central High School. Because he is an active alumni, the school assigned him an email address, JoeSchmoe@CentralHS.edu. Joe also got his bachelor's degree from Big State University, so he is also SchmJ2134@BSU.edu. When he started his master's degree, he got JSCHM112@HigherLearning.edu.

His local ISP is Verizon, so he also has Schmoe.Joe@verizon.net, and his work email is JosephSchmoe@GIManufacturing.com. When he was looking for a job he used BigMan@MSN.com, and when he joined a group about weight lifting (one of his hobbies), he got BigLifter@Yahoo.com. One of his suppliers put him on a work team that is using joint documents on Google, so he also has Engineer@Gmail.com. He is also very active in his church, which assigned him the email LeadUsher@SaintJohn.org. When he became the Treasurer of his hunting lodge, he also became Treasurer@HuntersRus.org.

Joe can handle this proliferation in a number of ways: 1) manually checking each one, 2) forwarding each one, or 3) automatically checking each one.

**Manually Checking.** Joe can check the messages at each one of the ten different addresses. At work, the IT people set up Outlook for him, so he goes into that to check his GIManufacturing messages. The people at his church gave him the instructions for using Thunderbird to check his SaintJohn messages, so that's the program he uses to check that email. At home he uses the Verizon website to check his Verizon email. Each of the others has set up a web client, so he goes to the web page of each of the other six addresses to check his messages.

The problem with this method is that checking so many different email servers is a lot of work. Joe usually doesn't keep up, and weeks can go by before he gets a message at one of the more remote email servers. Eventually he will probably stop checking them altogether. He may end up asking everyone to send him messages at BigMan@MSN.net because he checks that one every day or so. But even that is a problem. The other email servers are holding onto his email messages, but since he never goes in and checks them, the mailbox fills up until it consumes all of the space he is allowed (his quota) and then starts bouncing (which means that anyone who sends a message gets a return message saying that the message couldn't be delivered because the mailbox was full - which leaves a very bad impression and conveys laziness).

**Forwarding.** Alternatively, Joe can forward (which means to issue the instruction to each of the email servers to send any messages they receive to another email address) to his "main" email address. Let's say, in this case, most people know his address as BigMan@MSN.net, so he sets all the other addresses to forward to that one. This method is better than the first, although it doesn't always work. AOL email, for example, does not have a forwarding feature. Furthermore, people send to one email address, but get a reply from a different email address, which is confusing for them.

**Automatically Check.** For most business people, the best method is to set up a mailbox to go out to all the other mailboxes and retrieve the mail. Although this method takes a little more time and knowledge to set up, it will facilitate *for others* communicating through email with you. We will go over in detail how to do this in the next section.

***Setting up an Email Client for Multiple Email Addresses***

Joe could handle the problem in a different way. Instead of going to each of the ten different email servers using the email client the technical people described, Joe could decide which one email client he wants to use and set it up to go to all the different email servers and pick up his mail. To do this he would use an email client such as Outlook, (or Thunderbird, or Zumbra) that has this capability (not all email clients do). In the Tools menu he would set up a different account for each of his ten email addresses. He enters in the setup screen for each email account the name of the email server, his email name, and whatever password he uses to access that email. Then, when he clicks on "Send/Receive", Outlook goes out to each of the email servers and downloads the messages. To set up all his email accounts, he would need to understand the four different ways that email clients can connect to email servers.

Often, the IT people who set up the email servers provide choices for the people who are connecting their email clients. (If they haven't, you must choose the one recommended by the IT people who set up the email server.) POP3 provides the most flexibility in terms of storing and retrieving messages from a single personal computer. The messages get downloaded and you can put them into folders, read them even when you are not connected, easily move them to another location, and make sure they are backed up.

On the other hand, if you travel a lot or prefer to be able to see all your messages from any machine at any location, IMAP or HTTP are better choices because the messages stay on the server. The disadvantage is that you can't read the messages if you are not connected to the Internet. Also - you are relying upon the provider to back up the messages, and it is very difficult to move them to another service provider because they are physically stored on a server you don't own or have administrative access to.

Additionally, reading email with a web email client is not as flexible as reading email with a PC-based email client. For example, if you have 90,000 messages over the past year and want to find one that was sent to you from BigGirl@yahoo.com on January 20th, you can easily set up those search parameters in Outlook for the messages that are stored on your hard drive. You can't search that specifically on a web client. You can also more easily set up folders and rules on messages stored on your PC.

A more and more common configuration is to have your email download to your cell phone. Configuring your cell phone to get email is similar to setting up an email client on your PC. Alternatively, a person can simply use the browser on their cell phone and go to the websites that hold their email such as gmail.com or yahoo.com.

Regardless of how many email addresses you have, it makes sense for you to learn how to effectively manage email[*].

## Groupware

Groupware does much more than email. Groupware can provide an entire subsystem for application development. The term *groupware* has been applied to systems that are designed to enable groups of people to work together "virtually." Groupware has been around long enough to be considered legacy software, and it is not a growing category[†]. Newer cloud capabilities with mobile systems have outstripped the features of groupware and decreased the need for its collaboration capabilities.

Groupware can be thought of as add-on to email including forms and shared folders. Novell Groupwise, Microsoft Sharepoint (the groupware portion of Exchange/Outlook) and IBM Domino (the server portion of Notes) are the main legacy systems used by larger companies. The shared folders that enable people to log into the system to place or retrieve files so that people didn't have to attach files to email messages in order to share information. Sometimes this type of software is called portal software because it forms the "entryway" for employees to get into a corporate website (often called an intranet).

## Community Webs, Wikis & Blogs

A Community Web is any website that provides tools for people with a common interest to get together and share ideas or collaborate. Early on websites such as Yahoo Groups, Topica, and Google Groups provided some of the functionality of groupware such as shared file space, instant messaging, listserv, even web conferencing. Some community webs provide a needed service, such as Kayak, Youtube, Angieslist, Craigslist, Digg, StumbleUpon, etc. Many community webs are provided by a business who have an interest in building up loyal clientele, so they provide opportunities for users of the website to share their opinions. Amazon, Ebay, Taobao, Pandora would all fall under this category (along with hundreds of thousands of other sites) because in addition to providing a way for people to buy things, they enable people to connect with each other; providing reviews, answering questions on the products, publishing favorite lists, etc. There are literally thousands of community webs utilizing hundreds of different types of web tools.

Community Webs became even more popular after a certain kind of software known as *wikis* and *weblogs* became available, which made starting a running a community site much easier. The popularization of wiki software started with Wikipedia.org, a free world-wide encyclopedia that anyone in the world can edit. Wikipedia proved itself to be highly structured and relatively accurate, with the advantage of being able to adapt to new information much more rapidly than

---

[*] You can review resources in effective email management https://www.marshaegan.com/resources/articles/#email-prod, last accessed on September 25, 2020.

[†] The use of "groupware" is going down due to the growth of "cloud services" (which provides file sharing as well as basic applications as well as email). The term, and the need for this type of software, is disappearing. However, large companies with a large investment in groupware still use it extensively, which is why it is included in this textbook. Smaller or newer companies would have no need for it.

any printed encyclopedia. Started in January 2001, Wikipedia has grown to over fifty million articles in over 250 languages edited by hundreds of thousands of active contributors.

The type of web application software used by Wikipedia is available on many other sites (or available for download on your own website). But though Wikipedia is an unvarnished success, smaller Wikis never really panned out, though some say the concept of knowledge representation underlying them is just as valid today as always[*]. While the heyday of wikis is past, there is still a lot of wiki software in use. Wikimatrix.org is a website that compares over a hundred different wikis which are available. The most popular wikis are MediaWiki, DoKuWiki, PmWiki, and TWiki.

A blog (which originally stood for weblog) is a web page where a person can post their opinion. Unlike online discussion boards where many people both post and read what is posted, a blog is a web page where one primary person posts (though complicated by the fact that most blogging software includes a reply capability so that readers can read and comment on the blog). A discussion board is where everyone chimes in, whereas a blog is more like one person "holding court" and responding to each comment.

At this point there are hundreds of thousands of blogs, and the novelty has worn thin. Successful busy people rarely have the time necessary to post frequently, so most blogs are done by independent consultants trying to attract clients or people with waaayy too much time on their hands. (A marketing consultant may talk up blogging, but there is little evidence that blogging leads to real sales except in some very specific situations such as independent consultants or personal services.) Furthermore, for anyone who reads many blogs, it becomes quickly apparent that there is a role for editing even in cyberspace. Most newspapers, magazines, and news media have tried adding blogging as one of their tools for getting and keeping readers. Many have stopped trying. As a separate entity blogs are likely to disappear in the next few years as they are already fading out and have been supplanted mostly by social media.

## Social Media

### Social Networking Apps

Social networking is another form of collaboration software, but one which has expanded far beyond its initial beginnings. Social networking sites encourage virtual communities by allowing individuals to post information about themselves, and then invite other people they know to connect to their online profile to see their posts. LinkedIn.com is a business networking site with more than 660 million users in more than 200 countries. In 2016 it was acquired by Microsoft which at first threw its future influence into question, but which seems to have solidified into a positive as its 2020 revenues were more than eight billion dollars[†].

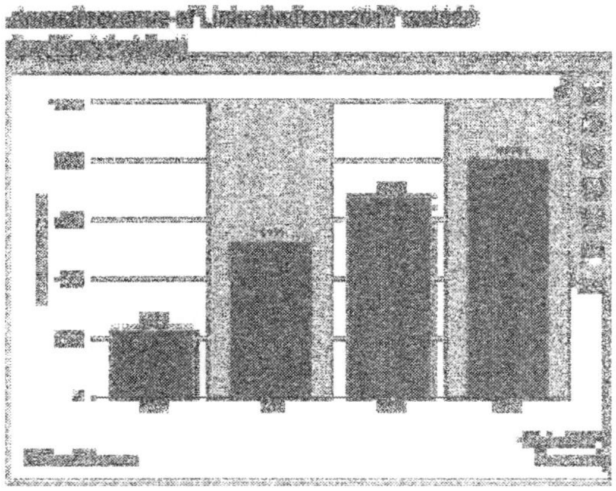

[*] ***Why are we still talking about wikis?*** By Sabine Melnicki, published on September 5, 2019. https://sabinemelnicki.at/blog/why-are-we-still-talking-about-wikis/ Last accessed September 25, 2020.

[†] How LinkedIn failed by Casey Newton posted on Jun 14, 2016. https://www.theverge.com/2016/6/14/11933166/linkedin-microsoft-acquisition-thirsty-email-company. Last accessed September 26, 2020.

**Figure 45. Linkedin Revenues***

Facebook.com rose from a small college website in 2004 to over 2.7 billion monthly active users as of the second quarter of 2020. Facebook IPO'd (Initial Public Offering) in May of 2012, and has grown from $35 a share to over $254.82 a share (in September 2020). Its popularity has been amazing (almost half of all Internet users as seen in Figure 46), but not without cost. Concerns over privacy, hacking, and undue influence have dogged Facebook and Mark Zuckerburg, its CEO and controlling shareholder. Zuckerburg has mostly dodged the bullet, promising more control by users and fewer scandals by third parties given access to the data.

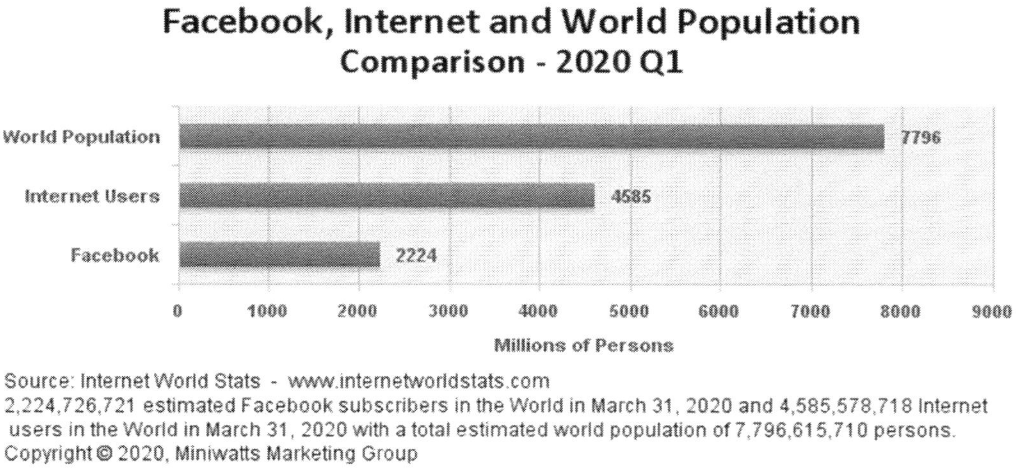


**Figure 46. Facebook popularity**

Facebook notwithstanding, previous social media sites did not do so well. Several precursors of Facebook (Six Apart, LiveJournal, MySpace.com, Friendster) all bit the dust in a relatively short time frame.[†,‡] Facebook purchased Instagram. WhatsApp and 86 other competitors. Google tried to take over the social networking space with Google+ (Google Plus), but did not prosper and Google+ was shut down in 2019.

Twitter appears financially successful, though the stock itself ($43.84 per share on Sept 27, 2020) languishes below its 2014 peak of $69 a share. Twitter has become very popular with politicos, journalist, and other people who are focused on the latest political news and celebrity activities. Pinterest also seems to be holding its own. WeChat is also quite popular, albeit in China and other countries where it has over a billion users. Newer social media sites include Snapchat, TikTok, Clubhouse, Caffeine, and Houseparty.

### *Scandals*

One of the devastating side effects of hugely popular social media sites is the proliferation of opportunities for unwanted sharing of private information. In October of 2010, Facebook admitted that third-party developers (such as those who created Farmville and Mafia Wars, two popular multi-user games played through Facebook) were transmitting private user ID data. It was

* Annual revenue of LinkedIn from 2017 to 2020. By J. Clement, Sep 2, 2020 https://www.statista.com/statistics/976194/annual-revenue-of-linkedin/ Last accessed September 26, 2020.

† Why Facebook Triumphed Over All Other Social Networks, by Gil Press, posted Apr 8, 2018. https://www.forbes.com/sites/gilpress/2018/04/08/why-facebook-triumphed-over-all-other-social-networks/#1ca7defc6e91 Accessed September 26, 2020.

‡ Friendster relaunched as a gaming site with 115 million users in 2011, but in June of 2015 suspended all services.

against policy, but technically possible, so many Facebook users' private information was compromised. In April of 2018 Facebook again faced controversy as it was reported that Cambridge Analytica harvested private data from over 87 million Facebook users and used them for a disinformation campaign that may have influenced the 2016 presidential election*.

### *Single Sign-On*

When one is surfing many different sites on the web, one gets tired of creating hundreds of different usernames and passwords. ***Single Sign On*** is the concept of being able to access many different sites without having to login over and over. One useful side-effect of hugely popular social media sites where people post their public and private profiles is the ability to use a shared login. Nowadays, many other sites are allowing people to use their Facebook or Google profiles instead of creating a whole new username and password. Since single sign-on has been a holy grail for decades, this would seem to be a good thing for password-creation-weary surfers. The question remains, however, as to how secure it is to use a social media login for other websites. It may be that users are opening their passwords up to compromise by allowing shared logins. Only time will tell.

Additionally, browsers have started to save login information (username and passwords) and credit card information. Whether this is a safe convenience or a risky activity is still to be determined as well.

## Search

With the explosion of information available on the web, search has gone from a librarian's back-end skill to a major interface for most people. When the web first began, there were dozens of contenders for search engines. Not any more.

### *Search Engines*

A *search engine* is a site where people go to find other sites (a more generic form of an aggregating site). The most popular search engine site today is Google.com (unless you are Chinese, in which case Baidu.com is the most popular). Because of its focus on the human context of the information, Google became the dominating search engine, quickly outpacing Lycos, Webcrawler, Ask, Excite, Ask Jeeves, AltaVista, Yahoo, and MSN. In 2010 Bing, a new search engine by Microsoft, was incorporated as the default search engine in all new Windows installations, and replaced both MSN and Yahoo by 2012. Bing started to catch up and got to about 30% usage, but just as quickly went back down to less than 3% (as of 2019) despite its Microsoft Windows advantage of default status in Windows Internet Explorer and Edge. In the end, Bing just isn't a very good search engine and, like all the precursors of Google, fell below the obviously superior Google search application.†

* Cambridge Analytica and Facebook: The Scandal and the Fallout So Far, by Nicholas Confessore, posted April 4, 2018. https://www.nytimes.com/2018/04/04/us/politics/cambridge-analytica-scandal-fallout.html Last accessed September 27, 2020.

† *I Used Only Bing for 3 Months. Here's What I Found—and What I Didn't* by Brian Barrett posted on October 17, 2018 https://www.wired.com/story/tried-bing-search-google-microsoft/ and last accessed on September 27, 2020. An interesting experiment that provides insight on why people would go out of their way to change their default search engine from Bing to Google.

***Google Paradigm Shift***

Sometimes it takes a shift in thinking to see the power of the public Internet and the World Wide Web. In the year 2001, I was working as a technology consultant, and I was working on a technical glitch on my own computer. My system kept telling me it was out of memory, though I knew for a fact it had plenty of memory. I spent hours reading through my system documentation, looking at the virtual memory settings in my control panel, trying one thing, then another, then another. I was getting frustrated and beginning to make noises of exasperation when my husband wandered into my office to find out what all the pounding and swearing was about.

"I can't figure out this %$&* problem," I said with a tone of disgust. "I've been working on it for hours."

"Did you Google it?" my level-headed, practical, but computer-illiterate husband asked.

"What do you mean?" I responded, not quite understanding, and a bit annoyed that he thought he could help me. I'd been on the public Internet since 1992! Of course I'd been on all my technical resource sites, the Microsoft support site, and read everything I could about memory problems. What did he think I was, an idiot?

He ignored my patronizing tone and calmly sat down at my computer, wrote down the exact wording of the error message on a piece of paper, switched over to Google.com, typed in the wording exactly as it appeared on screen, and got a whole list of troubleshooting sites that gave the exact right diagnosis and fix for my problem. Yes, I had been an idiot. It wasn't the memory at all, of course, it was a virus for which my anti-virus software did not yet have a fix. I downloaded a file that deleted the virus and my system was fine 10 minutes later.

I didn't grow up with the capability of going to a single source to get, within seconds, any type of information that I needed, so it didn't come naturally, But like everyone else, I now can't imagine doing anything, or writing anything, without my ever-present-all-knowing adviser at my beck and call.

### *Search Engine Optimization*

How do search engines list their sites? Well, the answer is constantly changing and is a whole subject onto itself. The work involved in getting a high profile on the search engine listings led to a whole new industry—search engine optimization (SEO). Anyone considering putting a new Web site up may get inundated with sales pitches from SEO consultants. SEOs keep up with the latest changes in how search engines work.

For example, in the past, the more keyword meta tags (an invisible word embedded in the code on a site), the higher the chance that the site would appear in the search list. At time went by, however, search engines took notice of people focusing only on keyword meta tags. They changed the rules so that sites that had too many keywords on each page became disqualified.

***Be Careful When Hiring SEOs***

SEO consultants sometimes do more harm than good. For example, in the past, the when the more keyword meta tags embedded in a site, the higher the chance that the site would appear in the search list, SEO consultants focused exclusively on loading up on keywords, throwing everything but the kitchen sink into the keyword meta tags. But when search engines took notice of this method of getting listed, they changed the rules so that sites that had too many keywords on each page became disqualified. (This does not apply to all meta tags, just *keyword* meta tags. The title meta tag and the description meta tag are well supported by search engines and should definitely be entered on every page on your site.) The abuse by those early search engine optimization consultants made keyword meta tags useless.

Another early abuse that led to search engine algorithm change was the number of links. A friend of mine, Charles, spent $60,000 on a supposedly sure-fire search optimization scheme proposed by a consultant. At the time, the search engines counted how many times each Web page on a site linked to each other Web page. Through special coding, Charles managed to get his site listed first on all the search engines every time someone entered any of the words on his site—until one day when suddenly his site didn't show up at all. The search engine rules disqualified sites that had too many internal links.

Although the rules are constantly changing, some basics seem to stay the same. The following items tend to influence placement on Google.[*]

- Search term use in title tag, headings, and body text.
- Relationship of body text content to search terms (topic analysis).
- Global link popularity of site (how many other sites link to yours).
- Age of site and document (the older the better).
- Amount of index able text content (more text, less graphics).
- Frequency of updates to page (weekly is better than monthly).
- Documents not too deep (less than four folders in the path).
- Accuracy of spelling and grammar (no misspellings or errors).
- Well-designed code (no coding errors).

## World Wide Web: Under the Covers

While not technically a software package, the World Wide Web is the software application that runs on top of the Internet. Let's delve a bit into exactly how websites work – technically.

### World Wide Web Webservers and Browsers

A website is a server application (software) that runs on a physical server (hardware) attached to the Internet. The web server interacts with the client software called a web browser.

A list of web server vendors and their market share over the past fifteen years can be found in Figure 47.[†] The major webservers are Open Source Apache, Open Source nginx (pronounced "engine X"), Microsoft IIS, or Google GWS. The web server waits for a request from a browser program and then serves up the web page requested.

The browser program, the client side of the client/server transaction, is what does the requesting. The programs include Google Chrome (clearly now the market leader), Microsoft Internet Explorer or Edge, Mozilla Firefox, Open Source Opera, or Apple Safari. (Web browser market share can be found in Figure 48).[‡]

You might note that the web server marketshare had remained relatively stable until about 2014 when the extra capabilities of nginx (load balancing) made it the front line favorite, taking over the top spot previously held by Apache. The browser market has also done a complete turnaround. In 2009, Microsoft Explorer had the majority share of the market, and Google Chrome just 5%. Now Google Chrome is the largest share, dwarfing everyone else.

## Application Servers

Also on the world wide web are application servers, which run the applications that the webservers provide through the website to the web browser. If the web page contains programming such as a list of products or services to pick, a shopping cart, or some kind of game to play or update to post, then the webserver connects to the application server to run the program. Common application server software includes Geronimo, JBoss, Tomcat, Glassfish; IBM WebSphere, BEA Weblogic.

---

[*] I originally got these items (2016) from Google Developer's website which published a guide for developers that explained what influences ranking, but Google has since removed that guide. However, another web page that mirrors much of what is said here and may be more up to date is How to Rank Higher On Google In 2020 by Brian Dean posted on February 24, 2020 https://backlinko.com/rank-high-on-google and last accessed on September 27, 2020.

[†] Netcraft, an Internet research firm in Bath, England. posted as of March 2016. Accessed in June of 2016. http://news.netcraft.com/archives/2016/03/18/march-2016-web-server-survey.html

[‡] *Browser Market Share Worldwide - August 2020* posted on https://gs.statcounter.com/browser-market-share#monthly-200908-202008. Graph was dynamically generated on September 27, 2020.

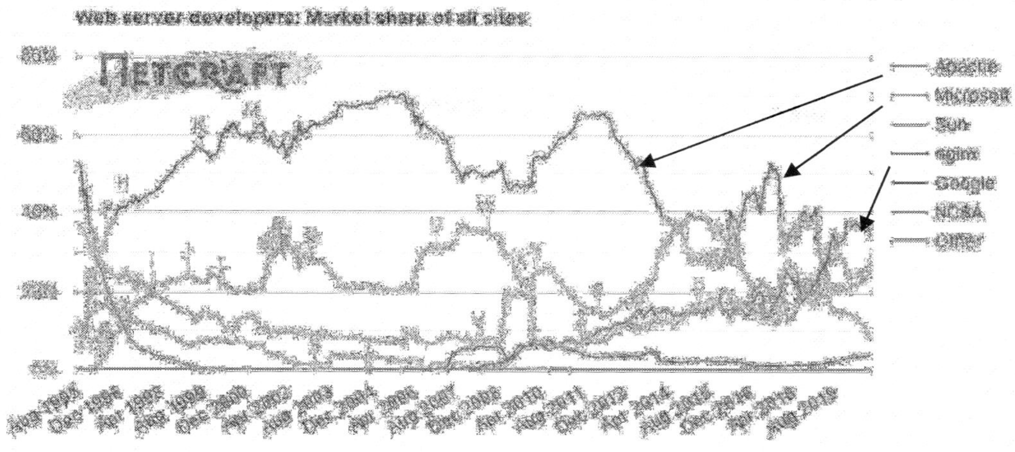


**Figure 47. Market share for top web server software over time.**

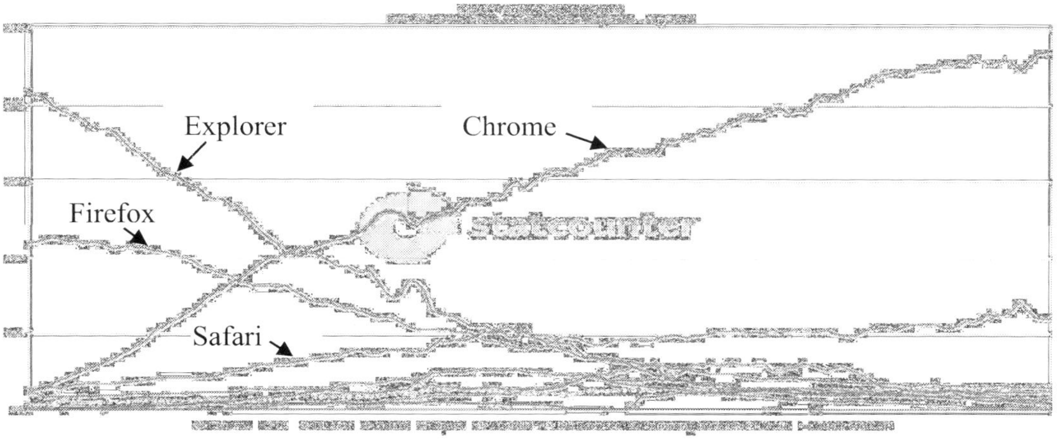


**Figure 48. Relative market share of web browsers.**

## *How a Website Works*

When clicking on a hyperlink, the browser issues an HTTP command (hypertext transfer protocol)* to request the display of a web page at a URL (universal resource locator)†. The webserver application responds by displaying the HTML (hypertext markup language) stored in the file on the website at that URL‡. Every click of the mouse on a hyperlink causes the same cycle to occur.

HTML is not really a full computer language because it does not have flow control command such as IF..NEXT or WHILE...DO. The only thing HTML can do is place text and graphics and other objects (such as movies) on a web page to be rendered by a browser upon request. You can see the HTML code by looking at any webpage in your browser and choosing to view the source code ([Ctrl-U]. Notice that HTML uses tags, identifiers surrounded with the less than (<) and greater

---

* We are simplifying here and ignoring the difference between HTTP and HTTPS. The latter adds some encryption and validates that the owner of the website is who is supposed to be the owner of the website.

† The browser does not know how to find the URL, but the Internet device called a router can ask another type of server (a DNS, Domain Name Server) for the IP address associated with the URL and so send the HTTP request to the right place. More details on this will be found in *Chapter 3: Network*: Internet.

‡ If the URL requested does not have a webfile name on it, but only a domain, then the web server automatically sends the file called index.html or default.html to the browser.

than (>) symbols. Tags normally have open tags and close tags, such as `<head>` and `</head>` or `<body>` `</body>`. To boldface a word, for example, it would have the opening bold tag and then a closing bold tag to turn off the bold, as in `<bold>` **word** `</bold>`.

The HTML tag to hyperlink something is:

```
<a href ="http://webdomain/subfolder/filename" >Hyperlink name</a>
```

For example, if we wanted to put the words *Click Here to See the Picture* on a webpage, and have someone click on those words to see a picture named Myfriends.jpg in the folder called Images on the website called AllMyFriends.com, we would use the following command:

```
<a href ="http://AllMyFriends.com/Images/Myfriends.jpg" >Click Here to See the Picture</a>
```

HTML has no commands to interface with a database such as MySQL or Oracle. For that you need middleware such as PhP, Python or ASP (discussed in more detail in the Chapter 5: System Software).

Many websites are not pure HTML code anymore. For one thing, most of them use CSS (Cascading Style Sheets) to help standardize the formatting. For another, they often store the text and graphics in databases and then use middleware to produce the text and graphics in a format that is readable by the web server.

If the URL does not seem to have a specific filename, or if the URL contains symbols like question marks, chances are the website is using middleware instead of pure HTML to manage the content. For example, the URL for a club website that talks about the benefits of the program might look like this:

```
https://taiji.clubexpress.com/content.aspx?page_id=22&club_id=335754&module_id=85614
```

The middleware (in this case, asp) dynamically writes the HTML code when requested. Such sites are using a content management system (CMS). Many sites use a content manager such as Wordpress instead of native HTML. This makes it easier for the owners and authors of the website to make changes. CMSs also add the capability of allowing the reader to login so that access to different parts of the site can be controlled. However, a CMS has two disadvantages: it slows down the website so that it takes longer for the browser to display the web page, and it often means that the web page isn't static, so a Search Engine like Google cannot find it. In order to make a WordPress site findable, for example, you would install an add-on software in order to feed the tags and meta tags to the search engine.

Despite the extra trouble, CMSs make the website much more manageable, and so they are the most popular way to put a website. While WordPress is the most common CMS in use today, others include Microsoft SharePoint, Oracle Portal, and dozens of open-source systems including Mambo, Drupal, Joomla, DotNetNuke, and WebGUI.

The next two pages, in Figure 49, shows the html code of a popular website by interface guru Jakob Nielsen. See if you can identify some of the tags in the code and then go to the website and see if you can identify the content.

When you see the symbol below, it means that there is code that was removed so that we could give you extracts from various sections of the website. You can see the full source code of any website by pressing **[Ctrl-U].**

* Curly line graphic denotes where text has been deleted so that you can see the beginning, middle, and end of the html page. Meta names are not displayed on the webpage, but are read by search engines. The banner shows at the top of the page and the footer shows at the bottom of the page.

```
<!DOCTYPE html>
<!--[if lt IE 7]>    <html class="no-js lt-ie9 lt-ie8 lt-ie7"> <![endif]-->
<head>
      <meta charset="utf-8">
      <title>Nielsen Norman Group:  UX Training, Consulting, &amp; Research</title>
       <meta property="og:title" content="Nielsen Norman Group:  UX Training, Consulting, &amp; Research" />
      <link rel="stylesheet" href="https://media.nngroup.com/static/css/d009767e7722.css" type="text/css" />
      <link rel="stylesheet" href="//fast.fonts.com/t/1.css?apiType=css&projectid=c5846260-9def-46ec-9407-0f1349bd2d64" type="text/css">
      <meta name="description" content="A leader in the user experience field, NN/g conducts groundbreaking research, trains and certifies UX
practitioners, and provides UX consulting to clients.">
      <meta property="og:description" content="A leader in the user experience field, NN/g conducts groundbreaking research, trains and
certifies UX practitioners, and provides UX consulting to clients." />
      <!--[if lt IE 7]>
         <p class="chromeframe">You are using an <strong>outdated</strong> browser. Please <a href="http://browsehappy.com/">upgrade
your browser</a> or <a href="http://www.google.com/chromeframe/?redirect=true">activate Google Chrome Frame</a> to improve your
experience.</p>
      <![endif]-->
    <header id="page-header">
      <div id="cart" class="shopping-cart mobile-cart show-for-small-down"></div>
      <div class="row">
         <div id="masthead" class="masthead small-12 medium-7 column">
```

```
            <!-- The button to toggles menu on small screen -->
            <a id="toggle-menu" class="show-for-small-down">menu</a>
            <h1><a href="/"><img src="//media.nngroup.com/static/img/logo-wordmark.png" width="74" height="30"> Nielsen <span
class="norman">Norman</span> Group</a></h1>
            <p class="tagline show-for-medium-up">Evidence-Based User Experience Research, Training, and Consulting</p>
            <!-- The button to toggles form search on small screen -->
            <a id="toggle-search-form" class="show-for-small-down"><i class="fa fa-search fa-flip-horizontal"></i></a>
```

```
                  <li><a href="/reports/">Reports</a></li>
                  <li><a href="/articles/">Articles</a></li>
                  <li class="has-submenu"><a>About NN/G</a><i class="arrow-down right"></i>
                     <ul class="submenu-content">
                        <li><a href="/about/">Overview</a></li>
                        <li><a href="/people/">People</a></li>
                        <li><a href="/about/why-nng/">Why NN/g?</a></li>
                        <li><a href="/about/contact/">Contact</a></li>
                        <li><a href="/news/">News</a></li>
                        <li><a href="/about/history/">History</a></li>
                        <li><a href="/books/">Books</a></li>
                     </ul>
                  </li>
               </ul>
            </div>
            <hr class="header-divider show-for-small-only"/>

            <div id="form-search" class="column small-12 medium-5" style="display:none">
               <div id="cart" class="shopping-cart show-for-medium-up"></div>
               <form name="searchform" method="GET" action="/search/" class="row collapse">
                  <div class="masthead-actions row collapse">
                     <div class="column small-12">
                        <div class="row collapse">
                           <div class="small-9 columns">
                              <input name="q" type="text">
                           </div>
                           <div class="small-3 columns">
                              <a class="button search" href="javascript: document.searchform.submit();">Search</a>
                           </div>
                        </div>
                     </div>
                  </div>
               </form>
            </div>
         </div>
      </div>
```

```
<footer id="page-footer">
   <div class="row hide-for-print">
      <div class="column small-12 medium-6">
         <section class="footer-popular">
            <h1>Popular Topics</h1>
            <div style="float: left;  width: 15em; display: block;">
```

```
<ul class="no-bullet">
        <li style="padding-bottom: .5em;"><a href="/topic/applications/">Application Design </a></li>
        <li style="padding-bottom: .5em;"><a href="/topic/credibility-persuasion/">Credibility &amp; Persuasion </a></li>
        <li style="padding-bottom: .5em;"><a href="/topic/e-commerce/">E-Commerce Design </a></li>
        <li style="padding-bottom: .5em;"><a href="/topic/email/">E-Mail Newsletter Design </a></li>
        <li style="padding-bottom: .5em;"><a href="/topic/eyetracking/">Eyetracking </a></li>
        <li style="padding-bottom: .5em;"><a href="/topic/human-computer-interaction/">Human Computer Interaction </a></li>
        <li><a href="/topic/information-architecture/">Information Architecture </a></li>
</ul>
</div>
<div style="float: left; width: 15em; display: block; list-style: none;">
<ul class="no-bullet">
        <li style="padding-bottom: .5em;"><a href="/topic/intranets/">Intranet Design </a></li>
        <li style="padding-bottom: .5em;"><a href="/topic/interaction-design/">Interaction Design </a></li>
        <li style="padding-bottom: .5em;"><a href="/topic/mobile-and-tablet-design/">Mobile and Tablet Design </a></li>
        <li style="padding-bottom: .5em;"><a href="/topic/user-testing/">User Testing </a></li>
        <li style="padding-bottom: .5em;"><a href="/topic/web-usability/">Web Usability </a></li>
        <li style="padding-bottom: .5em;"><a href="/topic/writing-web/">Web Writing &amp; Content Strategy </a></li>
</ul>
</div>
<p> </p>
     </section>
    </div>
    <div class="column small-12 medium-6">
      <div class="row">
        <div class="column small-6">
          <section class="footer-about">
            <h1>Follow Us</h1>
            <ul>
        <li><a href="/articles/subscribe/">Subscribe to our weekly UX newsletter</a>, Jakob Nielsen&#39;s Alertbox</li>
        <li>Visit Don Norman&#39;s website, <a href="http://www.jnd.org"> www.jnd.org </a></li>
        <li>Visit Bruce Tognazzini&#39;s website, <a href="http://www.asktog.com"> www.asktog.com </a></li>
        <li style="padding-top: 8px;"><a class="twitter-follow-button" data-show-count="false" data-size="large"
href="https://twitter.com/nngroup"><img alt="Follow @nngroup on Twitter"
src="//s3.amazonaws.com/media.nngroup.com/media/editor/2013/12/18/twitter%20follow%20button.png" style="width: 117px; height: 20px;"
/></a></li>
</ul>
          </section>
        </div>
        <div class="column small-6">
          <section class="footer-about right-column">
            <h1>About</h1>
            <ul class="no-bullet">
        <li><a href="/about/contact/">Contact information</a></li>
        <li><a href="/about/">About Us</a></li>
        <li><a href="/about/why-nng/">Why NN/g? </a></li>
        <li><a href="/return-policy/">Return Policy</a></li>
</ul>
          </section>
        </div>
      </div>
    </div>
  </div>
  <div class="row">
    <div class="copyright column small-12">
      <p><a href="/copyright-and-reprint-info/">Copyright</a> &copy; 1998-2016 Nielsen Norman Group, All Rights Reserved.</p>
    </div>
  </div>
</footer>
</body>
</html>
```

**Figure 49. HTML selected source code from beginning, middle, and end of https://www.nngroup.com (Jacob Nielsen is the father of webpage interface design).**

* Curly line graphic denotes where text has been deleted so that you can see the beginning, middle, and end of the html page. Meta names are not displayed on the webpage, but are read by search engines. The banner shows at the top of the page and the footer shows at the bottom of the page.

## 6-1. Interactive

1. **Class Activity:** Go to several web pages and view the source code. For each one, identify the following:

    Is the site written in pure HTML, or does it use a content manager?

    Can you tell what kind of content manager it runs?

    Can you identify, specifically, at least three elements that you see on the screen (text, graphics, or both) and then find those specific elements in the source code?

2. **Independent Activity:** How many email addresses do you have? If you only have one, set up at least three different email addresses on three different domains. Then, choose one of the email servers to become the "aggregator" of the email being sent to all four. Send out test messages to all four addresses. Watch them all come into the single email server. What ways do you think people use to manage their email?
3. **Independent Activity:** Look at the last 20 emails that you have sent. How many followed formal business email etiquette? How many followed social friends etiquette? Look at the last 20 emails you received. How many followed formal business email etiquette?
4. **Class Discussion:** Many students rely on texting much more than emailing. What are the advantages? What are the disadvantages? What do you think texting will look like ten years from now? What about email - will that change over the next ten years?
5. **Class Discussion:** What sort of business model do you see as the most viable for social media sites over the next ten years? When Google first began, many said that it would never have a viable business model, and yet it has become a very profitable company with long-term viability. How are the social media sites similar? How are they different?

## 6-2. Self Study Test Questions:

1. What is the difference between the Internet and the World Wide Web?
2. What are different ways that businesses use the World Wide Web?
3. Why is collaboration such a strange category?
4. Why would businesses need to know all the different ways to collaborate using the World Wide Web?
5. What is an intranet? Who would use it?
6. What is an extranet? Who would use it?
7. Why is email called the "killer app"?
8. What is the difference between IMAP and POP3?
9. List at least 6 of the rules of email etiquette.
10. Why is it important to use the rules of formal business email?
11. What is HTML? Give a few examples of HTML tags.
12. How would you look at the source code of a web page?
13. How does a web site work? What is a webserver? What is a web browser?
14. What are the guidelines that govern how high on the listing a website will appear on the Google search engine?

7

# Chapter 7: Application Software

Overview: In this chapter, readers will learn the following topics:

- ➪ Financial Software (ERP, Pre-ERP financial management, Personal financial software)
- ➪ Specialized Use Software (financial, ERPs, CAD/CAM, vertical market, etc.)
- ➪ Media Management Software (graphics/publishing, music management and video editing)
- ➪ Office Software (word processing, spreadsheet, database, presentation.

Application software is the software that an end user would typically use at home or in an office. Application Software breaks down into several useful subcategories. Examples of each category can be found in Table 16.

**Table 16. Technology Map Application Software Categories**

| MAJOR | MINOR | CATEGORIES |
|---|---|---|
| Application Software | Financial | ERP (Sage Business Cloud, Global Shop Solutions, One System ERP, SAP Business One, Oracle NetSuite, IQMS EnterpriseIQ, Workday ERP, MS Dynamics 365, Infor CloudSuite, Exact Macola 10, Infor LN, VAI S2K Enterprise, Oracle HCM, SYSPRO ERP) Pre-ERP: Oracle (Hyperion, Siebel, Retek, PeopleSoft, JD Edwards, Vantive); SAP (OutlookSoft, Triversity, TopTier, B1, Lighthammer); Infor Global (SSA Apps, Mapics, Geac, SCT Adage, FACTS, Lawson); Epicor (Dataworks, ROI, Scala, CRS); Microsoft Business Dynamics (Great Plains, Real World, Solomon, FRX, Navision, Axapta); Sage (ACCPAC, SBT, ACT!, Best, BusinessVision, BusinessWorks, FAS, MAS90/2000, Peachtree, Platinum, SalesLogix, SimplyAccounting, Tetra, Timberline), Consona (Compiere, Made2Manage, Capri); Exact (Macola, Vanguard); Intuit (QuickBooks, Mint); GnuCash, Yodlee, Personal Capital, Money Patrol, Pocketsmith, Empower, CountAbout, YNAB, Goodbudget, Dollarbird, Wally, Spendee, Bottom Line. |
| | Specialized | CAD/CAM (Computer Aided Design/Manufacturing) TurboCAD, Google SketchUp, Autodesk AutoCAD, BOBCAD-CAM, Dassault Systèmes CATIA, C & G Systems CAM-Tool, Cimatron and GibbsCAM, Nihon Unisys Dynavista, Planit Edgecam, DP Technology Esprit, Open Mind HyperMill, CNC Software Mastercam, Siemens NX) |
| | | Logistics: PathGuide Latitude WMS, 3PL Central Warehouse Manager, RAMP Enterprise WMS, Slingshot Enterprise Business Suite, LogFire, Sage X3, and JDA |

| MAJOR | MINOR | CATEGORIES |
|---|---|---|
| | | Vertical Market (Medical, Attorney, Retail, Hair Salons, Healthcare Providers, Massage Therapists, etc.) |
| | Media Management and Design | Graphics & Publishing (Adobe Creative Suite*, CorelDRAW, Microsoft Paint or Visio, Quark Author) |
| | | Streaming Music Managers (Spotify, Apple Music, Tidal, Amazon Music, YouTube Music and Pandora Premium) & Music Editors (Adobe Audition CC, Audacity, ocenaudio, Acoustica Standard Edition, Amadeus Pro, Fission, Hindenburg Journalist, SOUND FORGE Audio Studio 12, TwistedWave, WaveLab Elements) |
| | | Video (Sony Vegas Movie Studio, Apple Final Cut and iMovie, Avid Media Composer, Microsoft MovieMaker, CyberLink PowerDirector, Corel VideoStudio Pro, Adobe Premiere Elements, MAGIX Movie Edit Pro, Roxio Creator/VideoLab, NCHsoftware VideoPad, Pinnacle Studio HD) |
| Application Software | Office Applications | Wordprocessing (Microsoft Word, Google Docs, Open Office Writer, Page) |
| | | Spreadsheet (Microsoft Excel, Open Office Calc, Google Sheets) |
| | | Database (Microsoft Access, Filemaker, Open Source MySQL, Microsoft SQL, Oracle RDBMS, IBM DB2 Google Forms) |
| | | Presentation (Open Office Impress, Google Slides, Microsoft Powerpoint) |

## Financial

Financial software is of special interest to people in business. Financial software has been developed to track all the financial transactions of a person or company.

For individuals and families, money management programs like Intuit Mint† are used by individuals to track their income and expenses. These days, financial packages are more likely front ends for banking and investments, connecting directly with financial institutions so that people don't need to enter their transactions, but rather download them automatically from the website. They also contain reports to calculate net worth and graph expenses.

---

* Adobe has purchased and integrated many other graphics and design packages into the Creative Suite including Illustrator, Photoshop, Flash, InDesign, Acrobat, Fireworks, Dreamweaver, Contribute, Encore, and Audition. This is why Graphics and Publishing had to be combined as a category whereas in all previous years they were separate. Because both print and web publishing have become so graphics oriented, bundling these programs makes sense.

† Quicken was the Intuit program used by most individuals and families until 2016 when it was sold to H.I.G. Capital. *Intuit sells Quicken to private equity firm in management buyout*, by Gregg Keizer, posted March 4, 2016. https://www.computerworld.com/article/3041032/intuit-sells-quicken-to-private-equity-firm-in-management-buyout.html. Last accessed September 30, 2020. Since then it has mostly disappeared. *5 Quicken Alternatives Now that Quicken Online is Dead* by Sarah Li Cain posted September 24, 2020. https://www.consumerismcommentary.com/quicken-alternatives/. Last accessed September 30, 2020.

However, corporations need more sophisticated financial software than individuals need. One difference between corporate and individual financial software is the ability to do "accrual" accounting. Individuals primarily use "cash" accounting; in which income is recorded when the check is deposited, and expenses are recorded when the check is written. Accrual accounting, instead, needs to record the income when it is **earned**, and record the expense when it is **committed.** The rules identifying proper accounting treatment such as recording income, expenses, assets or liabilities are known as GAAP (Generally Accepted Accounting Principles). In accrual accounting, besides accurately recording income and expenses, companies need to record accounts payable (money that needs to be paid in the future) and accounts receivable (money that will be received in the future), as well as fixed assets (buildings and equipment owned by the company), liabilities (obligations of the company such as money borrowed that must be paid back) and equity (how much the investors in the business have contributed to the value of the business).

## *ERP (Enterprise Resource Planning)*

When financial software first started to be used by small businesses in the eighties, it focused on the general ledger and was known as accounting software. In time, however, modules were added for all the other functions of a business that required transactions that eventually posted to the general ledger. Along with the concept that the financial software could help leaders of an organization plan and manage the organization came a new term, Enterprise Resource Planning (ERP). Initially focused only on huge corporations, the functionality of ERP has reached down into the financial software used by even the smallest companies. ERP systems generally have multiple modules available so that corporations that use them can identify specifically which modules they wish to use. The more common modules such as General Ledger (GL) Accounts Payable (AP), Accounts Receivable (AR), Sales Orders (SO), Purchase Orders (PO) and Human Resources (HR) are often found in all the financial packages, whereas the rest of the modules may or may not be available with that particular ERP system. Customer Management Software (CMS) is another module that is now often combined into a financial package or ERP.

The combination of modules is designed to span the entire supply chain. In the early days of computers there were hundreds of financial software vendors, but a great deal of consolidation has occurred over the past thirty years. Now only a handful remain, although the pre-ERP software has not gone away as many businesses still use that financial software, and switching to a new financial package is not very easy to do. A list of all the software packages and were found on Ned Lilly's website, *http://www.erpgraveyard.com/tombs.html*. The first page of the October 2020 version is shown in Figure 50.

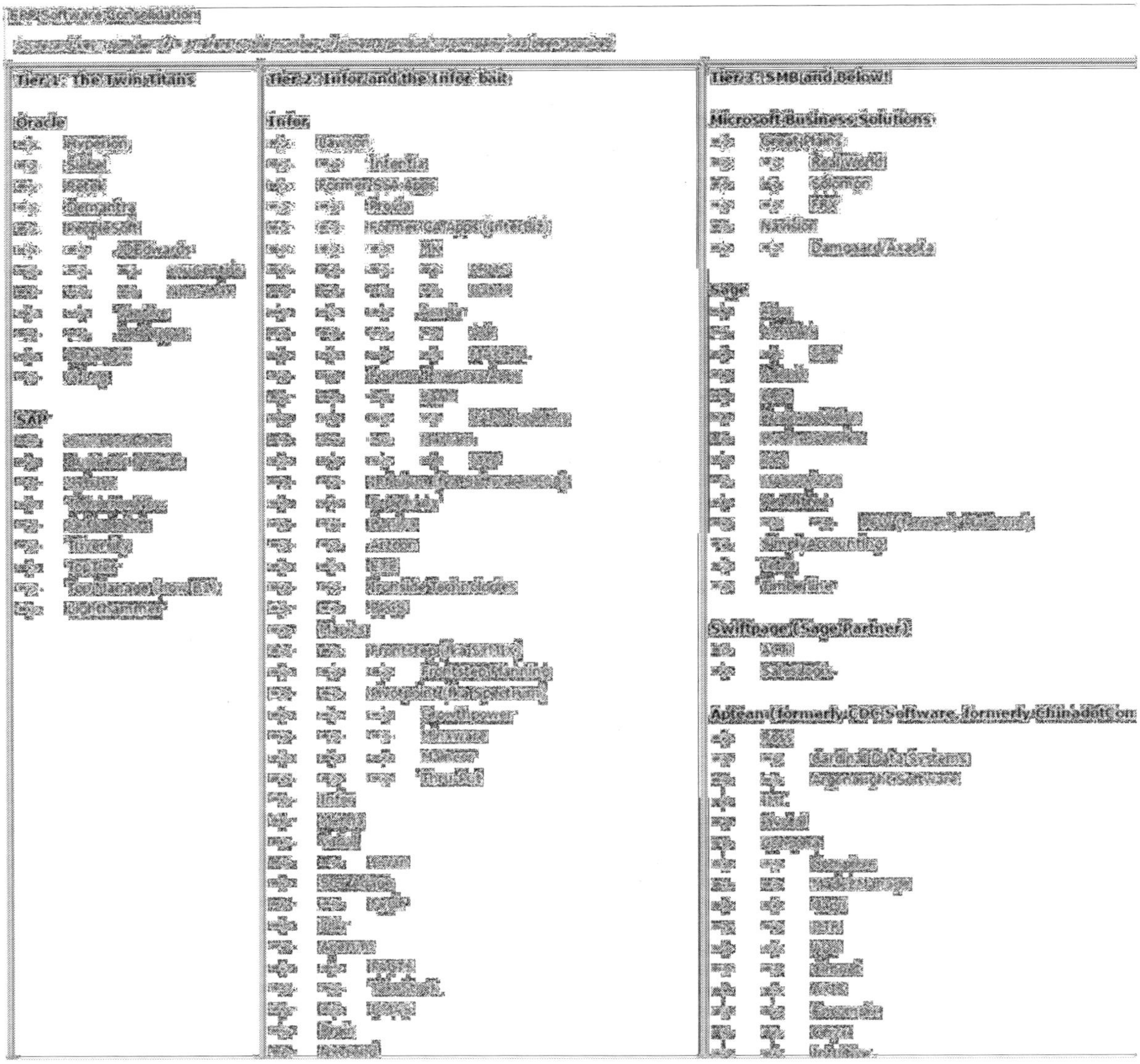

**Figure 50. Evidence of financial software consolidation: The ERP graveyard.**

The market share can be seen in the graph in Figure 51*. A few years ago, most large companies had a financial package from either SAP or Oracle (the Two Titans of ERP). Medium or large companies usually have a financial package from either Microsoft, Infor, or Sage. The most popular software for small companies is Intuit's QuickBooks, which isn't often included in ERP software statistics despite having all the modules that an ERP system would have. Both SAP and Oracle had been losing market share to online-oriented newcomers to the ERP market who work in the cloud rather than installing hardware and software on the company premises. However, both SAP and Oracle now offer their own cloud-based alternatives. Workday, Workforce Software, Cornerstone OnDemand and NetSuite are competing ERP vendors worldwide, that are all cloud-based.

* *Top 10 ERP Software Vendors, Market Size and Market Forecast 2018-2023* by Albert Pang, Misho Markovski and Andrej Micik. Published on December 6, 2019. https://www.appsruntheworld.com/top-10-erp-software-vendors-and-market-forecast/. Last accessed on October 2, 2020.

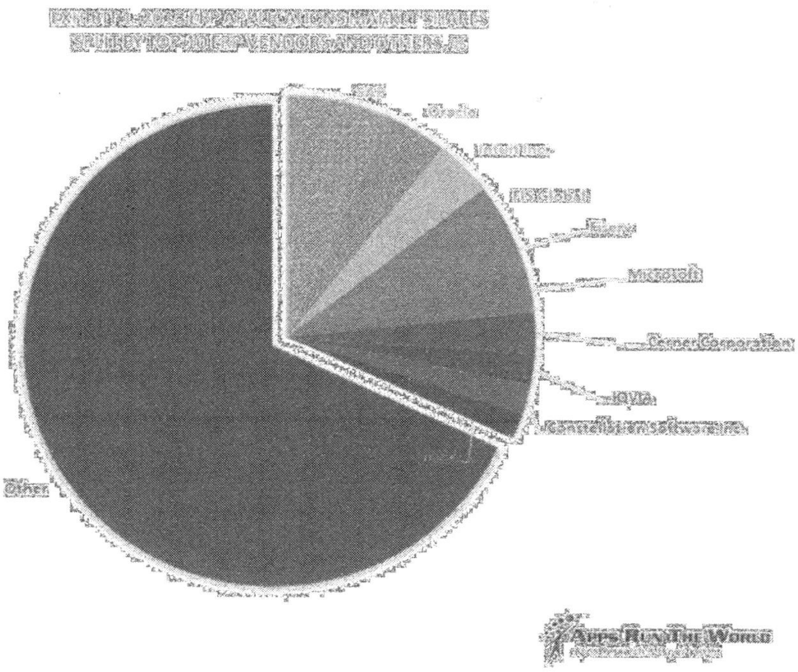


**Figure 51. Financial software market share.**

Furthermore, the ERP market has encompassed other previously separate markets of Customer Relationship Management (CRM), Human Capital Management (HRM), and Applicant Tracking Systems (ATS). The largest global companies and which packages they use for these functions can be found in Table 17*.

**Table 17. Financial Packages by Global Customer**

| Company | ERP | HCM | ATS |
|---|---|---|---|
| Walmart | SAP S/4 HANA | Workday HCM | Workday Recruiting |
| Exxon Mobil | SAP S/4 HANA | SAP HCM (HR) | SuccessFactors Recruiting |
| Apple | SAP S/4 HANA | SAP HCM (HR) | In-House ATS |
| Berkshire Hathaway | Oracle ERP Cloud | Oracle HCM Cloud | Oracle Taleo |
| Amazon.com | In-House ERP | Workday HCM | In-House ATS |
| UnitedHealth Group | Oracle E-Business Suite | Oracle HCM Cloud | Oracle Taleo |
| McKesson | SAP S/4 HANA | Workday HCM | Workday Recruiting |
| CVS Health | SAP S/4 HANA | Workday HCM | IBM Kenexa BrassRing |
| AT&T | SAP ERP ECC 6.0 | Workday HCM | Oracle Taleo |
| AmerisourceBergen | SAP S/4 HANA | Workday HCM | Oracle Taleo |
| Chevron | SAP S/4 HANA | Workday HCM | SuccessFactors Recruiting |
| Ford Motor | SAP S/4 HANA | Oracle HCM Cloud | IBM Kenexa BrassRing |
| General Motors | SAP S/4 HANA | Workday HCM | Oracle Taleo |
| Costco Wholesale | SAP S/4 HANA | SAP SuccessFactors HCM | Oracle Taleo |
| Alphabet | SAP S/4 HANA | Workday HCM | Google Hire |

Almost all financial software is built on top of a database, so often in addition to choosing the financial software, a company can also choose to run the software on one of several databases

* *Top 10 ERP Software Vendors, Market Size and Market Forecast 2018-2023* by Albert Pang, Misho Markovski and Andrej Micik. Published on December 6, 2019. https://www.appsruntheworld.com/top-10-erp-software-vendors-and-market-forecast/ . Last accessed on October 2, 2020.

(though Oracle Financials, JD Edwards, and PeopleSoft run best on Oracle database, while SAP runs best on HANA*, and Microsoft Business Dynamics run best on SQL). .

One important fact to keep in mind is that financial software for larger companies does not really exist as "off the shelf" packages. Rather, large companies purchase the software "by the seat" (as if it were off the shelf) but are actually purchasing the development time of the consultants who "customize" the package. Systems like SAP, JD Edwards, Hyperion, Oracle Financials, and PeopleSoft are closer to development environments than already-programmed software packages†.

### *XBRL (eXtensible Business Reporting Language)*

Although not considered a package, one of the protocols often used in conjunction with financial systems is XBRL (eXtensible Business Reporting Language). The Securities and Exchange Commission (SEC) has required companies to submit financial data using the XBRL schemas (list of fields) since 2009. According to the Financial Accounting Standards Board (FASB): "*The U.S. GAAP Financial Reporting Taxonomy is a list of computer-readable tags in XBRL that allows companies to tag precisely the thousands of pieces of financial data that are included in typical long-form financial statements and related footnote disclosures. The tags allow computers to automatically search for, assemble, and process data so it can be readily accessed and analyzed by investors, analysts, journalists, and regulators.*"‡ The standardization possibilities due to the use of a common formatting language enabled a much more rigorous review of financial data for public companies by the SEC and other entities.

Many problems occurred, however, because of the newness and constant revisions of XBRL at the time. In 2009 few (if any) financial software systems had been programmed to publish in XBRL, therefore companies had to take the old reports coming from their systems and re-code them with XBRL tags or purchase add-on software to do the translation. Hitachi Xinba XBRL, Altova MapForce, Rivet Software's Crossfire and CrossView, and Calcbench were some examples of add-on programs that took financial data and encoded it in the required XBRL format. Most financial systems these days publish in XBRL directly, and add-ons probably aren't necessary. However, the SEC just recently made a new requirement that companies use iXBRL (Inline eXtensible Markup Language) which formats the XBRL code into a human-readable format. There may be a similar gap currently in ERP systems with iXBRL reporting for the next few years.

## Special Use Software

There are several categories of software that are only of interest to certain industries or types of businesses. This section is split into manufacturing, logistics, and service industry software.

### *CAD/CAM/CAE Software*

Many manufacturing and engineering businesses need software that is specifically developed to design and manufacture products, interfacing with the robotics or equipment on the factory floor. These systems are called Computer Aided Design (CAD) or Computer Aided Manufacturing

---

* *Why did SAP buy Sybase?* Published on January 15, 2013. https://blogs.sap.com/2013/01/15/why-did-sap-buy-sybase/ Last accessed on October 2, 2020. The previous edition of this textbook said that SAP ran best on Sybase and that SAP had just purchased Sybase, which was the #1 database for Wall Street applications. But apparently SAP changed its mind and went with HANA instead.

† The reasons for treating custom programming like an off-the-shelf package have to do with the rules of capitalization under Generally Accepted Accounting Principles, discussed in further detail in a later chapter.

‡ As reported on https://irblog.prnewswire.com/2013/05/22/mind-the-gaap-for-xbrl/, written May 22, 2013 by PR Newswire. Last accessed July 5, 2016.

(CAM) or Computer Aided Engineering (CAE) systems. In the early days of large computers (mainframes), companies such as General Motors and Lockheed had dedicated computers with proprietary software to help engineers design. But when personal computers started to proliferate, the capabilities migrated as well. One of the earliest pioneers was AutoCAD , a program published by AutoDesk in 1982, followed within the next decade by SolidWorks published by Dessault Systems, and TurboCAD by IMSI/Design. Nowadays the top CAD programs are progeCAD, sxCAD, AutoDCR, CADISON.

CAD systems initially just printed out two-dimensional blueprints or drawings on a large plotter. However, as the field evolved, CAD systems went to three dimensions, and started linking to CAM systems that cut, bend, or connect materials in order to manufacture the finished product. Popular CAD/CAM systems are Solid Edge CAM Pro, GibbsCAM, Fusion 360, SOLIDWORKS CAM, and SolidCAM.*

One of the most recent innovations for CAD/CAM is to link with other systems such as Building Management Systems (BMS), Building Automation Systems (BAS), Orthotics and Prosthetics, or even Restoration Dentistry systems. The BMS can use the CAD designs, developed as part of the architectural design of a building, in order to manage the heat, light, air conditioning, plumbing, etc. BMS systems are just now developing, but the software is mostly coming from companies that manufacture sensors and controls like Johnson Controls, Siemens Building Technology, Delta, Distech, Circon and KMC controls. Replacement limbs can be produced to fit specific needs that are more comfortable and customized for the individual. The Dentistry systems can reproduce teeth to fit specific mouth spaces and configurations.

## *Logistics Software*

Because "supply chain and logistics" has become an important service for manufacturers, there are some software packages that have been developed specifically for those activities. Logistic activities include the management of goods and resources as they move through the process from raw material to finished product. PathGuide Latitude WMS, 3PL Central Warehouse Manager, RAMP Enterprise WMS, Slingshot Enterprise Business Suite, LogFire, Sage X3, and JDA Software are all packages that are developed specifically for logistics by third party companies (which means outside of the normal ERP systems discussed in this chapter, which usually have logistics modules).

## *Service Industry Vertical Market Software*

Vertical market software is an application that is designed to meet the multiple software needs of a particular service industry. For example, a dentist's office might purchase DentiMax, StarByte Dental or EasyDental. An insurance agent might purchase *Your Insurance Office*, and a Massage Therapist might buy *Your Massage Office*. Generally these packages include: a scheduling program, a word processor, a customer database, a human resources module, and financial software. For someone who has limited experience with computers, these integrated packages can help automate simple office tasks very easily and provide much-needed technical support.

As more and more people became familiar with general use programs such as Microsoft Office, Google Apps, or Open Office, vertical market software has been decreasing in popularity due to the high cost and limited capabilities of vertical market software.

* CAD-CAM Software: List of the Best Systems For 2020 by Khaleel Hayes. https://www.selecthub.com/manufacturing/cad-cam-software/ Last accessed on October 2, 2020.

# Media Management and Design Software

## *Graphics/Publishing**

There have been, in the past, many independent programs to do graphics. In recent years, most independent graphics companies have been bought up by companies selling web and print publishing packages so that the graphics can be more easily incorporated into the publication. The term *graphic* includes line art, cartoon drawings, pictures, and photographs.

It is deceptively difficult to combine text and graphics on paper or on the computer screen despite the many years that people have been doing it. Publishing is the art of laying out (on a printed page or a computer screen) the graphics and the text. Generally, the text is still written with a word processor. The publishing software deals with issues such as styles, hyperlinks, headers, footers, stories continuing on other pages, hyphenation, placement of graphics, etc. The top publishing program in the field of print publishing is InDesign, part of the Adobe Creative Suite†.

Graphics packages come in two different types: raster graphics or vector graphics. Raster graphics focuses on each individual bit and its color, such as Microsoft Paint or Photoshop. Vector graphics, on the other hand, focus on objects rather than each individual bit, such as Illustrator or CorelDraw. While graphics may originate from different sources or media, once digitized, they generally conform to one of the major graphic formats.

You can digitize a graphic by either creating it with a computer program initially, creating it with digital device (such as a camera), or by scanning it with a scanner to transform it into digital media. As noted earlier, unlike text (which is relatively easy to store digitally and take up very little space), graphics are difficult to store digitally and take up relatively large amounts of space. The more complex the graphic, the more difficult the storage and the larger the space needs. A color drawing takes up more space than a black-and-white line drawing, and a photograph takes up more space than a color drawing. Over the years various compression formats have developed to minimize the amount of space necessary to store graphics. Common graphic formats are:

- **GIF** - Graphic Interchange Format. One of the first graphic formats, used most commonly for color drawings that allows transparency, or the ability to see something under a certain area of the drawing. GIF files may also be animated - showing several pictures in the same spot in succession giving the perception of movement.
- **BMP** - Bitmapped. A format developed by Microsoft for Windows graphics that doesn't normally include compression algorithms, so the files are extremely large. BMP is similar to the RAW format that some cameras allow to be stored (albeit with huge files).
- **JPEG** - Joint Photographic Experts Group. A format developed specifically for photographs, allowing the user to choose the amount of compression. More compression means lower quality, but for screen-applications such as the web, lower quality is not necessarily a problem.
- **PNG** - Portable Network Graphics. Developed specifically for web applications. Because it is newer than all the other formats and in the early days, browsers did not always

* This category has been moving all over the place in recent years. Originally it was lumped into Office and was usually included as a free accessory with the operating system (Microsoft Paint or MacPaint). But as graphics became more sophisticated more advanced packages were published (Corel Draw, Adobe Illustrator). In the last version of this book graphics and desktop publishing were combined because the software began to integrate the two. Probably next version of this book the graphics/publishing will be moved to Development Software because except for the simple capabilities built into the camera software of your phone, the people who need graphics capability are web developers. Print graphics will probably start taking a backseat to web graphics in the near future.

† Previously, the dominant force in professional desktop publishing had been Quark, but between 2002 and 2008 it lost nearly all its market share by not updating the program to keep up with new OSs.

support PNG (though they all do now) it is not always welcome by developers (as demonstrated in Figure 52[*]).

- **TIFF** - Tag Interchange File Format. A format developed by Adobe that stores all the source information of a bitmap but can be compressed (though still always a very large file).

**Figure 52. PNG vs. JPEG**

Of course, every graphics program such as Photoshop, ClarisDraw, PaintShop, and Illustrator has its own proprietary format as well, and other graphics programs can understand some of those proprietary formats. Word can import over two dozen different graphic formats. However, for transferring data from program to program, using one of these exchange formats is recommended.

Compressing or converting from one format to the other usually necessitates the loss of some information, lowering the quality of the graphic (known as lossy in digital media parlance). JPG is lossy, while PNG and TIFF are lossless. A lossless format means a format that doesn't lose much information when compressed or converted.

Many graphics programs allow you to choose the amount of compression when you save a file with a JPEG (JPG) format. Figure 53 shows a high-resolution photo (1600 pixels per inch, the measure of picture resolution on a screen) that takes 4 megs to store. A higher resolution photo looks better than a lower resolution photo (up to a certain point). The perceived quality of a graphic is a combination of the media in which it is presented (i.e., print or screen) and the ability of the human eye to discern fine lines and shadings. A higher quality photo does not pixelate. To *pixelate* means that you can visually see individual pixels - an indication of poor quality.

If the photo shown in were loaded on a web page, it would take 7 minutes to see the photo at slower connection speeds. Figure 55 shows a highly compressed low-resolution version of that photo, which would only take four seconds to download. Research shows that most people will only wait eight seconds for a web page to download, so understanding the time it takes for a graphic to load is important. Web designers think very carefully, therefore, before using photos, and they almost never use high-resolution photos. You will also note that in Figure 54 that same low quality jpeg is shown much smaller, where the poor quality is not an issue. Hence the size of the display of a graphic is highly related to its appearance. Show a graphic larger than the dimensions were designed for and the picture will not look good.

**Figure 53. Uncompressed JPEG photo - 4,277 Kilobyte file**

[*] Courtesy of DLPNG.com.

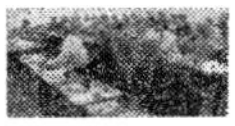

**Figure 54. Low quality JPEG shown smaller.**

**Figure 55. Low-quality JPEG (96 ppi) at 70% compression - 14 kilobyte file.**

Until the recent push for extremely high-resolution monitors, graphics on websites were usually extremely low: less than 96 pixels-per-inch (ppi), because early computer screens couldn't show higher resolution. So if someone tries to print a photo that comes from a web page, usually the quality is terrible because there isn't enough information about the colors of each dot for the printer in a photo stored at 96ppi. A printer needs an image of at least 200ppi to print correctly at 1200 dots per inch.

## Music Managers and Editors

Music Management Software is designed to enable people to manage all the songs and albums on their systems. Many also connect to online music stores and allow people to purchase songs and albums (even videos). The top program for iPods, iPads, iPhones, and Mac computers is Apple Music App (formerly called iTunes), though it also runs on Windows. Newer music services available with their own music management include Spotify, Apple Music, Tidal, Amazon Music, YouTube Music and Pandora Premium. Additionally, there are Music Editors which enable people to edit the music itself, including Adobe Audition CC, Audacity, ocenaudio, Acoustica Standard Edition, Amadeus Pro, Fission, Hindenburg Journalist, SOUND FORGE Audio Studio 12, TwistedWave, and WaveLab Elements. While mainly used by musicians, these programs may also be used in conjunction with the next type of software, video editors.

## Video Editors

Never before has it been so easy to edit video. In the early days of film, editing was done, physically cutting (with a real blade) and splicing together (with epoxy) different parts of a film reel. During the eighties and nineties software made it possible to digitally choose the points to splice and then have a computer system control the copying and pasting (virtual) from one set of tapes to another. The only issue was that each edit degraded the quality of the video, so editing was still a highly skilled art. Today, however, with most of the video in digital rather than tape format, video editing is available to anyone. Top programs to do editing are Adobe Premiere Elements, Avid Media Composer, Apple Final Cut Pro, CyberLink PowerDirector, HitFilm Express, Apple iMovie, Corel VideoStudio Ultimate, NCHsoftware VideoPad, Windows MovieMaker, and DaVinci Resolve. Some packages, such as Final Cut and Avid Media Composer (though Adobe Premiere is trying to horn in) are professional level video programs that are used by major filmmakers*. Other video editors range from free with the OS and very easy to use (like iMovie and Windows MovieMaker) or relatively sophisticated and inexpensive like VideoPad and Resolve. Due to the popularity of video-sharing on social media, many people are becoming more interested in learning how to edit and upload videos. Most video editing packages

* *Why I Switched to Final Cut Pro X After 25 Years of Working on Avid* by Michael Yanovich posted on August 19, 2019. https://nofilmschool.com/switched-final-cut-pro-x-after-25-years. Last accessed October 5, 2020.

these days (including the rudimentary editing functions built into phones and cameras) have a "Youtube upload" or "Facebook upload" button that enables quick sharing of videos recorded on webcams, phones, or cameras.

## Office Software

Office software is the general-purpose software that you would see and use every day. General application office software is like a multi-purpose room; it provides a basic configuration or design for doing something, but the actual content and structure is left up to the user. The main types of general office software are used in offices and homes for a variety of computer tasks. Office software is categorized by its purpose: word processing, presentation, database, and spreadsheet. In this section, we will highlight some of the more useful features of word processing, spreadsheet, database, and presentation. While a little more "hands on" than the rest of this textbook, this section is not designed to teach you how to use these programs; just to recognize features that you may find extremely useful (and perhaps are not already using). Some of the features and functions discussed in this section are:

- Word: Search and Replace
- Word: Simple Formatting, Saving, Printing
- Word: Styles and Templates
- Word: Headers, Footers, and Page numbers
- Word: Sections
- Excel: What If Analysis and Goal Seek
- Excel: Graphing
- Excel: Arguments
- Excel: Advanced Functions (PMT, FV, VLOOKUP/HLOOKUP)
- Excel: Sorting, Subgrouping, Autofiltering
- Excel: Protecting Cells and Worksheets
- Excel: Pivot Tables
- Access: How Databases Differ from Other Applications
- Access: Creating a Database
- Access: Designing a Table Structure
- Access: Entering and Editing Data
- Access: Simple Form
- Access: Simple Query
- Access: Joining Tables
- Access: Advanced Queries
- Access: Simple Report
- Access: Advanced Reports
- Powerpoint: Rules for a good presentation
- Powerpoint: Creating an outline for a presentation
- Powerpoint: Adding clip art and photos to a presentation
- Powerpoint: Setting up the master design or theme of a presentation
- Powerpoint: Adding animation and customization to a presentation
- Powerpoint: Printing presentations, note pages, and handouts

### *Suites of Applications*

Typically, word processing, spreadsheet, presentation, and/or database programs are packaged and sold together in *suites* such as Microsoft Office, OpenOffice, Google G Suite, WPS Office and

LibreOffice[*]. Microsoft Office is the most popular, traditionally used by over 90% of those using computers. However, Microsoft Office has gotten more and more expensive in recent years, and users complain that it is too complex to understand. Many technologists refer to Microsoft Office as "bloated," because it is chock-full of features and functions, many of which are not often used (or don't work the way they should).

As a result, many people have been searching for simpler alternatives. OpenOffice and LibreOffice are similar to each other and to Microsoft office, but they are open source (which means they are free, but the level of technical expertise needed to install and maintain them is much higher).

With the exception of WordPerfect Office, most office software comes in an online office suites version. Google Apps and Microsoft 365 are both online versions of the familiar desktop-version of Office. The software resides on a webserver on the Internet rather than installed specifically on someone's desktop computer. That limits its features and leads to security and reliability risks. However, it also opens up those documents for groups of people to work collaboratively on the same document (a capability which is considered "cloud computing"). In most cases, however, the online office suites are lacking some of the most powerful capabilities of the desktop version of the software. Features such as mail merge and advanced book formatting capabilities are not available in the online versions of the word processing software, and advanced features like goal seek and multi-level graphing are not available in the online versions of the spreadsheet.

With applications that can be either online or desktop, it is even more important to know exactly where the application resides, and where the data is stored. The next section goes over file location information.

## Applications vs. Data

Office applications are instructions for the computer to enable you to create, save, and modify your *data*. An important concept to understand is the difference between your data and the application that created it.

Applications usually consist of hundreds of files that work in concert together. The main file usually has an *extension* (three or four letters that appear after the dot in a filename) of **.exe,** and it works with many other files, some of which will have a **.dll** extension. [†]. (If an application icon disappears from your desktop or menu, for example, you can sometimes start the program anyway by going to the *Program Files* folder and double clicking the program name with the .exe extension inside its program folder.)

Some applications (most, actually) have a native format (usually proprietary) to save files. You can usually tell what type of application created a file because it has the extension normally associated with that program. For example, word processing files may have a **.doc** or **.docx** or **.odt** or **.rtf** or **.wpf** extension. Unfortunately, Windows and Macintosh operating systems hide the file extension by default, so to see the extensions (a very useful thing to do) you must find the place where you can show the extensions. In Windows, go to **MyPC--View** and check the box for **file name extensions.** In Macintosh, go to **Finder--Preferences--Advanced** and click on **Show all file extensions.**

Desktop applications are usually installed on one computer and cannot be used when on another computer. But the *data* you create can often be read, converted, saved, and modified in many

[*] For an excellent comparison of the three Office Suites, read http://www.techsoup.org/support/articles-and-how-tos/ms-office-vs-openoffice, written by Dan Rivas on May 11, 2016. Accessed July 4, 2016.

[†] The term *extension* is also applied to add-on programs or features for web applications. These are two different things and should not be confused.

different formats. Some formats are not proprietary* such as RTF and TXT†, but are limited in some ways. For example, a TXT file cannot store graphics.

Another important aspect of data files is the difference between a *source* file and an *output* file. A source file contains all the information necessary to make whatever changes or modifications might be needed to reproduce the original file with changes. Usually a source file must be stored in the native format of the program which originally created the file. A file may also exist in an output format, however. An output format is a format designed to be easily read or viewed by other people using other programs. For example, if you publish a Word document on a website, the result is an HTM file (written in html) which can be read using a web browser. If you publish a Word document to a PDF format, the resulting file can be read by anyone with Adobe Reader, a free add-on that most web users have already downloaded since so many people publish to a PDF format on the web. An output format can be read but usually cannot be modified. (If it can be modified, it is usually in a very limited way, such as adding annotations or fill-in fields). Think of an output format more as an *electronic printed version* of the file rather than the original source file. (Indeed, many people using free PDF creators actually *print to the printer* [**PDF writer** or **PrimoPDF]** in order to create the file in a PDF format. That way they can create a PDF from any program, even if the original application doesn't have a Save As PDF option on its menu.)

### *Folders and Files*

One of the most important concepts to understand is where your data files are being stored. Windows 10 systems and the Macintosh have a *Documents* folder. Older versions of Windows has a *My Documents* folder. The expectation is that you will store your files in these folders as they have been set up for you to store your data.

There are a few issues, however. Not all programs store data in these folders. Some programs still store data files in the Program folders (which makes it extremely difficult to backup or copy). Additionally, the *Documents (*or *My Documents)* folder is an *alias*. The folder (which looks as if it is at the top level of the directory) actually points to another folder that physically exists somewhere else (usually much deeper in the file structure). If your username is *aaaa111*, on a Windows 10 system, for example, (as can be seen in Figure 56) then the *Documents* folder would actually be located on *C:\Users \aaaa1\My Documents*. That would be fine if all programs understood the alias, but not all do. So sometimes you have to know where the *real* location of the file is. To make matters worse, Microsoft and Apple tend to change the default location of the *Documents* folder each time they come out with a new version, making it that much more difficult to know where your data is so that you can copy and/or back it up.

* Many standardized file formats of today were at one time proprietary, but either were released by the original vendor as a standard or taken over by a standards organization or committee. For example, RTF, stands for Rich Text Format, and was originally an IBM wordprocessing format. Now, however, most wordprocessors can read and write in RTF format, making it one of the standard formats easily transferrable from program to program.

† RTF stands for Rich Text Format, which can store simple formats such as bold and italics and limited graphics. TXT stands for Text, which is simply text – no formatting, no graphics.

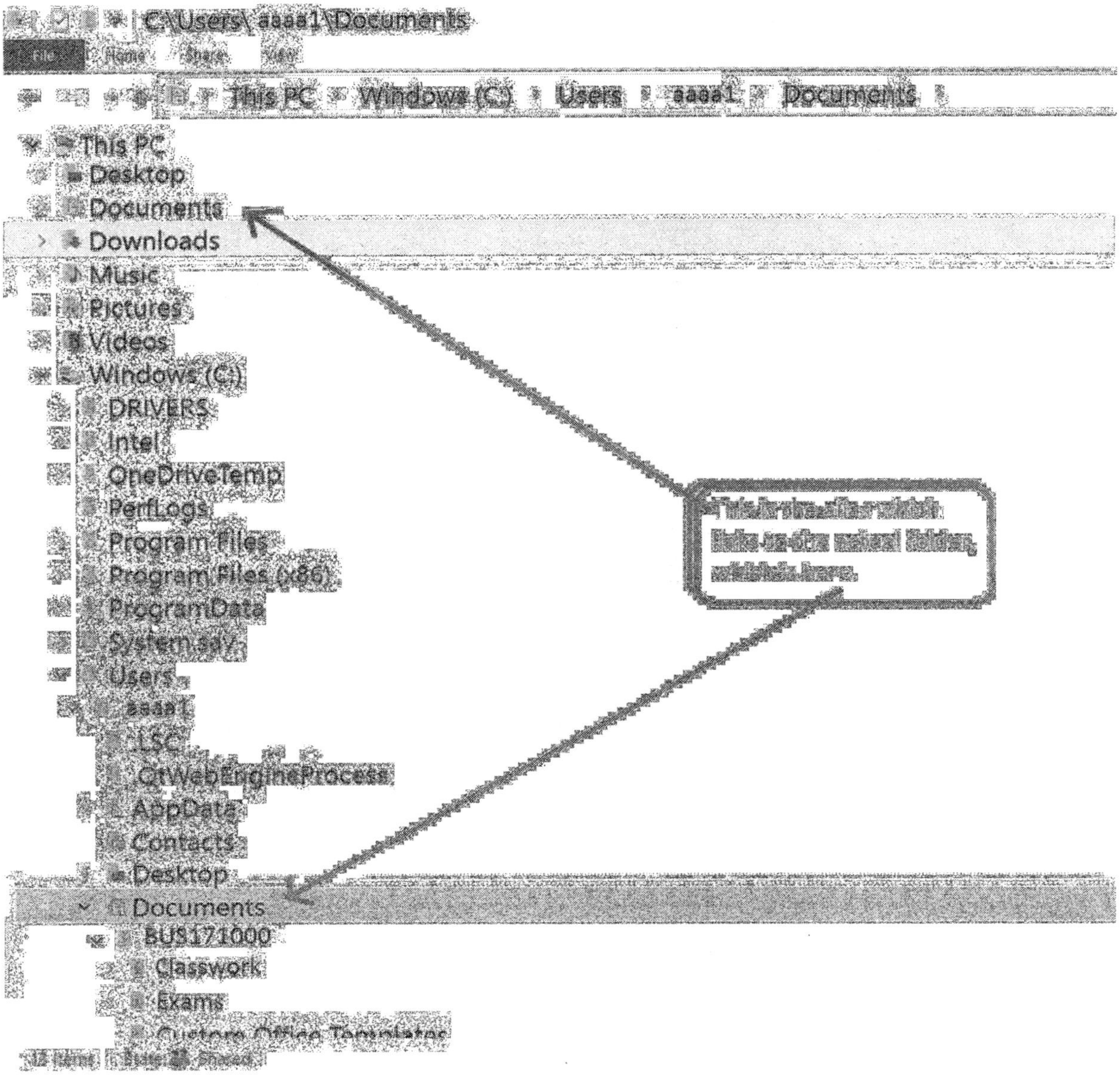


**Figure 56. Documents alias.**

Also keep in mind that just because you are looking at a folder in the file structure does not mean that it is the current *active* folder, or the folder into which the application will store the file. The currently open active folder is usually indicated with shading of some type. (Previous versions of Windows used a tiny open folder icon, but it was difficult to see.)

You can create folders in order to better organize your files by right mousing on a blank area of the right file panel of the file structure and choosing **New--Folder** and giving the folder a name.

The upshot is that everyone should become familiar with the hierarchy of folders on their system and should take the time to organize their folders by creating subfolders and specifically identifying into which folder each of their files should be stored. Always know where your data is located, as well as where you can find the backup copy. **(You *do* have a backup copy, don't you?)**

### *Network Storage vs. Local Storage*

Usually there are several options for storing your files. In most cases, you can store them locally on the hard drive or a flash drive or on a CD or DVD disk. You may also be able to store your files on a folder on a network drive. Network drives can appear in several different ways, but one of the most convenient is as a *mapped drive*[*]. When you map a drive, it appears to all your programs just like another disk drive connected to the computer, and all programs have some way of letting you change the drive when you work on a file. This way, the application does not need to have any special understanding of the network paths in order to save or open a file on the network drive. But mapping drives is becoming less common as cloud drives become more common.

Due to the proliferation of cloud drives, almost all of the major operating systems have started *encouraging* users to use the cloud instead of local storage. The reasons for this are many – and mostly to the benefit of the vendor, not the user. The biggest advantage of cloud storage for the user is the ability to access the files from other computers and devices, which is a big advantage. The problem appears because users don't understand the difference between local storage and network storage, so they often get confused. Vendors have tried to sway users by making the cloud storage the default – sometimes going so far as to make it difficult-or-impossible to store anything on a local drive. That's not that big deal for users with only a few hundred megabytes of data. But as pictures and videos proliferate, storage needs of users are exploding – which suits the vendors just fine since they charge for more storage. Users need to become diligent in understanding where their data is, and how much space it is taking up so they don't end up paying thousands of dollars each year to store data on a cloud which they could do on a local hard drive for free[†].

## Saving Files

Throughout the time you are typing into a document and especially when you are finished with it (before you print), you should save your document [Ctrl-S]. Normally Microsoft Office saves all documents in the *Documents* folder which might be in your user folder on your hard drive but might be on Microsoft's cloud, OneDrive. Google Apps normally saves in *My Folder*, which is a folder on Google's website, and OpenOffice saves in the last folder which was open; none of which is always a good thing. You might want to save the document on a network drive or a flash drive (also called a USB drive or memory stick), or maybe even to a CD or DVD disk. Normally you can change the location where the file will save. In Google Docs, you must download a file to change its location.

Another issue is downloading files. It is best if you choose the file which downloaded files will be stored, but typically the default folder is either the Downloads folder or the Desktop folder – neither of which is a good idea. To change the location of where files are downloaded, go to your browser settings and choose the option to allow you to choose the folder each time a file downloads.

---

[*] Mapping a drive instructions are found in *Chapter 5: System Software*.

[†] My storage needs are far beyond a typical user, and so I've become an expert on the process of keeping track of all my files Even I screw up sometimes. I wrote about 30 instructional guides on various software programs back in 1988 which I carefully backed up onto a tape. Recently I paid a data-restoration company to restore those books, but they were unable to get them from the tape. And last summer I lost several hours of video that I had carefully recorded and several hundred pictures that I took on my phone. Now I have a 10 Terabyte External hard drive to which I backup up my computer, my phone, my laptop, and all my cloud drives every night.

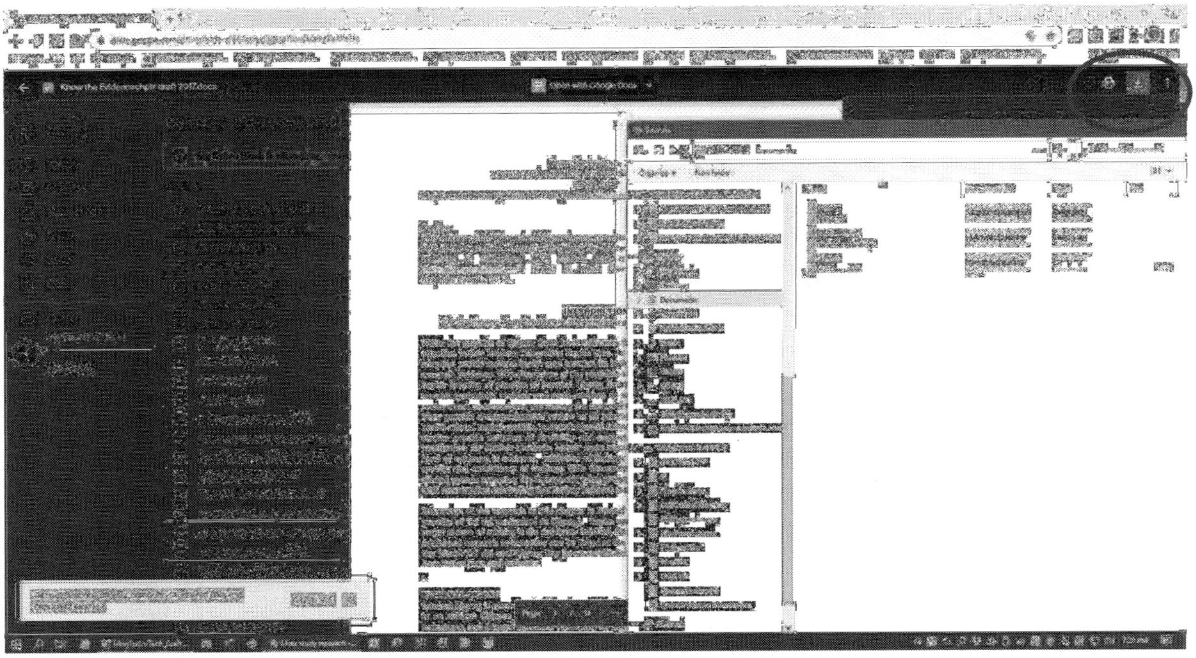

**Figure 57. Changing the location of a file in Google Docs by downloading file.**

To Download the file you choose **Download** icon and then choose the format to save it in. Then you would choose to **Save** and then choose the proper drive and folder, as seen in Figure 57.

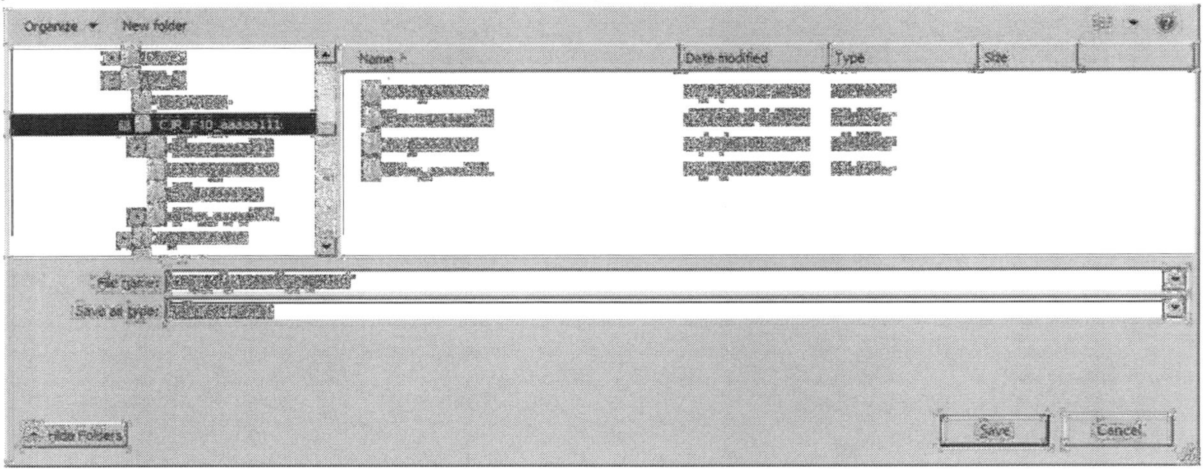


**Figure 58. Changing the location of a file in Google Apps.**

In earlier versions of Word, to change the location or filename, click the list option box to the right of the *My Documents* at the top of the dialog box as you can see in Figure 59.

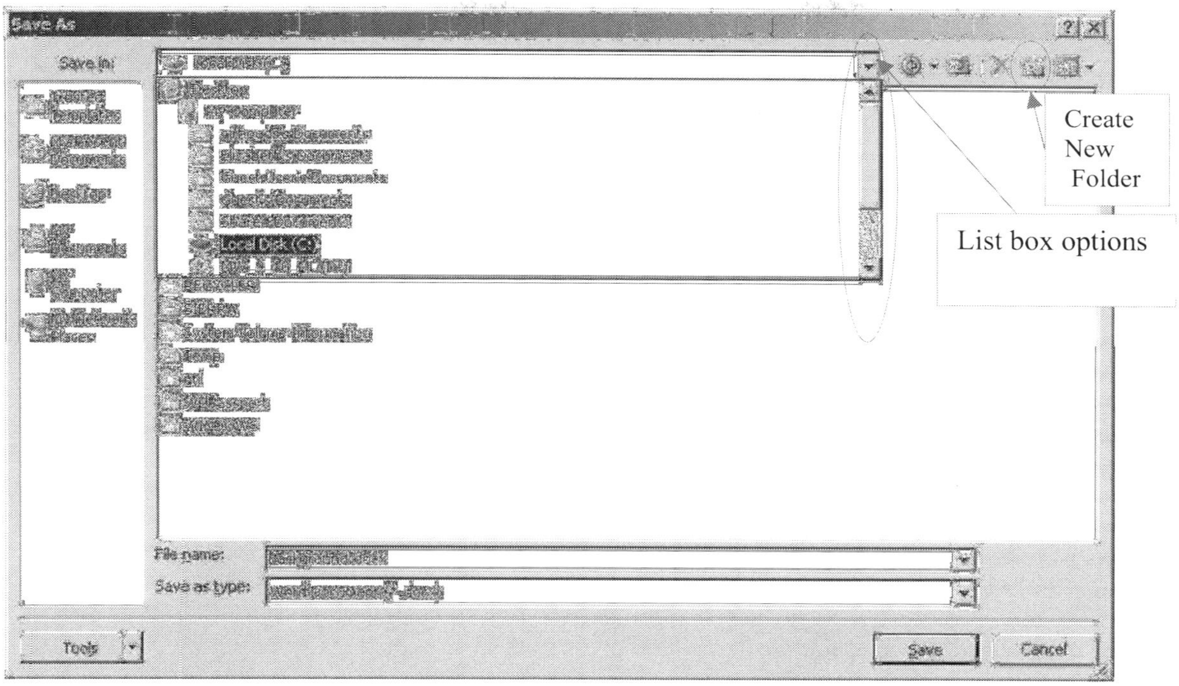


**Figure 59. Changing the location of a file.**

In Word, the Save As dialog box looks very different, as can be seen in Figure 60. But the concept is the same; select the folder to save the file, and change the name if necessary.

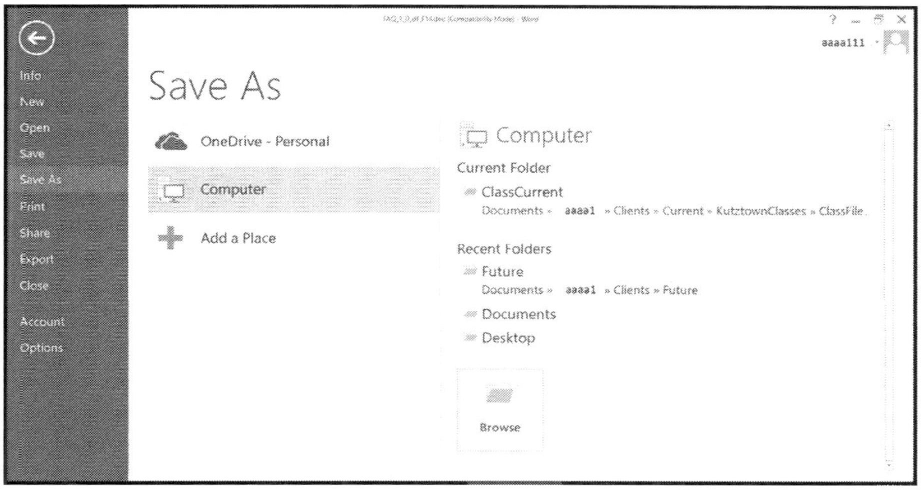


**Figure 60. Word Save As dialog box.**

In all cases, you can save the document in a variety of places:

- Choosing a folder from the C: drive saves the file somewhere on the hard drive inside the computer you are presently using.
- To save to a USB drive, choose D: or E: (whichever says "removable drive" or lists the brand name of the type of USB drive you inserted into the USB port of the computer).
- To save to a network drive, choose F: or G: or any other letter of the alphabet.
- To save to a cloud folder such as Google Drive, Drop Box, or OneDrive, select it from the list or click Browse to browse all the drives and folders.
- Alternatively, you might need to choose **Network Places** and choose the folder from the list that appears.

***Common Mistake Not Watching Save Location***

Watch which folder your file is being saved into. One of the most common (and most frightening) mistakes is to save the document, but not be aware of the *location* where it was saved. Word keeps track of the last folder a file is opened from or saved to, so if you aren't watching, your file will get saved in a different folder than you expect without your realizing it. When you go back to open the file again, it is gone (or you think it is). As long as the machine has not been rebooted, the answer is to look in the recently opened file list. If you open the file from there, you can do a **Save As** to save the document in the proper location. If the computer has been rebooted, then you can use the **Search** command to enter our filename and search the entire disk for the file*.

Some computer labs (like those at Kutztown University) completely erase the hard drive every time the system is rebooted. This prevents students from installing programs or making any changes to the systems, but it has the side effect; if you accidentally save your file to the C: drive, your file will disappear when the system is rebooted. This makes it especially important for you to know where your files are saved at all times.

## Search and Replace

Another useful feature of Office programs is Search and Replace. All programs have a way of searching for text, and it can be very useful to be able to do it. In most Office software, you can use the shortcut [Ctrl-F] and type in the text you wish to find. In addition to finding text, you can also replace what you find with anything that you type in [Ctrl-H]. † For example, suppose you were a consultant who is modifying a previously written proposal for a different client. In this example, every occasion of Walrus Associates would be replaced with Johnson Materials, Inc.

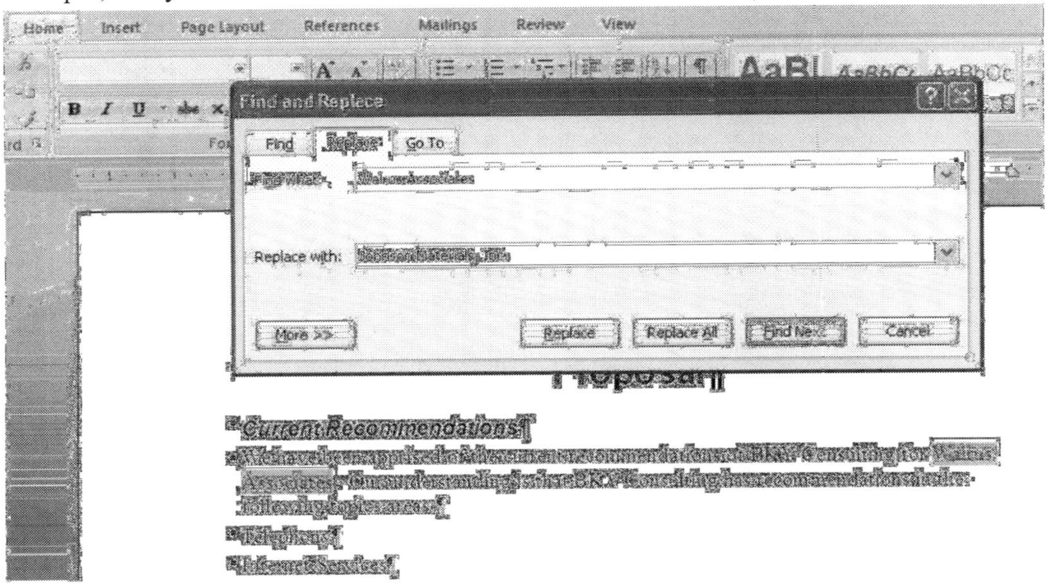


**Figure 61. Search and replace text.**

Some search and replace functions (such as that found in Google Docs) are simple. Others, such as the one found in Microsoft Office, has very powerful modifiers that allow you to use it in ways that you might not have thought of. Some options available might be:

* Keep in mind that Windows 10 search is terrible, and often doesn't find files even though they are there. Use another search utility such as *Search Anywhere.*

† Unfortunately, in Office 2013 Microsoft has decided to put a Navigation panel up in front of the search when you press [Ctrl-F]. Just click on the option box on the magnifying glass and choose Advanced Find to get back to the find and replace dialog box. You can, however, still choose [Ctrl-H] that takes you directly to the dialog box for both Find and Replace.

- Match case, which would match the capitalization that you enter in the Find what: box. If you type in Apology, it won't find apology.
- Find whole words only will not find the word or letters within a word. If you try to find you it will not find your.
- Use wildcards will enable you to replace one or more of the letters with special characters known as wildcards - *?* for a single letter and * for multiple letters.

If you use the ?, then the letter represented by the ? will be highlighted. If you use the * it will find any number of characters, but it will not highlight or select them unless you have actual text on both sides of the wildcard. Then it will highlight or select all of the text between the two letters, including spaces. For example, if you search for *y*u,* then ***you*** will get highlighted, but so will *an**y one for u**nited.*

- ***Sounds like*** will find words with similar sounds. If you type in *il* you will also find *ill, el,* and *all.*
- ***Find all word forms* (or *find similarity)*** will find all the tenses of a verb, and both singular and plural forms of a noun. If you type in *swim,* it will also find ***swam*** and ***swum***. If you type in *knife* it will also find ***knives***. If you type in *lie*, it will also find ***lying*** and ***lay***.
- To find a word that begins with the letters, look for a choice such as ***Match prefix***.
- To find a word that ends with the letters, look for a choice such as ***Match suffix***.
- You may also be able to ***Ignore punctuation characters***, which will find ***it's*** and ***its*** when you just type in *its*.
- There might be a choice for you to ***Ignore white-space characters***, if you search for *you*, both ***you*** and ***by our** own* gets found.

Let's say, for example, that your supervisor has given you the feedback that you use the word "I" too much in your report. You realize that you can search for the word "I" and replace them all with the word "you." But when you search for I, you get 513 occurrences! You realize that the **Find** dialog box is finding every "i" in the document, even within a word, instead of just finding the complete word I. You can deal with that in two ways; either you can ***Match Case***, which would find **I** (as well as MIGHT), or you can **Find whole words only** which would not find MIGHT, but would find i. Or you could check both, and only find I.

### *Finding Invisible Codes*

Finding and replacing text is often very helpful, but not as useful as being able to find and replace spaces, styles and invisible codes. What do you do, for example, if someone (who didn't know about paragraph spacing) hit the **Enter** key and put an extra paragraph mark in between each paragraph, and you copy and paste those paragraphs into a document where you have already set the paragraph spacing on? You would get three spaces between each paragraph; in other words, much too much space.

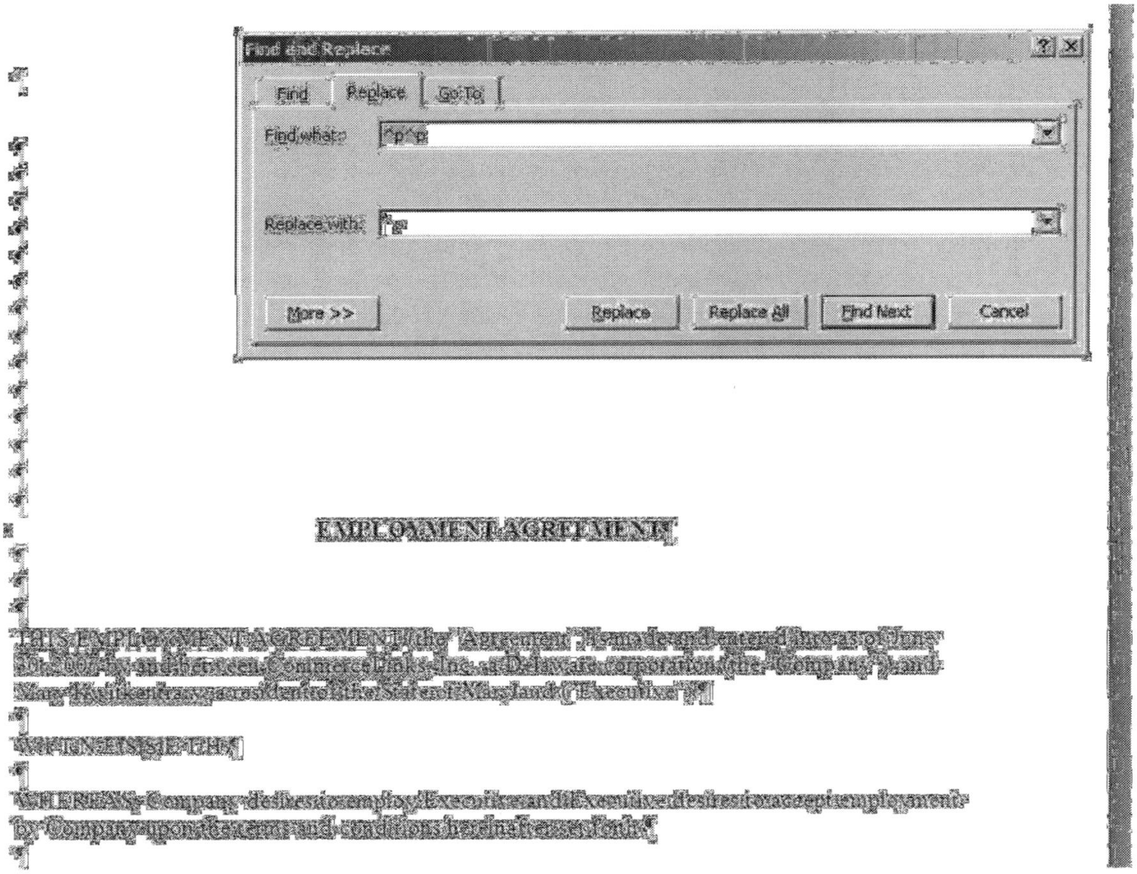


**Figure 62. Search and replace invisible characters.**

One way around that problem is to find every situation where there are two paragraph markers together and replace it with a single paragraph marker. Of course, you can't hit the **Enter** key to put the paragraph marker into the **Find** dialog box because pressing **Enter** *closes the box.*

In many cases, however, the invisible codes have visible equivalents. For example, a paragraph marker in Word has the visible equivalent of caret P (^P). If you look at the bottom of the **More>>** button in the search dialog box of Word, you will find another button called **Special**. Click on that **Special** button and you will see the options for entering invisible codes into the **Find and Replace** boxes (as seen in Figure 62).

In the example in Figure 62, if you click on ***Paragraph Mark*** twice in the **Find** box and **Paragraph Mark** once in the **Replace** box, you can replace every two paragraph marks with a single paragraph mark, eliminating the multiple spacing problem.

Another useful way to use this ability to replace a paragraph mark is to either merge or break up lines of text. This is commonly needed when converting information that has been downloaded from a mainframe computer or web page. For example, if you download a list that puts each word onto a different line, you can replace each paragraph mark with a comma or space to put them all on the same line as shown in Figure 63. To combine all the words from different lines onto the same line, you would replace every paragraph mark (^p) with a comma and a space (, ).

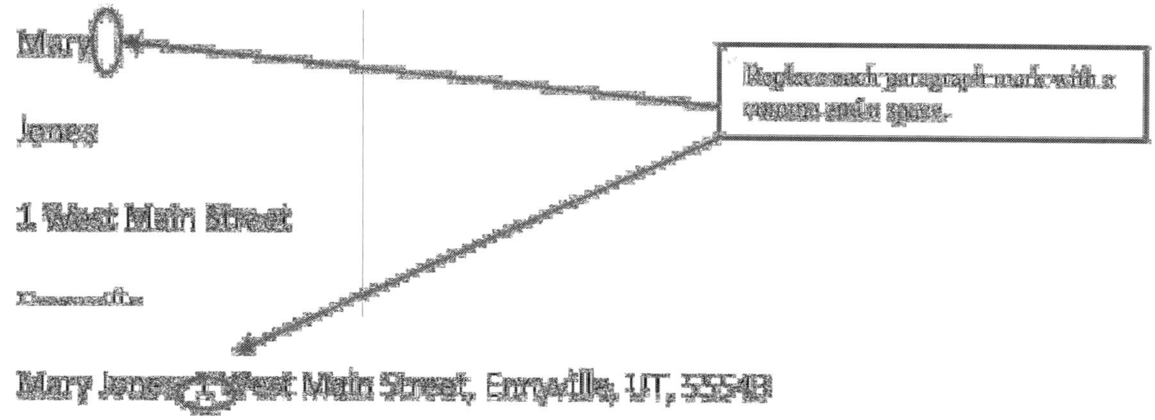


**Figure 63. Search and replace to clean up downloaded text.**

Now that we've covered features that are common to all Office software, it's time to go over some of the most useful features of word processing.

## Word Processing

A word processor is used to write letters and documents - the equivalent of a computerized typewriter ( See Figure 64). In the early days of computing (the '80s and '90s) there were dozens of word processing programs to choose from such as WordPerfect, Multimate, Displaywriter, WordStar, MacWrite, FullWrite, etc. For the last decade most people have used Microsoft Word because it was relatively easy to use while still having a lot of powerful features. Google Docs is becoming more popular due to its lower cost and availability, and WordPerfect still has a tiny loyal base of users who don't want to switch to Word. Many people have also started to use Pages, because it works on a Mac and an iPad. But still, the dominant word processor used everywhere is Microsoft Word.

**Figure 64. An electric typewriter (an IBM Selectric).**

While learning to use a word processor is relatively easy, getting the most out of it is relatively difficult. This section will highlight some of the features of Word that makes it powerful and useful.

### Styles

What do you do when you want to boldface a word? If you were typing in Word, most people would highlight the word and then click on the **Bold** button in the **Font** group of the **Home** tab on the Ribbon. And to indent it they would click on the **Indent** button on the **Paragraph** group of the **Home** tab on the Ribbon. And to space between paragraphs, they would click on the **Line spacing** button on the **Paragraph** of the **Home** tab on the Ribbon. But what if they want all three? And they want all three several times throughout the document?

That's where ***Styles*** come in. A Style is a combination of several formatting changes. The best thing about a Style is that you can change the formatting after the fact - and automatically update all the text that has that Style. For example, if you wanted to change all of the text formatted with *Heading 1* style from boldface to underline, you just modify the Style called *Heading 1* and they all automatically change. No need to go through and adjust each heading individually. While using styles may take more time to set up in the beginning, once you get used to it, you'll find that styles will save a lot of time.

Styles are important to use in long documents such as reports because it makes formatting consistent and editing easier. Therefore, it makes sense to apply almost all formatting with Styles, rather than just making manual changes.

Luckily, many word processors provide ***Built-in Styles*** that do not require any set up to use. Additionally, many of the built-in styles include automated tasks that will make your report even easier to organize - such as an automated table of contents and "keep with next" feature that will prevent your headings from separating across a page break and keep the heading with the first paragraph underneath it. You can also create a new style. In Word, for example, format some text the way you wish it to be formatted within the style, select it, and choose [Alt-H-L-S]*.

### *Invisible Style Codes*

Most word processors have a way of revealing the invisible codes like paragraph marks, tabs, and page breaks. These invisible codes, especially paragraph marks, because doubly important when you start to use Styles. If the style is a ***paragraph level style***, it is actually stored in that paragraph mark. For example, to see the invisible codes in Word, press [Alt-H-8]. To see the invisible code in Google Docs you choose **Edit--Edit HTML.** To see invisible codes in Open Office Writer, you choose **Tools--Options--OO Writer--Formatting aids.**

In most programs, as your cursor moves from paragraph to paragraph, if you look at the Styles, you will see the selected style change as your cursor moves, reflecting the style of the paragraph you are in.

### *Professional Appearance of Documents and Fonts*

It looks unprofessional to have a paragraph split in the middle at the bottom of a page. In a book proposal, for example, you wouldn't want an editor to be presented with something that looked unprofessional. To ensure that the paragraph cannot split across a page, you checked the **Keep Lines Together** option from the **Lines and Page Breaks** tab of the **Paragraph** dialog box [Alt-H-PG-AltP-Altk]. You can also add spacing between paragraphs (without double spacing the text within paragraphs, and without pressing the [Enter] key twice which may give you too much space and cause formatting problems later on).

The font list you have is dependent upon the list of fonts installed on the computer. Each program (like Word) adds fonts, so people who have installed a variety of different programs will have more fonts, and those with new computers or few programs will have fewer fonts. You can also purchase and install fonts. If you choose a font that few people have on their computer and you wish to share the file as a PDF, you can imbed the font in the file. Otherwise, if you choose a font other people don't have and you share the file, their system will randomly select a different font to substitute, but can lead to very unexpected (and unprofessional) results. For that reason, sticking with the most common fonts that most systems have already installed is best.

The font that you choose should match the level of formality you want to convey with the document. For business documents such as memos and reports, a more conventional font such as Times New Roman, Century Schoolbook, or Bookman Old would be a better choice than some of the progressive, more modern-looking fonts. Especially in conservative businesses such as banking, finance, manufacturing or retail, you want the content of the document to have the attention, not the font that it is printed in.

* In the interest of space, many of the instructions are given using the [Alt] keyboard options instead of the full instruction. For example, [Alt-H-L-S] means press the [Alt] key to access the keyboard shortcuts. You will notice letters and numbers appear by every menu option. H will bring you to the home ribbon. L will open up the Styles group, and S will choose Create Style.

A letter can have a "less formal" font than a report or memo. Fonts such as Comic Sans or Script Bold might be appropriate. For a brochure you might want to choose a font that is associated with the type of product or service. For example, a church might choose an Old English font, while a local theater might choose the Playbill font. A flyer for a haunted house might choose the Chiller font.

Newer fonts such as Avante Garde, Batang, Matisse, or Barnhard Fashion convey young, snazzy, or roguish characteristics. These fonts might be more appropriate in younger industries such as fashion, technology, or music.

Courier New is a different kind of font than most others. Courier New is a *fixed width* or *fixed space* font. That makes Courier New a good font to use when you want columns of numbers or text to line up - especially in a report or menu or any business document with a list. Since each letter takes up the exact same amount of space in fixed space fonts, (`an` **`a`** `for example, is no wider than an` **`i`**), fixed space fonts line up very easily. However, since it looks like a typewriter font, and since it takes up more room than a variable spaced font, Courier New isn't often used in reports or publications that require reading paragraphs of text. Fonts today are measured by their height in ***points***. There are 72 points in an inch, so a 72 point font will be one inch tall*.

## *The Importance of Professional Appearance*

You might be asking why anyone should spend time formatting a document. It doesn't matter how it looks, does it? What's important is what is said, right?

While it is true you can't judge a book by its cover, the fact is people get an impression by how a document looks as well as what it says. Just as you would trust a well-groomed doctor in a white coat more than one who looks slovenly and has dirty hands, you would also think that a financial analysis that crams tiny type onto a margin-less page is less credible than the one that separates out the paragraphs into headings, has a nice readable font, and plenty of margin space around the edges. The greatest marketing report in the world will get overlooked and ignored if presented in all capital letters, without breaks or headings. The "look" of the document provides that first, all-important impression to the reader - who might be your boss, your colleague, your client, your customer, or your supplier. Which impression would you like to make?

- Sloppy, inattention to detail, emotionless, boring?
- Bright, exciting, happy, artistic, "wow!"?
- Calm, professional, enthusiastic, knowledgeable?

While the first one may be an inadvertent or careless impression, and the second one may be appropriate in some limited industries, the third is probably what you will want to convey for most business documents.

Maintaining correct spelling and grammar throughout is a given, of course. Additionally, certain design "rules of thumb" exist that, without your having to write a single line to say so, convey that you are a highly competent, knowledgeable professional. In general, it can be said that:

**Serif fonts (fonts like Times New Roman and Century Schoolbook) of at least 12 points should be used for body text. A serif is the little edge at the ends of the letters as can be seen in**

Figure 65.

---

* Point measurement for fonts is a change from the "old way" when typewriters proliferated. On a typewriter, fonts were measured in ***pitch***, which is determined by how many letters could fit in an inch, a measure of width, not height. With a fixed-pitch font the larger the number the *smaller* the font (the exact opposite of points, where the larger the number, the *larger* the font). A Courier New *12 point* font is roughly equivalent to a Courier New *10 pitch* font. At the same time, the Courier New *12 pitch* font is roughly equivalent to a Courier New *10 point* font. (We use Courier New because we can only do a comparison such as this when the width of each letter is exactly the same.)

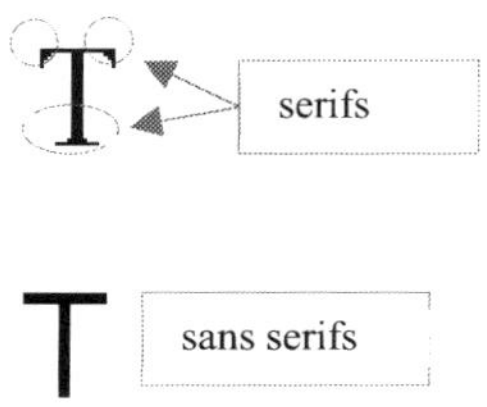


**Figure 65. Serif fonts vs. Sans serif fonts.**

Sans serif fonts (i.e., fonts like Arial and Helvetica) should be used for headings. Sans serif means "without serifs."*

A document should have, at most, two or three different fonts.

Use headings, subheadings, and subsubheadings as appropriate. Avoid presenting pages and pages of text without any headings.

**Headings** should be bigger than **subheadings**. **Subheadings** should be bigger than **subsubheadings**. All headings should be bigger than body text.

At least 30% (and up to 50%) of the page should be "white space."

Each heading should be connected to the text below (i.e., no page breaks between the heading and the first paragraph of a new topic).

There should be more space before each heading than between the heading and the paragraph below it.

Avoid all capital letters (they tend to convey INAPPROPRIATE SHOUTING THAT CAN BE VERY ANNOYING).

Avoid all lowercase as well and make sure you use appropriate punctuation (famous poets who do not capitalize or punctuate notwithstanding).

If you break the design rules in most businesses, no one will *say* anything - you will simply appear less professional and credible to your colleagues and superiors. This is especially true in conservative industries such as finance, banking, manufacturing, or retail.

There are a few noted industries where these rules can be (and in some cases should be) broken. Fashion, advertising and music are examples of industries that promote and expect creativity. Knowing how and when to break the normal design rules is encouraged in these types of businesses. But astute professionals in those industries *know* the rules that they are breaking, so even when being broken, the rules dominate. (You didn't think that e.e. cummings didn't *know* that names are normally capitalized, did you?)

### *Templates*

If a style is the first level of being able to create and edit documents quickly, the template is the second level. A template stores not only the styles, but also the page formatting, the headers and footers, the colors and themes, and any default text that you would like to see in that type of document. Once you learn how to use other's templates and create your own templates, you can save a great deal of time creating similar documents. As a student, for example, you could create a term paper template that includes the cover sheet, headers and footers, page numbers, and all the

* Yes, there are many many design-oriented people who will argue with me on this. I'm using the research on readability, not a sense of fashion. As noted, except for certain industries, readability and functionality trumps aesthetics.

fonts and designs that make the document look highly professional. Most applications have a way to create and use templates. For example, to save a document as a template in Word, simply create the document with all the changes that you would like to see, and then [Alt-F-A] and select Word Template as the file type. The next time you want to create a file with similar characteristics, open up the template instead of starting with a blank file*.

### Page Formatting

In a simple letter or document, normally the default margins don't need to change. Unlike letters, brochures, or other short documents, however, longer papers and reports generally have a few requirements like headers and footers, page numbers, and a cover page. In a short document, it is easy to "glance through" the document and find everything. With a longer document, the reader can't quickly glance through it to find whatever they are looking for. As the author, it is important for you to make it easy on the reader to find everything, and to understand exactly where the document came from, what page they are currently on, when it was written, and where to get it again if they need it.

### Page Margins

In the previous section, you learned how to modify the space around the text and paragraphs, change the indents of the paragraphs, and change the amount of space before and after paragraphs. However, in some situations, you might need to modify the margin of an entire document. For example, what if you want the document to have a 1.25" margin around the entire page? Or need to print on legal size paper (8½" X 14") or some other size paper?

Some business documents traditionally have large margins - such as letters and memos. A letter might have 1½ to 2 inch margins on each side - more if there is a logo in the header at the top. Longer reports and document like instructional manuals and policy guides may have slightly narrower margins - closer to 1 inch on each side. Brochure and flyers - especially folding documents - will need much narrower margins - probably ½ inch on each side. In most programs there is a tab or menu which contains page margin choices. There will also be a tab or menu where you can change the page orientation from ***Portrait*** to ***Landscape***.

In Microsoft Word and Open Office Writer, if you want to have some pages of your document Portrait and some Landscape (or different margins or different headers and footers) then you must insert a Section Break. A section break will allow you to set the page orientation, margins, headers, footer, and page numbers separately in different sections of a document. (Google Docs does not yet have this capability; the whole document must be either landscape or portrait, and must have the same headers and footers.)

If you want a cover page that does not appear as page 1 in your page numbering, then you need to put the cover page in a separate section. You can set a section to begin again on page 1. It is common to have the cover page and a table of contents in its own section using page numbering in roman numerals (i, ii, iii, iv, v, vi, etc.) and begin the page numbering again at Arabic number 1 when the actual content begins. This will also allow you to set wide margins on the first few pages and different headers and footers.

### Headers, Footers, and Page Numbers

When they are used in a document, ***headers*** appear at the top of each page, and ***footers*** appear at the bottom of each page. In all business documents, it is very helpful to have the name of the

* In reality, even a blank document is a template. The starting file of most programs is really just a default template. For example, in Word, the default template is called normal.dotx, and it is stored in a hidden template directory. If you modify that file, you can change the initial blank file that Word opens when it starts up. Also – a fair warning about templates that come with applications; often they are designed to show off advanced features of the application and are difficult to edit. They are great to use as-is, but if you want your own thing, do your own thing and then save it as a template rather than beating your head against a wall trying to make a minor change to an already existing template.

document on every page, as well as the author, the date the document was created or saved, and the path and filename of the document. (A ***path*** is the location of the document, such as *u:\aaaa111\Classwork* that precedes the document name.) It would be a lot of work to type these on every page, every time - not to mention futile because if you edit the document, it will repaginate in different places. With headers and footers you can type these in just once, and they will automatically repeat throughout the entire document.

Even better, both Word and Writer has ***fields*** (also called *Quickparts* in Word) that will allow you to automate many different types of information about the document:

- Path and filename,
- Date (the create or fixed date, and the dynamic or save/print date)
- Author name
- Document title
- Page number

If you use these fields, you won't have to edit any of that information if you were to open up the document from a different place or different time. They will automatically update.

Putting the path and filename in the footer of a document is an especially helpful business practice. Months or years later, when you pick up the paper document, you won't have to remember where you stored it; the location is at the bottom of the page. You might not think so right away, but getting into habit of putting the path and filename in your footer will save you much trouble and frustration as the years go on and the number of documents you need to track increases exponentially.

A filename and path can get very long, so choose a very small font, such as 6 point size (which you will have to type in as it is not normally a font size choice). It's also nice to place a line (paragraph border or horizontal line) between the footer and the text of the page for a very professional looking touch.

## Spreadsheets

The use of word processing was actually secondary to the primary use of personal computers in the early days. The very first "application" available for personal computers was VisiCalc, a spreadsheet program, which sold for $99 in 1979 and ran on an Apple IIe computer. Spreadsheets were the "killer app" of the personal computer.

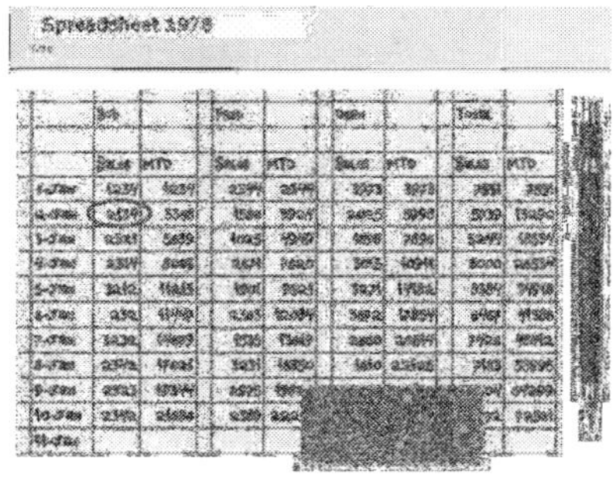

A spreadsheet enabled accountants to stop poring over paper lists of numbers where one mistake could cause hours of rework. (See Figure 66.*) Instead, they could type those columns of numbers into cells and have them automatically sum. If one of the numbers changed, the sum would automatically change.

**Figure 66. Manual spreadsheet.**

VisiCalc soon fell by the wayside as Lotus 123 became available on the IBM PC. Since IBM was the top computer vendor, they were already selling computers to all the large companies, so they could easily provide spreadsheet software to the finance divisions of those companies. Lotus 123 (eventually bought by IBM to become Lotus Notes) was followed quickly by Multiplan (eventually bought by Microsoft to become Excel). Excel is the most popular spreadsheet in use, today, and an essential skill for anyone in business.

* The graphic illustration of the manual spreadsheet is courtesy of Bill Jelen (aka "Mr. Excel") from http://www.mrexcel.com/tip076.shtml, originally published September 30, 2004 and accessed in May, 2008.

While again, most people know how to use spreadsheets in a basic way, there are many useful advanced features of spreadsheets that people would find helpful – but only if they know about them. In this section we will go over some of those useful features.

### *What-If Analysis*

A "what if" analysis is the ability to make a change to an assumption number of a spreadsheet, and observe the impact on all the formulas of the spreadsheet. All spreadsheets can do what if analysis – indeed, it is the whole purpose of entering formulas into a spreadsheet. For example, what if expenses were to rise by 3% next year? What if they were to fall by 2%? What if the interest rate was 6% instead of 5%? Once a spreadsheet is set up correctly with formulas and functions, changing one of the input values immediately reflects the results of a What If Analysis.

To do a what if analysis, you simply set up formulas and functions to use cells that contain the numbers. One potential point of confusion is that there is a **What If Analysis** icon on the Outline group of the Data ribbon of Excel - but it is not what is thought of as a traditional what if analysis, for which you need no menu.

But under the What If Analysis icon in Excel is **Goal Seek**, for which you do need the ribbon, and which is NOT available in all spreadsheets. While a what if analysis wants to find out what would happen if the value of an input number changed, whereas a *Goal Seek* allows you to specify what you want the answer to be based upon changing one of the inputs. A typical use would be to figure out what grade a student needs in a final exam in order to get an A in a course (shown in Figure 67), or what interest rate a borrower would need to have in order to get the loan payments on a new boat to $500 a month (Figure 68).

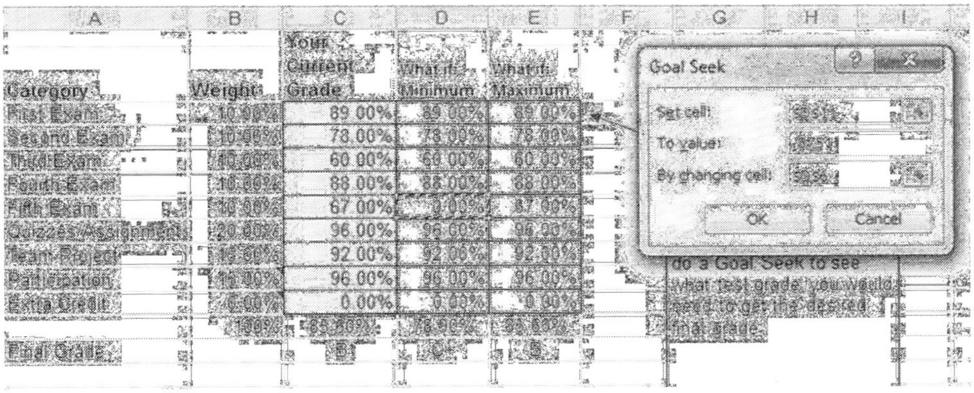


**Figure 67. Using Goal Seek to seek a final grade.**

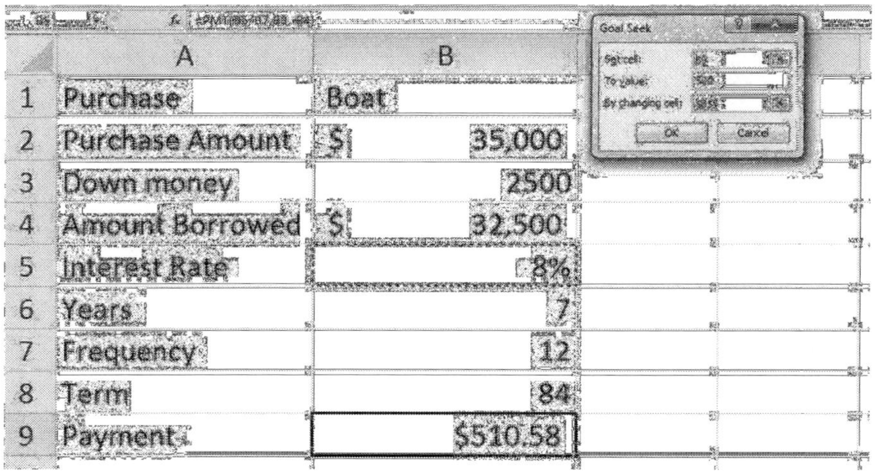


**Figure 68. Goal Seek for loan payment.**

### *Graphing*

Graphing (also known as charting) is almost always most easily done in a spreadsheet (though possible in a database, but not easy). The key to success is to make sure that the data is organized in a way that is expected for charting. Sometimes this means using the [Ctrl] select combination to select non-contiguous cells. Sometimes this means moving the data around. In general you want the X categories in the leftmost column, the Y series names at the top of each column, and the numbers to be plotted within the range selected.

Never include the details of a totals in the same chart as the totals. The totals will be many times larger than any of the detail numbers, making the graph useless for illustration purposes.

There are several kinds of graphs. A pie chart, as shown in Figure 69 can only have a single series. In this illustration, you can see that the Total for the year is being graphed, and that the X values are the different categories of expenses. To make this selection we needed to use the [Ctrl] key to make a non-contiguous selection.

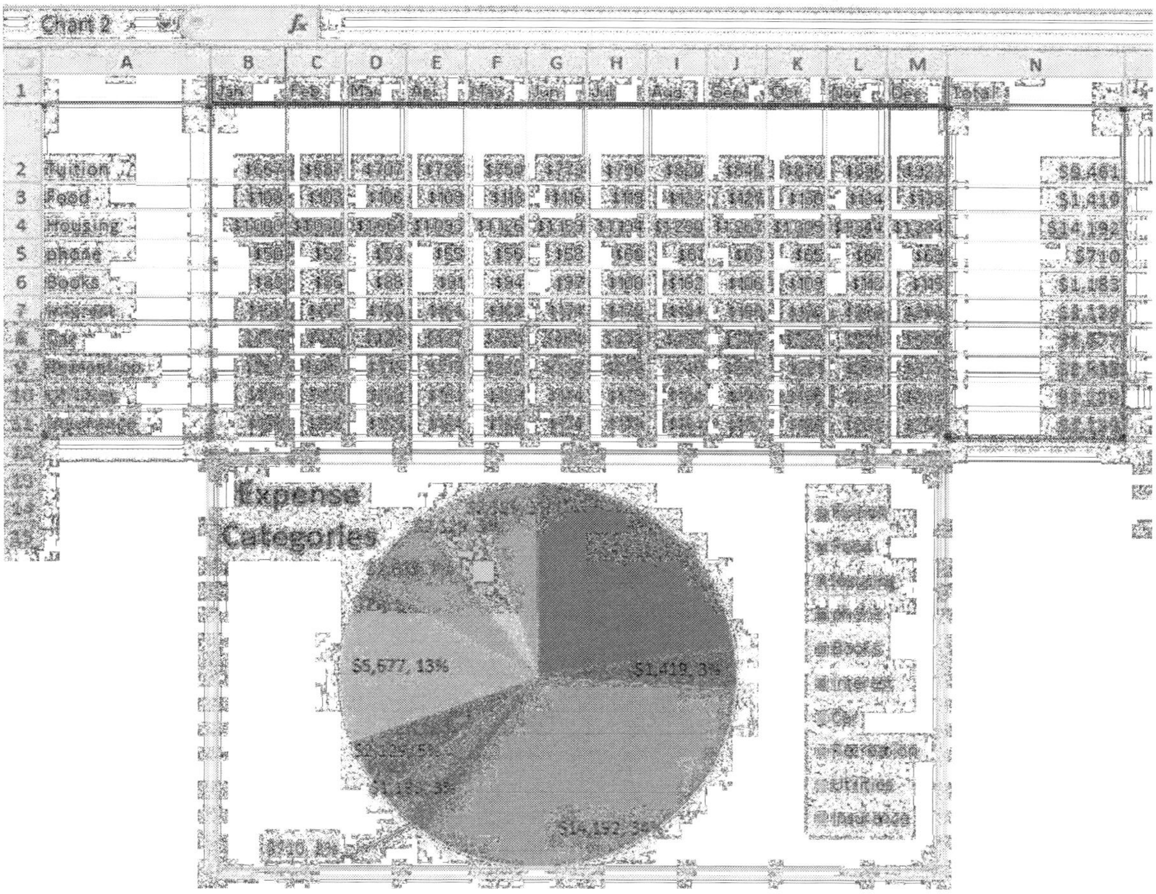


**Figure 69. A Pie Chart.**

Other types of charts are Column, Line, Bar, Area, Scatter, Surface, Stock, Doughnut, Bubble, and Radar. Charts can be 2-dimensional, or 3-dimensional. While the fancy graphs and the 3 dimensional graphs may be fun to play around with, in general they often obscure the point of the information a graph is meant to convey, so it is usually best to stick with the regular simple 2-dimensional Clustered Column, the Line, and the Pie. Once the data has been selected, and the graph has been inserted based upon the data, then each component within the chart can be selected and modified either using the context menus for each object, or using the Chart Tools ribbons (Design, Layout, and Format) which appear when the chart is selected.

A common problem is to try to print a spreadsheet with a chart on it but inadvertently **selecting** the chart first. If you select a chart, then only the chart will be printed, not the chart and the spreadsheet upon which it is based (even if the chart is imbedded in the spreadsheet, which is the default).

### *Advanced Functions*

While formulas, what if analysis, and goal seek are all very useful, the true power of a spreadsheet comes from the functions. Functions can take highly complex formulas that utilize advanced math like quadratic equations and calculus, and make them very simple so that people can use the power of the advanced math without becoming mathematicians themselves. Functions also allow us to use conditional logic in order to look up information or add or count data, which is a very useful capability. Many of the more useful functions will be discussed briefly.

While some spreadsheet programs have slightly different functions or slightly different inputs used within the function, all spreadsheets have these basic functions. They are the most common functions employed by typical people using spreadsheets (as opposed to highly specialized uses of spreadsheets such as finance or engineering). This discussion will cover both the use of the function within the spreadsheet, and the meaning of the arguments.

A function usually has one or more *arguments*. An argument is a variable, a holding place for a value to be determined. When you are entering a function, Excel usually displays the type of argument in a flag below the cell as you are entering it. Arguments are usually entered as the cell address of the location of the number. We enter cell addresses instead of hard coding the number so that we can do what if analysis and goal seeks. (If all the arguments of a function are hard coded, then you would have to edit the function within the cell to make a change. If you enter the number in a cell and then use the cell address of that cell as the argument, then to change the results you simply type a new number into the cell [much easier than editing a function]). You can see an example in Figure 70.

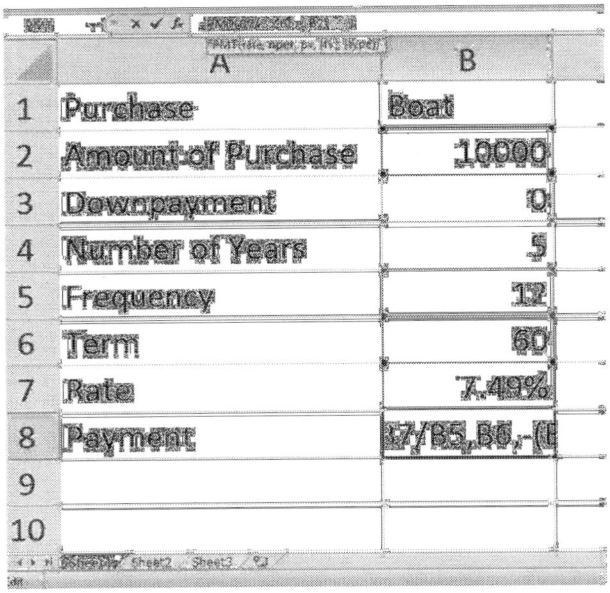


**Figure 70. Arguments appearing below function on formula bar.**

In this example, the arguments are rate (the interest rate of the loan for each payment), nper (the number of periods or number of payments of the loan), pv (the present value at the beginning of the loan). There may also be *optional arguments*, which are arguments that you don't need to enter, but may enter if you wish. You can tell they are optional arguments because they appear in square brackets []. In the example the optional arguments [fv] is the future value at the end of the loan, and [type] which is an *indicator*. An indicator (also known as a flag) must be explained in some way (usually by looking at the help screen for the function). In this case, the type indicator values are either 0 or 1. The value is 0 if the payment is made at the end of the month and 1 if paid at the beginning of the month. The *default* (the choice if no value is entered for the type) is 0, so if you omit this argument, then the assumption is made that the payment will be made at the end of the month. It makes a very small difference in the amount of interest for the term of the loan, so the payment will be a tiny bit different depending upon the type indicator value.

Also notice that one of the arguments in the illustration (**nper**) is darker (boldface) than the others. While you can't see it in the illustration, the blinking cursor is on the **B6** in the formula bar. That makes it the current argument, which is why it shows up bold. Knowing which argument is current is helpful if you are entering or editing the arguments on the formula bar.

### *=PMT(rate, nper, pv, [fv], [type])*

Understanding this function is easy when we think about how borrowing from banks works. When someone wants to buy something (like a house or a car) that they cannot afford to pay for all at once, they can borrow money from the bank (or financing company) to buy the item, and then pay back the bank a little at a time until the amount is satisfied. Banks, of course, don't lend us money from the goodness of their hearts; they lend us money because they are able to make a profit on it. When we pay them back, we pay them back **more** than we borrowed. How much more is dependent upon many things: what the interest rate is, how frequently we give them money, and how long it takes us to pay them back. The lower the interest rate, the more frequent the payments, and the shorter the time until they are paid back, the less we pay. Every day that we owe the bank money they charge us interest at an agreed-upon rate. "Another day, another dollar" is an accurate statement for a loan of $1675 at 4% interest for five years.

So when we borrow money, we pay back two different amounts; the amount that we borrowed (the principal), and the amount of interest we owe for the privilege of borrowing the money. The amount of interest we owe changes every day, so we cannot calculate a precise amount of interest we are paying back unless we know exactly which day in the term we are calculating it for. Every time you make a payment, some of it is for interest and some of it is for principal. Each time you pay off a little more of the principal, the amount of interest decreases, so over time the principal portion of your payment rises and the interest portion decreases. That's why, when you want to pay off a loan early, you have to tell them the "pay off" date (in the future), and they will calculate the "pay off balance" which is the principle plus interest you owe which you haven't yet paid, estimated up to the date that you have given them.

How much of your payment goes toward principal and how much goes toward interest takes a quadratic equation to solve. Until the proliferation of spreadsheets, unless they were really sharp mathematicians, borrowers had to go to the bank and ask for an amortization chart in order to estimate how much of each payment was interest and how much was principal.

Entering the argument for the Interest Rate is complicated because, by law, banks are not allowed to quote a monthly. They are only allowed to quote the annual rate (to avoid consumer confusion). However, the interest rate in the function must be the **per period** rate (usually but not always monthly). Regardless of the real term (the number of payments or the number of periods), the interest rate in the formula is divided by the number of payments per year. For example, if the interest rate charged by the bank is 6%, and our payments are monthly, then the interest rate we enter into the PMT function would be **.5%** (.005), which is the interest for one month.

Entering the principal is also complicated. From the bank's perspective, the principal is the positive amount, and the payments are the negative against the principal (at least the principal portion of the payment is subtracted from the original amount to arrive at the balance owed.) That means that under normal circumstances, the payment amount that we calculate with the PMT function will always be negative. Generally, of course, we don't want it to be negative. So we need to make the principal negative by putting a minus sign in front of the cell address of the principal inside the function. That way, our payment will show as positive. (We could also put a minus sign in front of the whole function in the formula bar, but that would make it difficult to use the function wizard, which makes it easier to enter the arguments. Plus, it would go against the principles of financial accounting about principal.)

For example, imagine that we stored the annual interest rate in cell A1, the number of payments per year (also known as the *frequency*) in cell A2, the number of years in A3, and the amount that we borrowed in A4. Then the payment function would look like this: =PMT(A1/A2,A2*A3,-A4).

Notice that we are doing a calculation inside one of the arguments. Since the nper (the number of periods, also known as the *term*) is the number of payments per year times the number of years, we can use the single cell inputs. (We might just as easily create a formula =A5 * A6, and put it in A7, and used that instead of doing the calculation within the argument of the function.)

You may also notice that we didn't enter the last two arguments, which are optional. If a balloon payment was made, (a large amount due at the end of the loan instead of paying off the loan slowly down to 0), then the FV would be the future value of the balloon payment. Additionally, because the bank charges us interest every single day, it would make a difference pay at the beginning or the end of the pay period. The type argument is True (or 1) if we pay at the beginning of the month and False (or 0) if we pay at the end. If we don't put anything the future value is assumed to be 0 and the payment is assumed to be paid at the end of the month.

In another example, the entire function is shown using the dialog box wizard (which comes up if we click on the ) for a boat purchase at 7.49% interest and monthly payments for five years. We did put the Term in a cell (which contains the formula `=B4 * B5`) but we did enter a formula to calculate the period interest directly into the argument (`B7/B5`). Notice that when using a formula in an argument, we do not include the equal sign.

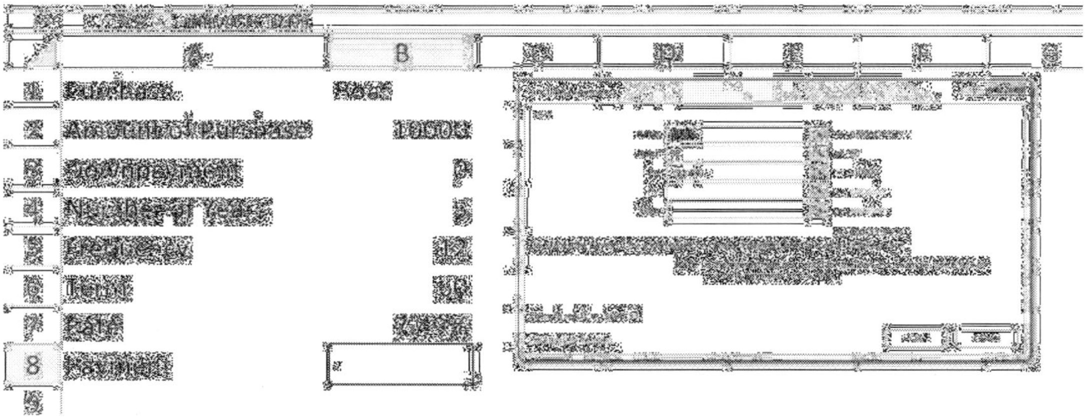

### *=IPMT(rate, per, nper, pv, [fv], [type])*

Related to PMT is IPMT, the interest portion of payment. This function is very useful to determine how much of a payment of a loan is interest. The difference between the interest and the payment amount is the amount paid on the principle of a loan (i.e. the amount of the present value of the loan that has not yet been paid). All the arguments are the same as PMT except for *per*, which identifies *which payment* the spreadsheet is calculating the interest. Since the principle changes every time you make a payment, the amount of interest changes each time as well. Therefore, we need to identify whether we want to know how much the interest portion is of the first, or second, or twenty-fifth, or two-hundredth payment because each one has a slightly different portion of interest.

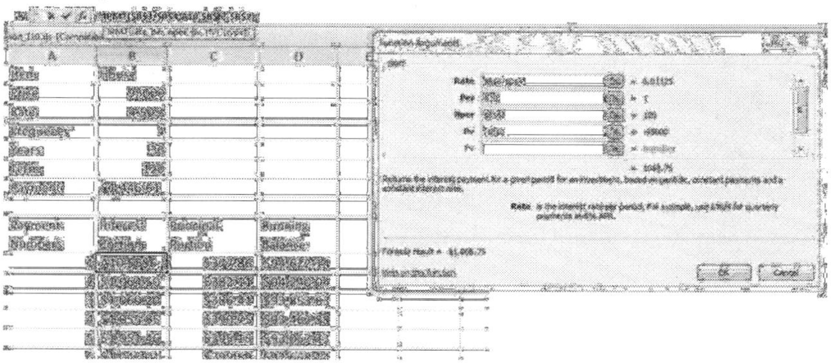

**Figure 71. IPMT function arguments.**

### *=FV(rate, nper, pmt, pv, type)*

Future value works exactly the opposite of payment. Instead of borrowing an amount and then paying it off in equal payments over time, future value allows us to put aside equal payments over time in an account with an interest rate, and it will calculate the future value of the account. For example, if you were to put $5 a week into a savings account with 2% interest for 20 years, you would be hard pressed to calculate the interest because it is compounded daily (which means that you get interest on the interest that has already gone into your account). But with Future Value, it would be easy. Assume you put the interest rate in A1, the frequency of payments per year in A2, the number of years in A3, and the amount of your payment in A4. In this case, let's also calculate the term in A5 (which is A2 * A3) and the amount of period interest in A6 (which is A1/A2). The future value function would be =FV(A6, A5, -A4). Again, we need to make the payment negative in order to show the principal as a positive number, and we started the account with 0 (our present value) so we can skip that argument, and we are going to pay at the end of the period so we can skip the type argument as well.

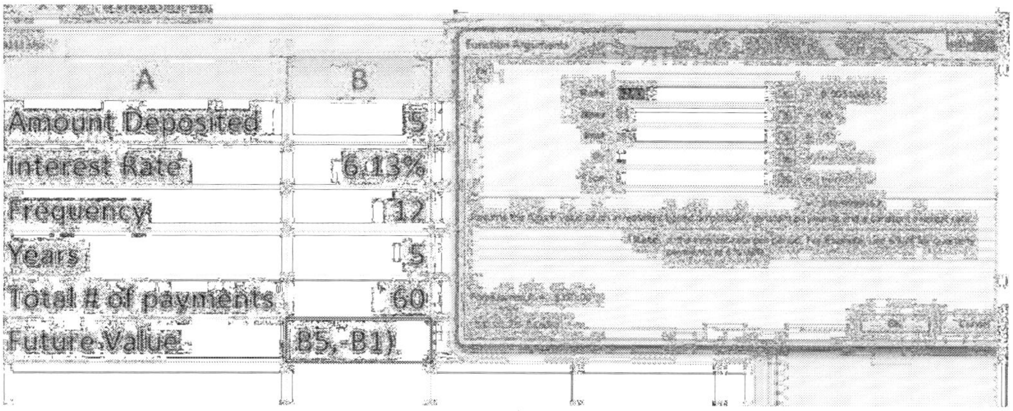


**Figure 72. FV function with arguments.**

### *=VLOOKUP(Lookup_value, Table_array, Col_index_num, [Range_lookup])*

Being able to look up information in a table is a very useful technique in spreadsheets. For example, a student might want to look up what grade they would get based upon their numerical percentage, or a businessperson might want to look up what tax rate they would pay based upon their income. If the table containing the information is vertical (i.e., in columns), then we use a Vlookup() as shown in Figure 73.

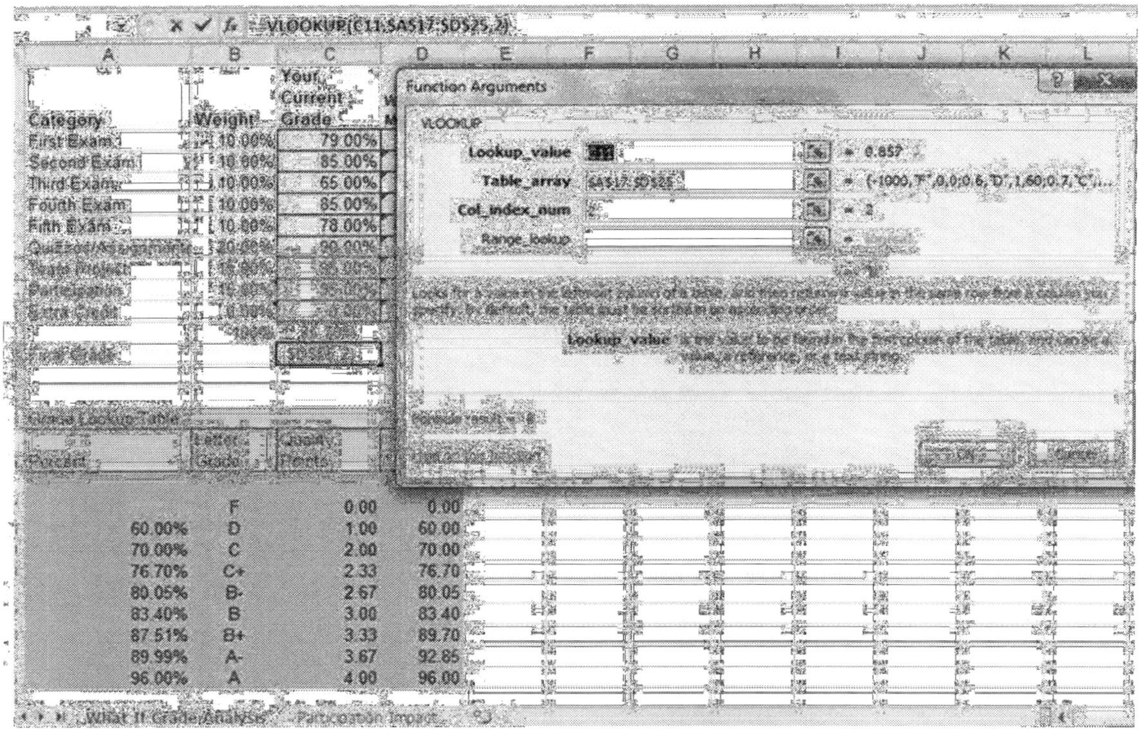


**Figure 73. Vlookup() function with arguments.**

The vlookup function has four arguments. The lookup value is usually the cell address where you enter the value that you want to look up. In the illustration, for example, the student is looking up the numerical grade 85.70% on the table to find out what the letter grad would be. Note that the lookup value is always a single cell address, never a range of cells.

The table_array is the argument that contains the range of the table. Do not include any column headers in the range. Furthermore, the left most column in the range (column #1) must be in alphabetical or numeric order. If the numbers are not in order, the function will try to interpret them as if they are in order, and you might not get a correct answer. You may notice also the dollar signs that appear in the table_array range. We can type them in manually or press the [F4] key to insert them in the cell address. They will make the cell addresses absolute rather than relative so that if we copied the vlookup function into another cell, the table range would not change.

The col_index_num argument is confusing, but simple once you understand. (It is not exactly intuitive. Intuitively we would expect the column index to be a column letter, like B, but it doesn't work like that.) The column index is the <u>number</u> of the column <u>within the table</u> that is expected to return a value. In the illustration, we want the letter grade (in this case, a "B") to appear in the cell as the result. Although there may be more columns in the table, we want the <u>second</u> column to return. So we would enter the numeric value "2" in the column index argument. If, for some reason, we were looking for, say, the Quality Points, then the column index argument would be "3". This argument allows us to use the same table for different vlookup commands so that we can get different values depending upon the column index.

The final argument is the Range_lookup, which is an optional indicator, but important. If we are looking up a number that might fall between any other two numbers, as is the case in the Figure 73 illustration, then we would want the Range_Lookup to be TRUE, which is the default (so we actually can skip it altogether). However, if the Range_lookup needed to be an exact match, such as looking up an identification number, then the indicator would have to be FALSE. FALSE is not the default value, so if we need an exact match, the value must be FALSE. Otherwise, we will get an incorrect result if someone were to enter something that was not in the lookup list at all.

Suppose you were trying to look up the name of a student based upon an identification number. The table looked like this:

| | |
|---|---|
| 1245678 | Joe Bloe |
| 1256785 | Mary Contrary |
| 1456746 | Gerry Geronimo |
| 1675678 | Suzie Sousman |

Suppose someone enters (incorrectly) the ID number 1354564. If the vlookup range_lookup indicator was FALSE, the answer of the function would be **#N/A**, which stands for Not Applicable because the ID number was not found. But if the vlookup range_lookup indicator was TRUE, then the function would result with the answer Mary Contrary. Even though the ID numbers didn't match exactly (1256785 is not the same at 1354564), because the value was BETWEEN two of the range values, the function would report it as if the ID was found, leading to mistake identity.

### *=HLOOKUP(Lookup_value, Table_array, Col_index_num, [Range_lookup])*

If the information is horizontal (i.e., in rows), then we use an Hlookup. Everything else is exactly the same as vlookup. You can see an example of Hlookup in

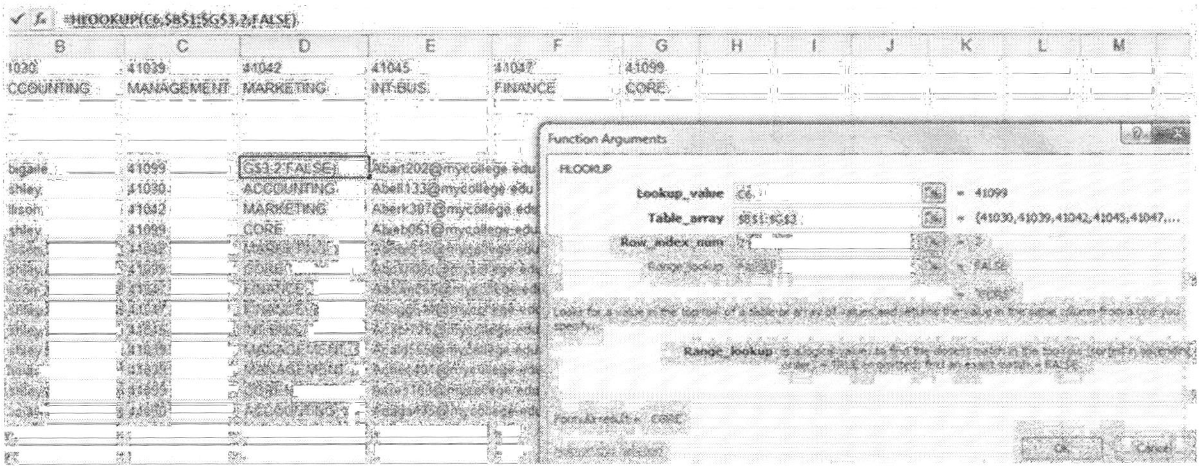


**Figure 74. Hlookup function with arguments.**

### *Sorting, Subgrouping, Autofiltering*

One thing that spreadsheets tend to do well is to quickly subgroup and subtotal a list of values. While this capability also exists (and is more advanced) in databases, if we just want a quick subtotal, using a spreadsheet to get the subtotal is quick and easy. Spreadsheets enable quick filtering, grouping, and subtotaling for any range of numbers if there are no blanks or formatting interfering with the range. ,

In Figure 75, you can see the results of a subtotal of the HighBalance on TypeOfClient. The little numbers over on the left can be clicked to change the level of detail included in the display.

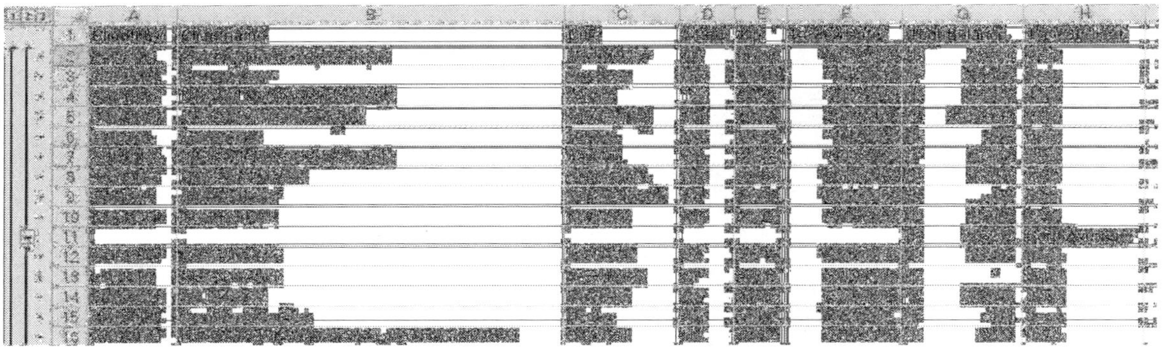

**Figure 75. Subtotaling in Excel.**

If we only wanted to see the totals, we would click on the little 2 to the left of the column headings, then the sheet would look like Figure 76.

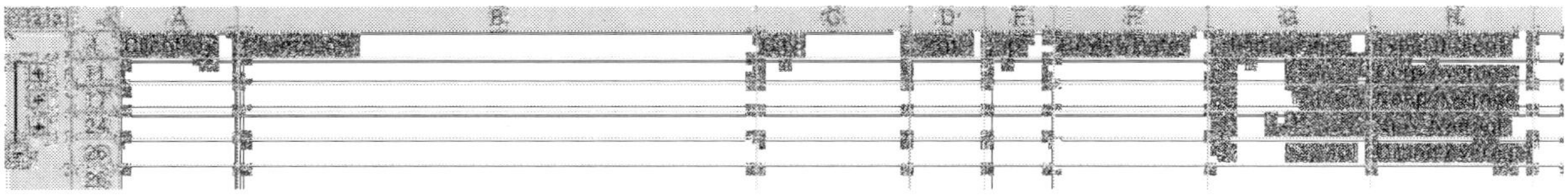

**Figure 76. Subtotals without detail level.**

It is important to remember to **sort** on the column which will be subtotaled before subtotaling it, . Otherwise, the subtotals will break up into sub sub totals based upon whatever column it was sorted on.

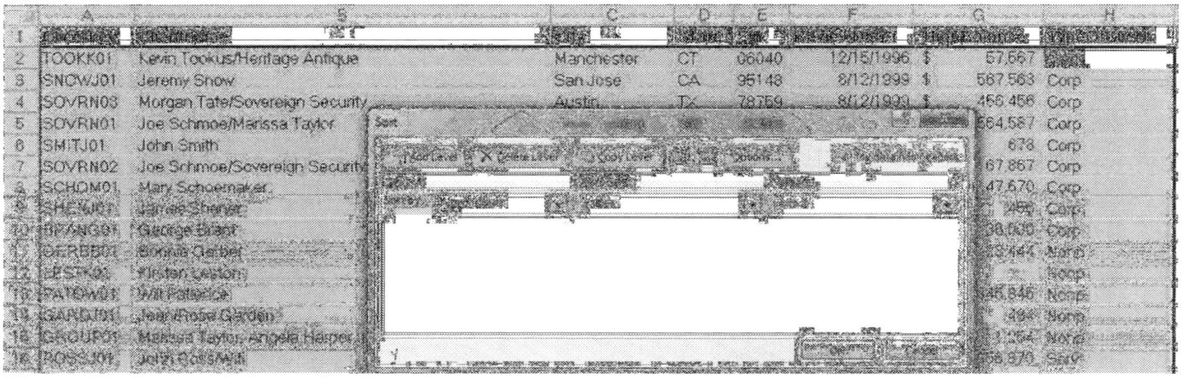

**Figure 77. Sorting information in a sheet.**

Notice in the illustration (Figure 77) that the entire data set is selected. Also notice that there are headings at the top of each column, and the Sort dialog box has the **My data has headers** choice checked. It is very important that the correct items are selected before we hit the Sort button. A common mistake is to try to select the column upon which we want to sort; that would be a mistake because it would sort **only** that column, not the rows of information related to the column. This mistake would permanently alter our data and, if we accidentally save it that way, cannot be undone.

Once we've sorted the data appropriately, we can choose the Subtotal from the Outline group on the Data ribbon, as shown in Figure 78. We list the column that we want the group subtotal for in the **At each change in,** the type of statistic we want (sum, max, average, count, etc.) in the **Use function**, and check the column we want to total. Notice also that this same dialog box is used to remove all the subtotals from a data set.

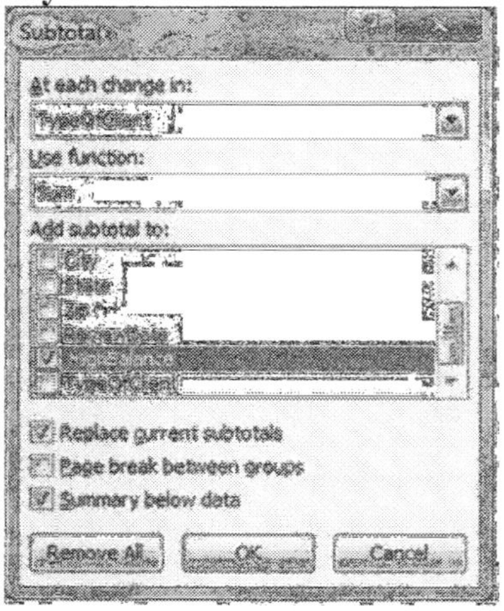


**Figure 78. Subtotal dialog box.**

Autofilter is also a very useful feature. When you click on the **Filter** icon on the **Sort & Filter** group on the **Data** ribbon, an option button appears at the top of each column. This autofilter allows us to easily choose criteria so that we can view subsets of the data without actually copying and pasting them elsewhere. An example filter to see only the rows where the client is from California or New York is shown in Figure 79.

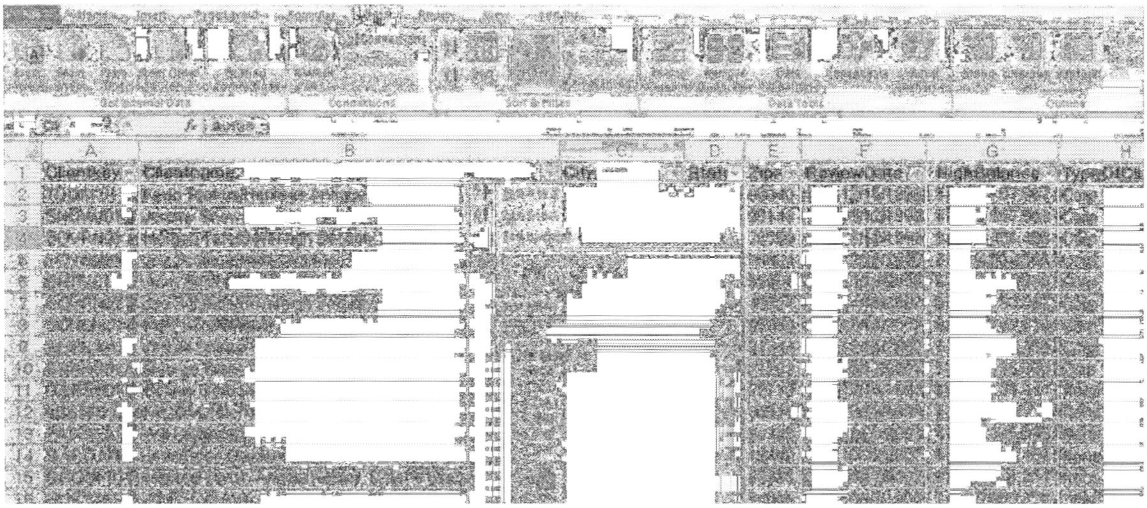

**Figure 79. Autofilter.**

## *Pivot Tables*

Another useful feature is Pivot Tables. Pivot tables allow you to easily create subsets of a dataset. Simply by using the menu items and dragging and dropping the appropriate fields into the table, a person can turn a datafile (as shown in Figure 80) into a more easily analyzed pivot table as shown in Figure 81.

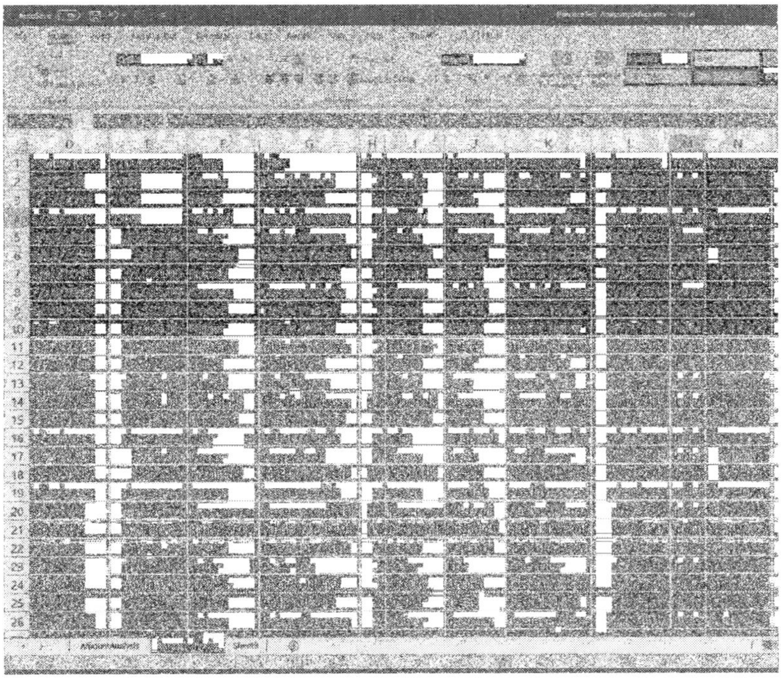

**Figure 80. Sample Datafile**

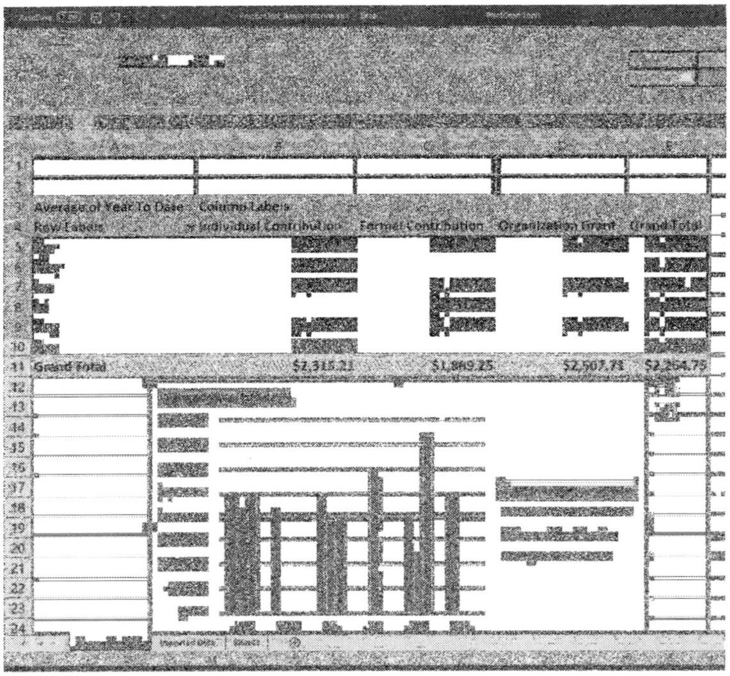


**Figure 81. Sample Pivot Table**

### *If Statement*

While it looks simple, the IF statement makes spreadsheets extremely powerful. With an IF statement, people can set conditions that allow them to quickly and easily see which records are which. For example, if I were a business owner, I could set up an IF statement that says "Great Work" if the value of the monthly sales was over a certain number, or I could use the if statement to calculate the amount of the bonus that an employee would get based upon whether or not they met their goals. If the net income is over a certain threshold we could calculate the tax on it, but if it were under the threshold, we could set the tax to zero. There are thousands of real-life, real-world examples of using the IF statement. The capabilities are unlimited - as long as we know how to use the function.

=IF(logical test, value_if_true, value_if_false)

The IF statement allows you to provide two different potential answers to a cell, but only shows one of them based upon criteria that you set. The generic form of the function is =IF(logical test, value_if_true, value_if_false).

The IF statement has three parts: the logical test (also known as a test criteria or condition), the value_if_true (also known as the value if true, what will appear as the answer if the logical test is true), the value_if_false (also known as the value if false, what will appear as the answer if the logical test is false).

The logical test itself also has three parts: the cell address, the operator (>,<,>=,<=,=,<>), and the comparison value or criteria which can be a number, a cell address, or text (in quotes). The operators are shown in Table 18.

For example, if we wanted to set a logical test to test if the value in A1 is less than 400, then the condition would be A1 < 400. If the number in A1 was 403, the condition would be met, and the resulting cell would show whatever was in the value_if_true argument. If the number in A1 was 398, the condition would not be met (i.e. it would be false), and the results of the cell would be whatever is in the value_if_false argument.

Note that there is a difference between equal and less than or equal to. If, for example, the number in A1 was exactly 400, then in this case, it would NOT meet the condition, because the number must be less than 400. If we wanted 400 to meet the condition, then we must use the operator <=, as in A1 <= 400.

**Table 18. Conditional Operators**

| | |
|---|---|
| > | Greater than |
| < | Less than |
| >= | Less than or equal to |
| <= | Less than or equal to |
| = | Equal to |
| <> | Not equal to |

To reiterate: a logical test as 3 parts:

- The cell address
- The operator (= <> > >= < <= )
- The comparison value

The comparison value may need to have quotes if the information is text, or hash tags if the information is a date. Numbers and cell addresses do not need anything around them. Here are several examples:

- A1 = "Hello"
- A1 >= #07/01/2016#
- B50 > 2000
- B50 <> A1

In Figure 82, we see the if statement that shows a simple condition: if the difference (the number in cell B20) greater than or equal to 50. If it is, then the cell shows the value 25. If not, then the cell shows the value 0 (zero). In this situation, our Grandma will give us a bonus if we manage to save $50 in our bank account by the end of the month. If we can, she will give us an additional $25. If not, we get nothing. The If statement (which you can see in the formula bar at the top of Figure 82) is =IF(B20 >=50,25,0). The wizard dialog box shows the IF function arguments. Notice that after we wrote the IF function for the first cell (B21), then we used the fill handle to copy the function to all the cells in row 21. Because we used relative cell addresses, all of the addresses adjusted correctly. Note that when you enter the text into the function wizard dialog box, then Excel puts the quotes in for you automatically.

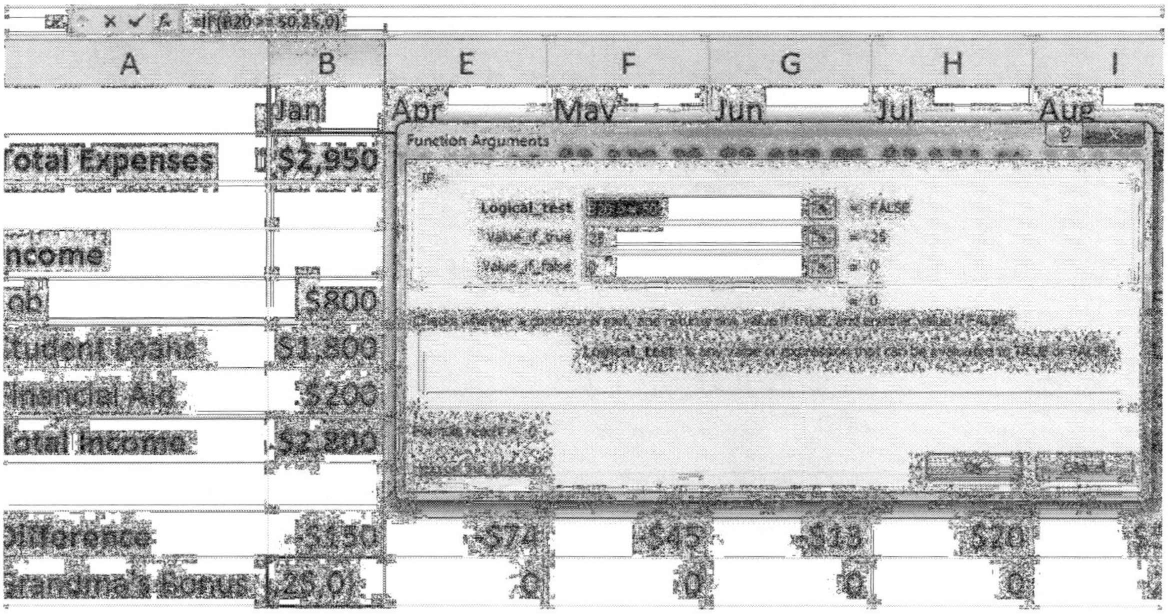

**Figure 82. Simple IF statement example.**

## Database

A database program is a collection of tables containing stored lists, usually accompanied by forms to enter information into the lists on the screen, reports to print the lists on a printer, and queries to specify which part of which list should (or should not) be included.

General purposes database management programs (usually called DBMS for Database Management Software) are available for standalone use, but for the most part, databases become part of the underlying structure of specialized software such as financial programs.

### What is a Relational Database?

The full name for the software that manages tables with related information is a relational database management software program. A database has the table as the "heart" of the system, while the forms, queries, and reports are all just different ways to see the information that is stored in the tables.

A table structure is composed of *fields* (or columns). For each field you must determine the name of the field, the type of information it stores, how much space to give the contents (i.e., what's the longest that can be stored in that field), and which field is the *primary key.* The primary key is the column containing information that is unique to each record in the database, such as a unique ID or a row number. The primary key becomes essential in sorting the contents of the table and joining two related tables together. The contents of a database are stored in *records* or *rows.*

Database design is a highly complex topic that requires a very long time to master, but you don't have to become a master of database design to create and maintain a database of information. In this module we will cover just a few basic rules of thumb and leave database design theory to the experts.

When you look at the data in the tables *through* a form, or a query, or a report, you are looking at ***the data,*** or the contents of the table. It is not a separate copy of the data, or another set of the data. A form or a query is like a window into a room of a house. Imagine standing inside a house and looking at the chair, the table and the lamp. Now go outside and look at the chair, table and lamp through a window. Now reach into the room, pick up the lamp, and smash it. When you go back into the room through the door, is the lamp smashed? In this way, if you change, or delete, the data while you are looking at it through a query, you are changing the data itself, in the table.

This is important to know so that you don't unwittingly change or delete your data just because you are looking at it through a form or a query.

In general, you can edit the data directly in the table. If you prefer to see the information for each individual record on the screen, you can create a form. Forms are generally used to enter data, especially if one person is creating the database and another person (a data entry operator) is going to enter the data. Forms can provide help, suggestions, pick lists, calculations, etc., so that there are fewer errors when entering data. While it is possible to print a form with the contents of the table, forms are usually designed for the screen, not the printer.

The query is a way of looking at just part of the contents of the table. Databases are designed to store large amounts of data, so you should not be able to see all the columns and all the rows for all the tables (and if you can, then it is probably not a good use of the database). Therefore, we need a way of saying which columns and which rows we want to look at or include in the form or report. We also use queries to create calculated fields, join tables, and create advanced subgrouping or totaling of the information in a table. Queries can be used to enter data, view the data on the screen, and to print, but they are most often used as *foundations* for reports. Queries have limited formatting capabilities, whereas forms and reports have formatting flexibility.

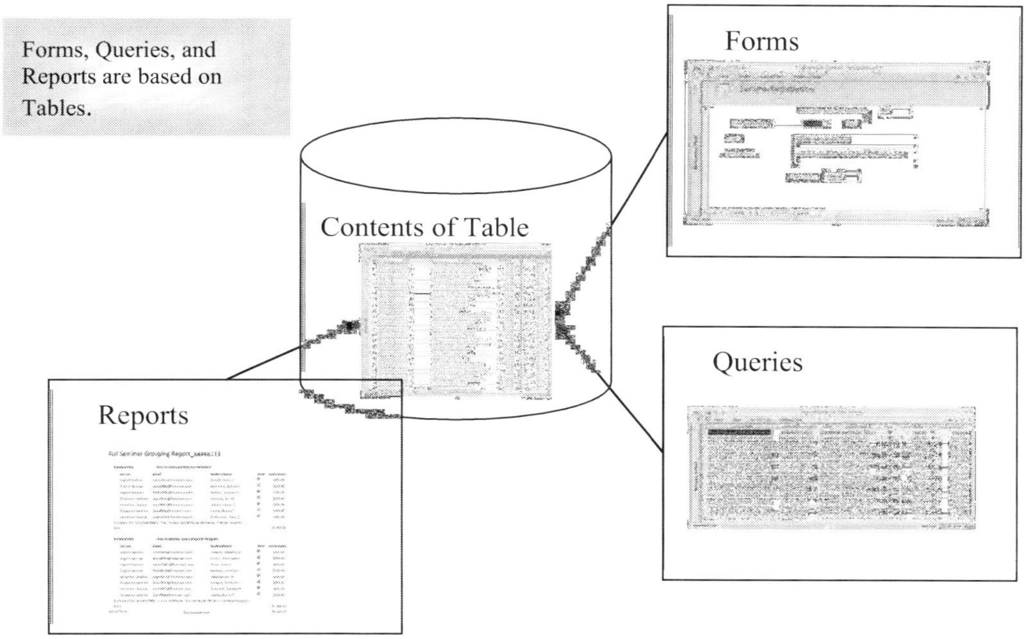


**Figure 83. Relationship of tables to forms, queries, and reports. Database tables are the main component.**

Reports are usually the end result that we want printed on a printer. While it is possible to look at a report on a screen, there is no way to enter or edit data on a report. Smart people base their reports on queries rather than raw tables. That way, you can include fields from joined tables and create calculated fields in a query. Furthermore, it is much easier to change the contents of a report by changing the query. You don't want to permanently change your table just to accommodate a single report.

### How Databases Differ from Other Applications

A database is designed to store large amounts of data. It wouldn't make sense to go to the trouble of setting up a database if the number of "rows" that you want to store is less than 200 or so. In that case it would make sense to use a spreadsheet. Anything that can be done with a spreadsheet can be done with a database, and anything that can be done with a database can be done with a spreadsheet. It is only the ease with which the task can be done that determines the difference. You wouldn't want to print a mailing list of 300 names and addresses with a spreadsheet, and you wouldn't want to create a pie chart of your expenses with a database.

While it was possible to store lists (such as names and addresses) in a word processor or a spreadsheet, neither was well designed for the purpose of managing lists. Using a program for a task for which it was not designed is like "fighting" the program; usually resulting in wasted time and effort, not to mention a great deal of frustration.

Some tasks are much easier to accomplish in a database than in a spreadsheet, so it makes sense for us to know which is which. Databases, however, by their very nature are a little more complicated than the other application software. They are great for storing large amounts of data, querying and reporting on that data, as well as reordering and totaling that data based upon criteria. Spreadsheets, on the other hand, are much better at calculating, analyzing, graphing, and displaying data.

When they first came out in the nineties, relational databases drastically changed the way programs stored and retrieved data. The relationship could be set up at the beginning and the data could be accessed anytime without programming. End users could create a database, store data in it, and report on it without learning how to program first.

dBASE II, published by Ashton Tate, was the name of the first relationship database management program widely available for the PC. As it became more and more recognized as useful, hundreds of other database programs became available encompassing a range of needs from cheap and easy to highly complex, expensive and robust. The most common databases used for new systems today (in order of complexity and cost) are Access, MySQL, SQL, PostgreSQL, Oracle, and DB2. Databases used commonly for systems in the past (but probably not for a lot of new systems today) are Sybase, Progress, Ingres, dBASE, rBase, Foxpro, Alpha4, Filemaker, Paradox, DataEase, ProFiler, Informix, Btrieve, and Watcom.

> Databases are much more complex than wordprocessors, spreadsheets, or presentation software, so it is quite a learning curve to use one. Although everyone should know how to create a database, it is even more important to learn how to *query* data from the tables and design *reports* for them. One important point to remember is that databases don't pull all their records into memory at one time. (Word processors and spreadsheets do - if you were to forget to save a word processing file or spreadsheet and your computer crashed, you would lose all your work). A database pulls into memory one record at a time. As soon as you move to the next record, the previous record is saved. However, because part of the database is in memory and part is on the hard drive (or server), the connection is *fragile*. For a non-robust database like Access, the system will corrupt the whole database if the connection gets broken.
>
> Furthermore, it is difficult to rename an *open* database. The file must be closed to rename or move or copy it. Finally, you cannot open an Access database through a browser on the web, or send it through email. The file might *look* like it opens, but since it doesn't actually get stored on a hard drive, you can't save any changes that you make to the file.

The database is often a component of an application. Instead of programming directly to store information, application programmers would build programming code on top of a relational database program. The database would handle the actual file-open, write data, and close file commands. The system that translated the programming language code into database commands is usually called *middleware* (discussed in more detail in the *Development Software* section of *Chapter 5: System Software.*

Access (by Microsoft) is probably the most common database available. It is generally easy to use (insofar as any database can be said to be easy to use. See inset: Relative Complexity of Databases.) Although easy to use, relatively inexpensive, and widely available, Access is not robust. Access databases often get corrupted or damaged so that they cannot be opened. It would not be wise to rely upon an Access database for business-critical data. Additionally, Access should be limited to relatively small databases. If the database gets much beyond 100 thousand or so records, it begins to get bogged down. It slows and is more easily corrupted. Access is also not a very good choice if you plan to put the data on a web server. While it is possible, web applications based upon Access databases "break" frequently, requiring frequent reboots of the web server. Despite its drawbacks, for an office to keep track of names, addresses, inventory, or lists of any sort, Access is the best choice because it is inexpensive and relatively easy to use.

Microsoft also sells a more robust database, MS SQL, (usually pronounced Sequel although sometime people spell out the letters, S-Q-L, standing for Structured Query Language. See the inset about multiple uses of the term SQL).

At the top of the spectrum is DB2, the only relational database management system that runs on large traditional computers. DB2 requires dedicated programmers and is very expensive. DB2 is not easily utilized via the web, so in most cases it is used in systems developed and designed by IBM for use on private networks within an organization.

***Multiple uses of the term: SQL***

One common point of confusion is that Microsoft sells a particular database called SQL, but the letters SQL also refer to a generic language developed in the '70s for IBM's relational database system (called System/R at the time) which was short for SEQUEL (Structured English QUEry Language). The query language called SQL is used with **all** relational database management systems, not just those with SQL in the name.

Oracle is the top database for both robustness and volume and is very capable of web transactions. Oracle can handle millions of records with ease, and hundreds of people accessing the data at the same time. Oracle doesn't get corrupted easily. However, Oracle is expensive and requires dedicated staff to install and maintain the database. It runs best on server systems utilizing one of the Unix-based operating systems.

For systems that require web access, the top inexpensive choice is MySQL, which is open source, and therefore doesn't cost anything to purchase. Most web hosting services include a copy of MySQL or Maria* already set up and configured for use on the website. A particular web programmer may prefer PostgreSQL (which is also Open Source), but that is not included for free as often as MySQL, so it is not as popular.

The term legacy is applied when we are talking about hardware or software or databases that has been used in the past, but which is no longer considered an up-to-date system. The phrase "nobody uses that anymore" applies to anything that is considered a legacy system, though what they actually mean is that no one would buy that anymore. Legacy systems are everywhere.

There are some industries that have had so many systems developed with legacy databases that they must still be maintained (though when developing a new system, few programmers would want to work in them as they don't have all the new features of web-accessible databases). One such database is Sybase, which was often used in financial firms on Wall Street. Progress was often used in the construction industry. Informix was a top choice in retail.

### *Where is the Data for a Database?*

There are some important differences between databases and other applications, however. Most applications like word processing, graphics, and spreadsheets, place the entire data file in memory

* Several years ago Oracle "took over" management of the code for open source mySQL, causing many to migrate to a look-alike open source database called **Maria.** As far as I can tell, there didn't appear any reason for this except distrust of a large organizations managing open source code.

while you are working on it. If your system should happen to crash while you are working on the file, you might lose whatever you just entered or whatever you had just changed, but you won't lose the original file, which is on the hard disk in storage, not memory. A database, however, is never entirely in memory and only completely in storage when it is closed (and for some databases, not even then as they are always "open" whenever the computer is on). Each time you pull up a record to read or edit it, that one record is retrieved from storage. When you move off the record, that record is put back into storage. That means that you can't really "save" a database table. The database automatically saves one record of the table at a time. (You can, however, save forms, queries, reports, and column settings as well as changes to the structure of the database. You can't "save" the data table because the records save automatically as you move through each one.)

This is a good thing in order to enable you to store large amounts of data; you are not limited to the amount of space in memory. This is a bad thing in order to maintain robustness of the data. If anything should happen to the connection between the computer system editing the data and the computer system storing the data while an edit is taking place, you wouldn't just lose the one record you changed, you could lose the **entire database.** It might get corrupted, and you would not be able to open it again. This happens all the time with small databases like Access (which is why they are not robust and should not be relied upon for mission critical data). Larger databases such as Oracle and DB2 have many technical features in place that reduce the chances that the database will get corrupted.

Understanding the nature of databases also explains why you cannot directly open a database on the Internet through a browser. While it may appear that the database is open because you see the portion you would normally see on the screen, the "back end" of the database (the part normally in storage) is actually being stored in a temporary location that disappears when you close the database, losing all of your work. Therefore, you should always download and save a database file, and then open up the file from a specific location on a hard drive or a flash drive when working with a desktop database like Access. It will not work on a cloud drive, which is accessed on the Internet through a browser.

### Creating a Database

Similarly, to create a database, you must first identify where it will be stored and you must have full access rights to the location. The structure of the database must be designed and created before you can enter any information into the tables. A database can only be moved or renamed when it is closed, not when it is open.

### Designing a Table Structure

In order to create a table, we need to identify the name of each field and the type of information to be stored in that field. Although different database programs have different names for the types of fields, these basic types should be available in all of them:

- Text or Alphanumeric character
- Number
    - Double (a regular number. This is the most common type of number field.)
    - Currency (a number with a currency symbol in front of it)
    - Integer (or short integer, a single digit, 1-9)
    - Long Integer (multiple digits, no decimal places)
    - Decimal (multiple digits, with many decimal places)
    - Single (a number without decimal places)
- Date or Date/Time
- Yes/No or Logic
- Binary
- Hyperlink
- Memo or Long Text

Additionally, one of the fields should be identified as a *primary key*. A primary key is a field in which the contents are unique to each record. We can't use name, for example, because two people might have the same name. It must be a field that would not be repeated no matter how many records we enter. Some designers use a field that automatically sequences to the next number. Microsoft Access is set up to use an automatically sequenced number as an ID if you don't set the primary key. It is usually better to set a primary key rather than use the default.

Additionally, we should be familiar with several rules of thumb regarding design:

1. For interoperability, it is best not to use spaces in field names.
2. Always set a primary key.
3. Always identify the length of a text field.
4. Separate first name, middle name, and last name into separate fields.
5. A number field is one in which the numbers might sometime be added, subtracted, multiplied, or divided.
6. If the field is a date, use a date field.
7. Whenever possible, add a lookup table or lookup value to a table structure.

While most databases allow you to place a space in the field name, when you do so the software will not be able to recognize the name as a field automatically anymore. That means you have to remember to put square brackets [] around the field name in queries and reports in order to get the program to recognize the item as a field.

Without a primary key, it is very difficult to join one table to another. A primary key also makes it easier to identify duplicates and eliminate the second one.

The default length is 255. That is a lot of characters to be blank, adding to the complexity and space requirements of a database unnecessarily.

Combining information together may seem like a good idea; until the first time you want to sort by the second component. For example, first name, middle name, and last name should be put into separate fields because otherwise you wouldn't be able to sort on last name. City, state, and zip should be put into different fields so that you can sort on zip code. It is always easier to put fields together using concatenation (combining text fields with a & or a +) than it is to separate data that's been combined into a single field.

Numbers like phone, zip, and ID are usually best stored as TEXT fields because they might have non-numeric characters and we would want them to sort based on digits, not value.

Dates that are stored as real dates have some capabilities they don't have if you store them as text. For example, if you store January, February, and March as text fields, they would sort by month name (February, January, March) instead of by actual date. You would also not be able to tell which date came before another date, nor could you say what 7 days later might be.

Lookup tables or lookup value lists enable someone who is entering data to pick from a list rather than enter the information directly. Choosing from a list greatly decreases the opportunities for errors and increases the effectiveness and efficiency of the database. GIGO is a phrase often used when dealing with data in computer databases; garbage in, garbage out. If we don't have strict guidance about what can and cannot be entered into a field, we will get data entries that do not make sense, cannot be sorted or grouped correctly, and cannot provide us with the answers we need from the database.

### Entering and Editing Data

Once data is entered into a table, it can be edited or deleted. One important concept to recognize when editing data is whether or not you are IN the text of the field or just ON the field. If you are **in** the text of the field, for example, if you press the [Home] key your cursor will jump to just before the first character of the field. If you are **on** the field and press the [Home] key, you will jump to the first column of the record.

Additionally, there are shortcuts that can help with data entry. In Access, for example, if you press [Ctrl-"] (i.e. control ditto) it will copy the contents of the field from the previous row.

You can look for a specific record with [Ctrl-F], and you can delete a record by selecting it and pressing the [Del] key on the keyboard. You can press [Tab] to move from field to field.

### *Simple Form*

A simple form will allow you to edit existing records or add a new record to the end of the database (*append*). Although you do not need to use a form to enter new data, if a table has many fields, it is often easier to enter the information through a form rather than try to scroll back and forth through the columns of table.

### *Simple Query*

Queries allow you to limit your view to just some of the fields, or some of the rows of a database. When you are in the Query Design view, you choose which fields you wish to see by double clicking on each one and it will get added to the query structure (usually shown below the table field lists in the query design view).

A query can also limit the rows or records that are included in the form or report or view by entering a *condition* which the records must match before they are included. A condition (also known as a logical test) is composed of three parts; the field, the operator, and the comparison value[*]. The comparison value is also known as the *criteria.* In the query screen, the fields included in the query are below the table fields. Below the names of the fields are several Criteria lines. For example, if you wanted to see only the people with the last name of *Jones*, in the column with the field was called **LastName** you would enter "Jones". The next time you *run* the query, only the records that met the criteria (the records whose last name was Jones) would appear in the query.

Each type of field has different rules for entering the criteria on the criteria line. A text field must have quotes around the criteria content. A number must be just the number, without quotes or currency symbols or commas. A number may also be combined with an operator such as greater than >. less than <, greater than or equal to >=, less than or equal to <=, not equal to <> or equal to =. A logic or Yes/No field must not have quotes either, and must be either yes or no, true or false, on or off, 1 or 0. A date field must have a date format that is understood by the software. Usually the software places a # around a date field as in #01/01/2012#. You cannot search or place criteria on a memo or long text field (a reason to avoid that type of field unless absolutely necessary).

### *Wild Characters*

It is also possible to use a wild character (*) in the criteria line of a query as long as it is prefaced with a LIKE operator. For example, to search for all the company names that start with the word **First** you could enter the criteria, LIKE "First*" in the criteria line, and you would get **First National Bank, First Department, First College, First Opportunity House,** etc. If you put the * at the beginning of the text, then you would also get any company where first is anywhere in the field, not just at the beginning, such as **Music First Company**. There are also more specific wild characters such as ?, which replaces only a single character.

[*] You will note a similarity with the logic test or condition of an IF statement.

### *Multiple Conditions: Boolean AND, Boolean OR*

Multiple conditions in queries are controlled by Boolean Logic*. Any condition on the **same** line is automatically considered an AND condition - i.e. both conditions must be met before the record is included. Any condition on **different** lines is an OR condition (you will note the OR to the left of the criteria row in the query structure). If the conditions are combined with OR, only one condition or the other must be met before the record is included. In Figure 84, an OR combination would give you everyone who has blonde hair, no matter what their eye color is and all the people with blue eyes no matter what color hair they have. An AND combination, on the other hand, would only give you those people who have both blonde hair AND blue eyes at the same time. One way to keep it straight is to remember that you always get MORE records with an OR combination.

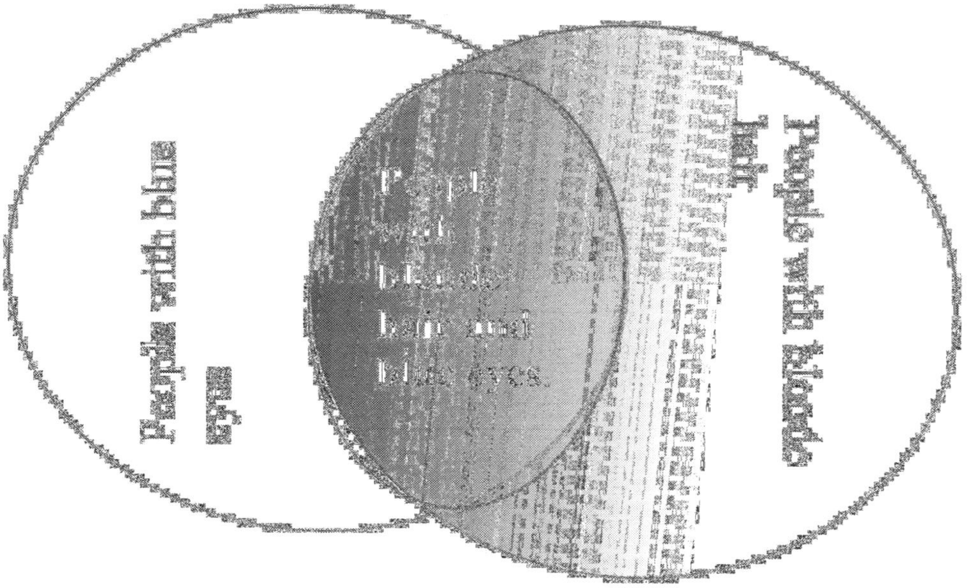


**Figure 84. Illustration of Boolean AND (middle section) and Boolean OR (both right and left sections).**

### *Joining Tables*

Relational tables are based upon the concept of ***normalization***. In the early days of flat file databases, if a customer purchased an item from a company twenty times, there were twenty different copies of their name, address, phone number, etc. In addition to taking up space, having so many duplicates caused problems when trying to determine which address was correct, especially if they changed. Normalization is the process of ferreting out duplicates and placing the information in a *lookup* table. In other words, have the order number and amount in one table, have the item purchased and all its information in another table, and have the customer and all contact information in another table. The tables are then *joined* in a query or view so that it looks like all the information is in one table, when in reality the information is coming from three different tables.

In order to join, each table must have a *linking field* or *foreign key*. A linking field is one where the information is the same in both tables; joined tables are said to be "related" on the linking field. Normally, the linking field is a primary key in one of the two tables. When the tables join, the table with the primary field as the key is known as the **one** side of the **one-to-many**

* George Boole was a mathematician who came up with the rules for combining multiple criteria.

relationship. The table that has duplicates of the information would be considered the **many** side of the one-to-many relationship. An illustration is in Figure 85.

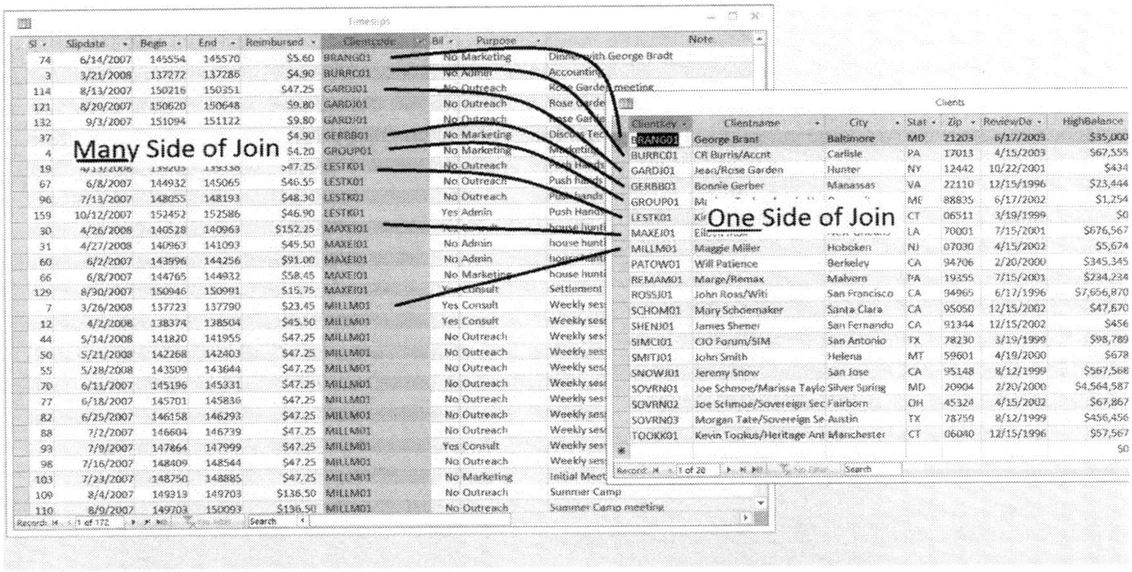


**Figure 85. One-to-many relationship.**

## *Advanced Queries*

All of the capabilities discussed so far (choosing fields, choosing criteria, combining criteria through Boolean Logic) is just the beginning. Another highly useful capability is creating new fields out of existing fields; *calculated fields*. A calculated field might be mathematical (as in `TotalPrice: [Price] * [Quantity]`) or it might be the combination of text (called *concatenation* as in `FullName: [Lastname] & ", " & [Firstname]` or `"Lastname" || ', ' || "Firstname".` In Access the concatenation symbol is an ampersand **&** and in Google Base the concatenation symbol is a double pipe ||.) In Figure `86,` you can see a query screen with both types of calculated fields as well as a join, and an example of Boolean logic. The query includes records where the (number of credits earned is more than 60 AND the non compliancy flag is set to Yes) OR (the level is "Seniors").

Queries can do more than just limit the rows and columns and create calculated fields. Queries enable you to subgroup and subtotal, create a crosstab query. At a more advanced level, you can also use queries to append rows from one table to another, delete rows, and update rows.

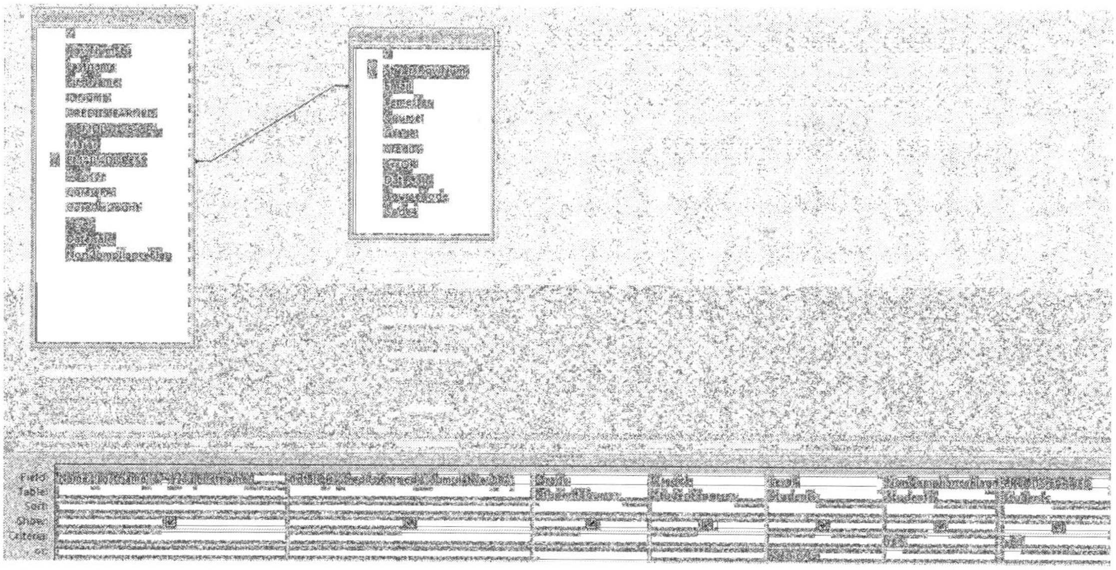

**Figure 86. Example query screen with calculated fields and Boolean AND/OR criteria.**

### *Simple Report*

Reports should always be built on queries rather than tables themselves so that you have more control over the information that is included in the report. Additionally, although it is possible to create calculated fields in the report itself, it is usually better to create the calculated fields in the queries and just include them in the report. Unlike queries, which have very limited formatting capabilities, with a report you can change the fonts, and add colors, etc.

A report design view has different components. A report header, a page header, a detail band, a page footer, and a report footer. Whatever is in the page header and footer appears at the top and bottom of every page. Whatever is in the report header shows up once, at the beginning of the first page of the report. Whatever is in the report footer appears once, at the bottom of the last page of the report. Whatever is in the detail band will appear once for every record included in the report. If the query has 257 records, there will be 257 rows in the detail band (even though in the design view, you only see one row.)

In Access, though it is possible to build a report from scratch using the design view, it makes very little sense to do so. It makes more sense to use the report wizard to create the report, and then edit the report in the design view. Also, you can select multiple fields in a report (you use the [Shift] key to do this) and change the formatting information or move them around the page either by dragging them, or using cut and paste. You can also include functions like SUM and COUNT for specific columns.

## *Presentation*

Although in the early days application software was limited to just text, eventually some applications became available that could show slides and graphs. Instead of laboring over transparencies or slide presentations, people realized they could connect the personal computer to a projector and show presentations much more easily than laboriously photographing, developing, organizing, and presenting slides. Harvard Graphics was one of the first presentation packages for the PC, followed by Freehand by Lotus. Both Harvard Graphics and Freehand fell by the wayside in a few years, however, as Microsoft released the easy-to-use PowerPoint program. The most popular presentation software today is Microsoft PowerPoint, Google Presentations, and Open Office Impress.

Nothing has caused more delight and more boredom at the same time than the proliferation of presentation software. Nothing is more boring or more useless than a presenter standing up and reading the items on a slide presentation. It's as if the ease with which a presentation can be done is counterproductive to the amount of thought that has gone into the presentation. Among all those who have been forced to sit through these boring presentations, a common phrase has arisen: *Death by PowerPoint*[*].

### Avoiding Death by PowerPoint

There are several simple easy-to-follow rules that will always allow your presentation to be of much higher quality than the typical presentation.

- Create the presentation by using an *outline view* with clearly delineated points and a hierarchy of information.
- Include graphics, illustrations, charts and other interesting objects in your presentation.
- Make sure *all* objects (graphics and text) can be easily seen and understood.
- Create short phrases of your points and do not use full sentences. Avoid putting more than six words on a line or more than six points on a slide.
- Look directly at the audience when giving a presentation. *Do not* look at the slides at all. Speak from notes that are in front of you, not by looking at the slides.
- Either use a clicker or have someone else switch from slide to slide for you.
- Do not "overanimate" Animation, if used at all, should be subtle and not time consuming.

### Using an Outline and Slide Sorter

It is usually very obvious to the audience when someone has put their "stream of consciousness" thoughts on a presentation. There is no hierarchy to the information. Every slide seems to be out of nowhere, with little connection to, the information before or after it.

Most good presenters know the importance of using an outline to develop a presentation. Generally, the outline can be typed into any program and imported into a presentation, but most presentation packages have an outline mode that allows you to enter the outline information directly, and then translates the outline into a slide structure.

Additionally, most presentation packages provide you with a slide sorter view so that you can easily see all the slides at one time in order to rearrange them as necessary. The slide sorter view is also the best view to use when merging different slide presentations (i.e., copying slides from one presentation into another).

### Using Graphics

All presentation programs allow you to import clip art and graphics into the presentation. Some also allow you to imbed video clips and sounds into the presentation. It is a good idea to utilize these capabilities; a presentation with just words can be awfully boring. However, use graphics judiciously. They must be relevant to the points being presented and add to understanding rather than confuse. Table, charts and graphs are especially helpful, but be sure that they are not too small to be seen by the typical audience member. Generally, anything less than a 20-point font is difficult to read on a slide show presented to a group of even a small size. The larger the group the larger the font must be. One common rule proposed by Guy Kawasaki is the 10/20/30 rule. Thirty point font, twenty minutes to present, only 10 slides.[†]

Also, when copying tables, charts, and graphs from spreadsheet or word processing programs, you want to use an Independent Bitmap or a Picture Metafile rather than a Microsoft object or a linked

---

[*] The phrase is not in any way limited to Microsoft's Powerpoint software but rather is a common on the concept of computerized slide show using any presentation software.

[†] More details can be found on http://guykawasaki.com/the_102030_rule/ written December 30, 2005 by Guy Kawasaki. Accessed October 9, 2020.

file. While there are advantages to inserting a linked object (like being able to update the source file from within the presentation package), in most cases the advantages are not worth the disadvantages (like not being able to resize or move the object without it changing drastically).

Also keep in mind that while photographs are eye-catching and often useful, they are also very large. If you plan to send your file through email or present it online, you want to have a smaller rather than larger file.

### Master Design or Theme of a Presentation

There are several levels of customization for presentations. One common mistake is to start making changes to individual text or headings in the presentation. This is usually a bad idea because there is usually a *master* design or *theme* that is controlling the attributes of the elements, such as fonts, bullets, spacing, colors, backgrounds, timing, animation, graphics, etc. The proper way to make a change to a presentation is to start with the top level (called *Master* in Powerpoint, and *Theme* in Impress). That way, changes made will automatically appear on all the slides linked to the master design or theme slide. For example, if you want your company name or logo to appear in the corner of every slide, you can insert it on the Master Design slide and it will appear on every page.

If a specific slide needs to be different, if it can't use the graphics of the master or needs to change the font color or bullet to be different on one particular slide, there is usually an option to unlink or suppress the master design or the graphics or colors or background on any particular slide. After you have made this choice, you can make specific changes to that slide since it is no longer controlled by the master design.

### Animation and Customization of a Presentation

Animations and customizations are ways to get "movement" in a presentation. You can have text appear or fade out, get bigger or smaller, fly in or jump around. Hundreds of choices are available. You can also customize the transitions between slides: open up like a shutter, wipe like a windshield wiper, or appear as a checkerboard, etc.

Using animation and transition customization is like cooking with spices; a little goes a long way. There is nothing worse than a speaker waiting for the animations to "finish" jumping around or forming on the slide so that the speaker can continue the presentation. This problem is caused, in part, by not following the rules that say presenters shouldn't be looking at the slides at all. Additionally, like fonts and colors, animations should be consistent. Don't do a spiral on one slide and a flash on another and a peek on yet another. You will confuse your audience rather than make a good impression.

Animations and transitions can usually be applied to the entire presentation through the Master Design, or applied to a single object or textbook by selecting it and applying the customization.

### Printing Presentations, Note Pages, and Handouts

All presentation packages provide space called a note page for authors to enter what they will say for each slide. It is always better to print out the note pages so that you don't have to turn around to read each slide so you can remember what you were planning to say. Since the note pages include pictures of the slides, you can see in front of you what the audience sees behind you. Usually, you can customize the note pages to print them out on smaller cards if you don't wish to have large pieces of paper for each slide.[*]

[*] Newer presentation software usually has a presentation view that relies upon the fact that there are two monitors. It puts the notes on one monitor and the presentation on another. I've yet to meet anyone who has mastered using this, or that has a system that works well with it. What happens most times is that both the slides and the notes appear, unwanted, on the presenting monitor.

Additionally, presentation packages allow you to print multiple slides on a single sheet - sometimes even with lines so that the audience can take notes. In general, it is a bad idea to give out a handout with all your slides on it before your presentation. This will decrease the interest and attention of the audience. But as an after-presentation follow-up, it is a great idea to provide a handout with small pictures of all the slides. Before the handout, it is best to provide a single sheet that provides just the highlights of your presentation.

# 7-1 Interactive

## A. *Financial Software*

1. **Independent Activity**: Interview a business owner or manager. What kind of financial software do they use? Who enters the data? Are there different reports for different levels of the organization?

## B. *Graphics Software*

1. **Independent Activity:** Go to a page on the world wide web with a photograph on it. Measure how long it takes for the photograph to fully appear. Download the photograph from the web page (right mouse on the photo, then choose either *Save Picture* or *Save Target as*, choose a location on your hard drive or flash drive to save the file).

   Print the photograph on a high-quality color printer. Does the printed version look the same as the web version? Repeat the experiment with another page where the photograph takes much less time to appear. When you print it out, is the quality higher or lower than the first one?

## C. *Wordprocessor*

1. **Independent Activity**: Write a letter to your favorite teacher asking for a referral for a job for which you are applying. Create at least three styles. Set it up so that the first page has your name and address in the center at the top (as in a letterhead) and a thin line at the bottom with the page number in the center. Structure the second page so that the upper right header has the name of the recipient and the bottom right has the pathname followed by the page number out of the number of pages. Save the letter as 7-1A_Letter_*userID*.docx. Where you see *userID,* put your own user ID (such as aaaaa111) so that your instructor can tell whose file it is and if you've completed the assignment.
2. **Independent Activity**: Create a template out of the letter and remove the items that will change, leaving the items that will always remain the same; save the template as 7-1A_LtrTmp_*userID*.dotx.*
3. **Independent Activity**: The document in this activity is designed to help a business manager communicate decisions that have been made to employees and shareholders. (There is a datafile for each section of this exercise, though in reality the datafile is just the completed version of the file before it, so you can just continue to work in the same file and just give it different names as you save at the end of each section.)
   - Open the 7-1B_SimpleFormat_df.docx document.
   - Examine the contents in each page. How many total pages are in this document?
   - On the second page of the document, the second paragraph (Margaret Malley) is not a hanging indent as the other paragraphs in that section are. Make the indents hanging so that the words wrap about half an inch on the second line.
   - Go to the fourth page of the document. The table title is currently aligned to the left, and you want to center it over the table. Make the necessary change.
   - In the third page, format the paragraph under the title as follows, and create a new style called SmallRedPara:
     a. Change the font size to 8 points.

---

* Note that to work AS a template, this .dotx file must be placed in the Word Template directory. You can find out where that directory is by going to **File—Options—Save.** The folder for the personal templates will be listed.

   b. Change the font color to red.
   c. Change the font to Arial.
   d. Change the spacing so that there are five points before each paragraph.
   e. Indent the paragraph on both the right and the left by half an inch.
   f. Set the alignment to Justified.
   g. Make sure the style is saved with all of the modifications.
- Go through and apply the SmallRedPara style to the first paragraph under each Header 2.
- Change the Header 2 style as follows:
   a. Change the font to a sans serif font.
   b. 15 points before each paragraph.
- Add the filename including the path name to the footer.
- Change the margins for the whole document to 1.5" on the left and right, 1" on the top and .75" on the bottom.
- Add a header with the Filename (without the path) centered on top and the Current Date underneath with a blue border line between them.
- Convert the information on page 1 to a Table so that it looks like the following:

| Name | Title | Organization |
|---|---|---|
| Maureen Nazinov | Campaign Director | Jewish Federation |
| Pam Jostelyn | Office Manager | Council for Chemical Abuse |
| Margaret Malley | Executive Director | Berks Counseling Center |
| Cheryl Love | Executive Director | Community Prevention Partnership |
| Nancy Singal | Program Director | Jewish Community Center |
| John Sogelman | Controller | Jewish Community Center |

- Add a header row to the table that starts on page 3 so that the titles appear on each page that contains information from the table.
- Fix the table on page 4 so that it instead of:

<table>
<tr><th>Old Way – Manual</th><th colspan="2">New Way - Using Technology</th></tr>
<tr><td rowspan="2">Type up and paste newsletter<br>(or pay a typesetter).</td><td>Desktop Publishing with Microsoft Publisher,</td><td>InHouse Digital Publishing</td></tr>
<tr><td>PagePlus Serif</td><td>Adobe Acrobat Professional</td></tr>
</table>

It looks like the following:

| Old Way – Manual | New Way - Using Technology |
|---|---|
| Type up and paste newsletter<br>(or pay a typesetter). | Desktop Publishing with Microsoft Publisher,<br>InHouse Digital Publishing<br>PagePlus Serif<br>Adobe Acrobat Professional |

- Find all the double spaces, and replace them with a single space.

- Cheryl Love has gotten married and would like you to use her new name, Cheryl Fienstein. Find the word Love and replace with the word Fienstein throughout the document.
- Insert a cover page at the beginning using the built-in Filigree cover page. Create a new page between the cover page and the rest of the document. Move the section About this Document onto a page by itself prior to The Decision: Should We Invest in Technology.
- Set up the document so that page #1 starts on the page with The Decision: Should We Invest in Technology.
- Have the page with the About this Document section show as page ii (roman numeral 1). Put the page number at the bottom in the center of the page.
- Change the footers so that the page number appears on the left of even pages and the filename and pathname appears on the right.
- Change the footers so that the author's name appears on the left of odd pages, and the page number appears on the right.
- In the Our Decision section at the end of the document, put the inventory table on a page by itself in landscape mode. Make sure there are no empty pages at the end of the document (though the last page will just have a title – but it should be portrait, and the headers and footers should be the same as previously).
- Put the paragraph about Peter Drucker in a text box on the right side of the page. Put a caption under it that identifies Peter Drucker as a management consulting guru.
- Save as 7-1B_userID.docx

## *D. Create a Simple Database*

1. Imagine that you have many family members in need. Since you have a lot of money, you freely lend it to them - but you want to keep track of them so that you can remind them to pay you back.
2. You decide to create a database with all the fields of information you might need to keep track of all the relatives who owe you money. What kinds of information should you store? What should you name the fields? What type of field should each be? Which field should be the primary key field? (Don't forget to follow the rules of thumb for database design.)
3. Name your database8-1E_FirstDB_*userID* and store it in your class folder on a hard drive or flash drive. (Remember that you cannot open or edit a database from a web or cloud drive.)
4. Create the table structure for the fields of information you need to store about your relatives. Call the table **Debtors**. Enter a couple of (hypothetical) records of information for relatives who owe you money.
5. Create a form to make it easier to enter the information. Enter a couple more records of information through the form. Call the form **Debtors_Form**.
6. Create a simple query to view just the name, phone, and amount. Call the query **Debtors_Info_Query**.
7. Create another query with all the fields from your table. Use a criteria to show just the two relatives who already paid back the money. Name your query **Paid**.
8. Create a simple report to see all of your data based on the **Debtors_Info_Query**. Call the report **Debtor Report** and show a total for the amount of money you have been owed.
9. Close the database entirely. Look at where it is located in the folder. Open it up again; is your data there? Are all your forms and queries and reports there?

10. Add another record. Look at your Debtors_Info_Query again. Is the new person there? Close the database.

## E. Create Simple Queries

1. Download the datafile 8-1F_MrktgDB_df. Rename it 8-1F_MrktgDB_*UserID*. Open up the database.
2. Review the information in each of the tables.
3. Create the following simple queries with the following names and criteria (the fields in parenthesis following the table name on the first line of each section).

   Category Sales: (Category, Sales Dollars, Sales Last Week)

   - Category 40
   - Categories more than 70
   - Category between 30 and 70
   - Items with more than $10,000 in daily sales

   Customer Loyalty (Department Name, Customer Name, Purchases Year To Date, Amount)

   - Customers who bought jewelry
   - Customers who bought jewelry or electronics
   - Customers who bought Women's Apparel
   - Customers who bought any type of apparel
   - Purchases from any Sovereign Customer
   - Customers who bought jewelry from California
   - Customer who bought jewelry from California or Connecticut
   - Purchases before 2008
   - Purchases between April 2008 and May 2009
   - Autoinvoiced Customers

   Company Financials (all fields)

   - Sort ascending price to book
   - Sort descending price to book
4. Close the database 8-1F_MrktgDB_*UserID*.

## F. Create Advanced Queries

1. Download the datafile 8-1G_MrktgDB_df. Rename it 8-1G_MrktgDB_*UserID*. Open the database 8-1G_MrktgDB_*UserID*.
2. Create the following advanced queries with the following names and criteria (the fields in parenthesis following the table name on the first line of each section).

   Customer History Table (Age, State of Residence)

   - Average Age by State

   Category Sales: (Category,Item, Sales)

   - Sum sales by item
   - Sum sales by category
   - Sum sales by category by item

3. Create a Crosstab Query queries with the Sales Day (the day of the week) across the top and the items down the side and the sum of the sales for each item for each day inside the grid. Make sure there is a total for each row. Call the query **SumSalesByDayByItem_Crosstab**
4. Close the database 8-1G_MrktgDB_*UserID*.

## G. Create Advanced Reports

1. Download the datafile 8-1H_MrktgDB_df. Rename it 8-1H_MrktgDB_*UserID*. Open the database 8-1H_MrktgDB_*UserID*.
2. Create a query with the following fields:

   **Customer ID, Customer Name, State of Residence, Price to Book, LY Spend, Value of Last Purchase**.
3. In the query, add a field that calculates the number of orders per year by taking the LY Spend field and dividing it by the **Years Since Customer**. Name the calculated field **SpendDividedByLongevity**.
4. Save the query to be used in the report. (Call it whatever you like, but remember what you called it.)
5. Use the query to create a grouping report. Include all the fields in the report. Group on the **State of Residence** in the report. Total the **LY Spend**, and average the **Price to Book**, **Value of Last Purchase** and **SpendDividedByLongevity**. Set up the report to print landscape, and make sure all the columns are wide enough for the numbers to appear. Save the report and call it **AdvancedReport**.
6. Close the database 8-1H_MrktgDB_*UserID*.

## H. What-If and Goal Seek

1. Open the spreadsheet 10-1A_WhatIfGoalSk_df.
2. Edit the cell in A1 and put your name in it just before the word "Budget".
3. Position the view so that you can see the **Difference** at the end of the year at the same time as you can see the assumptions for income and expense changes.
4. Do a what if analysis. What if expenses were to increase by 5%. What EOY Total did you get? Write it here: _________
5. What if expenses were to decrease by 5%? What EOY Total did you get? Write it here: _________
6. Enter values until about half of the months have a positive difference and half of the months have a negative difference. What expense % change did you get? Write it here: _________
7. Do a goal seek. What would the expenses have to be in order to have exactly $500 difference between income and expense at the end of the year? What expense % change did you get? Write it here: _________
8. Do another goal seek. What would the income have to be in order to have exactly $1000 in difference at the end of the year? What income % change did you get? Write it here: _________
9. Write the numbers from the blanks in the appropriately labeled cell from B27 to B31.
10. Save the spreadsheet and call it 10-1A_WhatIfGoalSk_*userID*.

## *I. Graphing*

1. Open the spreadsheet 10-1B_Graphs_df.
2. Select the expense category names and the EOY totals for those categories. Create a pie chart using Layout 6. Add the chart title `Expense Categories` and make the percentages inside the pie pieces 11 points and bold. Change the color of the first category (Tuition) to black. Make the percentages in Tuition (34%) and Phone (12%) White so that they can be seen on the darker background of the pie piece.
3. Select the month names and the income categories and amounts. Create a stacked bar chart. Switch the rows and columns so that the income categories become the series and the months become the X axis. Add a title `Income Categories By Month`.
4. Save the spreadsheet and call it 10-1B_Graphs_*userID*.

## *J. Payment Function*

1. Open up a blank spreadsheet. In column A enter the following phrases:

```
Price
Down payment
Amount borrowed
Interest
Frequency
Years
Term
Payment
Total Paid
```

2. Imagine that you've decided to purchase a new car. You've found one for $12,000, and you have saved up $3000 for the down payment. The finance company is willing to extend a 4.25% annual interest rate if you are willing to pay in bi-weekly (every two weeks) installments for five years. Where do you think this information would go given the row headings we've entered?
3. Put the $12,000 in for the price, and the $3000 in for the Down Payment. Use a formula (=B2 - B3) for the amount borrowed. Enter the 4.25% (or .0425) for the Interest in B4, and =52/2 or 26 in Frequency. Enter 5 for the years in B6, and calculate the term (=B5 * B6) in the Term.
4. Use the PMT function to calculate the payment in B8. Don't forget that the interest rate must be divided by the frequency of payments per year and the Amount borrowed must be entered as a negative in order to make the payment positive.
5. For B9, enter a formula =B2 + (B7 * B8) that will calculate the entire amount paid.
6. Do a what if analysis. What would happen if you paid it back in three years instead of five? What if you paid monthly instead of bi-weekly?
7. Save the spreadsheet and call it 10-1C_PMT_*userID*.

## *K. Future Value Function*

1. Open up a blank spreadsheet. In column A enter the following phrases:

```
Amount saved
Interest rate
How often
```

```
Years
Number of Payments
Future Value
```

2. Imagine that you've decided to save up for a special vacation. You want to go to the Bahamas, and it will cost $6500 for the trip. Your bank will give you a 2.25% annual interest rate if you are willing to commit to putting at least $30 in every week for four years. Where do you think this information would go given the row headings we've entered?
3. Put the $30 for the amount, the 2.25% (or .0225) for the Interest Rate in B2, and 52 for How Often in cell B3. Enter 4 for the years in B4, and calculate the number of payments (=B3 * B4) in the Number of Payments.
4. Use the FV function to calculate the total amount in the account after four years in B5. Don't forget that the interest rate must be divided by the frequency of payments per year and the payment amount must be entered as a negative in order to make the future value positive.
5. Will you make enough to go to the Bahamas in four years?
6. Do a what if analysis. What would happen if you kept putting money in the account for seven years? What if you paid monthly instead of weekly?
7. Save the spreadsheet and call it 10-1D_FV_*userID*.

## L. Subgrouping and Subtotaling

1. Open the spreadsheet 10-1E_Subtotal_Stud_df
2. On the Students sheet, create a subtotal (hint - look at the Data Ribbon, the Outline section). Group by level (Freshman, etc) and calculate the sum of the credits earned for each level. Collapse the groups so that you only see the levels and the totals. If you got more that 4 or 5 levels, you might have forgotten to sort. If so, remove the subtotals, sort, and redo the subtotals.
3. Save the spreadsheet 10-1E_Subtotal_Stud_*UserID*.

## M. Vlookup and Hlookup Functions

1. Open up the spreadsheet 10-1F_LOOKUP_df. One sheet called IDlookup shows in the spreadsheet a list of names corresponding to various IDs. Another sheet called WageLookup shows the list of hourly wages corresponding to various names. Another sheet called 1GradeLookup shows a list of number grades and their corresponding letter grade. The final sheet is called MultigradeLookup and shows a number of grades to be looked up.
2. In the IDlookup sheet, in cell C1, enter a lookup function to look up the name based upon the ID. Which lookup should you use (Vertical or Horizontal)? Where is the value you are looking up in the table array? Where is the table array? Which column number should be returned? Should you require an exact match or not?
3. After you've entered the lookup function, change the lookup value to 977558421. What happened? Write the result here: _________ Change it again to 977558420. Write the result here: _________
4. In the Wagelookup sheet, in cell C1, enter a lookup function to look up the hourly rate based upon the name. Which lookup should you use (Vertical or Horizontal)? Where is the value you are looking up in the table array? Where is the table array? Which column number should be returned? Should you require an exact match or not?

5. After you've entered the lookup function, change the lookup value to your own last name. What happened? Write the result here: _________ Change it again to Easley. Write the result here: ________
6. In the 1Gradelookup sheet, in cell C1, enter a lookup function to look up the letter grade based upon the number grade. Which lookup should you use (Vertical or Horizontal)? Where is the value you are looking up in the table array? Where is the table array? Which column number should be returned? Should you require an exact match or not?
7. After you've entered the lookup function, change the lookup value to 70%. What happened? Write the result here: _________ Change it again to 69.99%. Write the result here: ________
8. In the MultiGradelookup sheet, in cell D2, enter a lookup function to look up the letter grade based upon the number grade. Which lookup should you use (Vertical or Horizontal)? Where is the value you are looking up in the table array? Where is the table array? Which column number should be returned? Should you require an exact match or not?
9. Copy the formula in D2 down into D3 to D10. Did it work? Why or why not? If it did not work, figure out what you need to do to fix it. (Hint - did the table array reference change? Did you want it to change?)
10. Save the spreadsheet and call it 10-1F_LOOKUP_*userID*.

## N. Review Graphs, Vlookup

1. Open the spreadsheet 10-1G_ReviewGrphVlkup_Stud_df.
2. First, look at the information on the Students sheet. Then go to the sheet called "Charts" and review the formulas on the sheet that dynamically chart the counts and averages from the Students sheet. (You will learn about some of these functions in the next chapter).
3. Create a pie chart that shows how many Freshman, Sophomores, Juniors, Seniors, and 5th Yr Seniors there are based on the Calculated Level in column O from the Students Sheet. (Hint – the functions to count them have already been done for you – you just need to create the chart.)
4. Have the chart show the title "Students By Level".
5. Have the number of students in each rank appear inside the pie pieces.
6. Make the pie bigger inside the box so that it fills the area. Make the numbers bold and bigger too.
7. Pull out the piece for 5th Yr Seniors. Make sure that the number is outside of the box and a leader line to it exists.
8. When finished, the chart should look exactly like this :

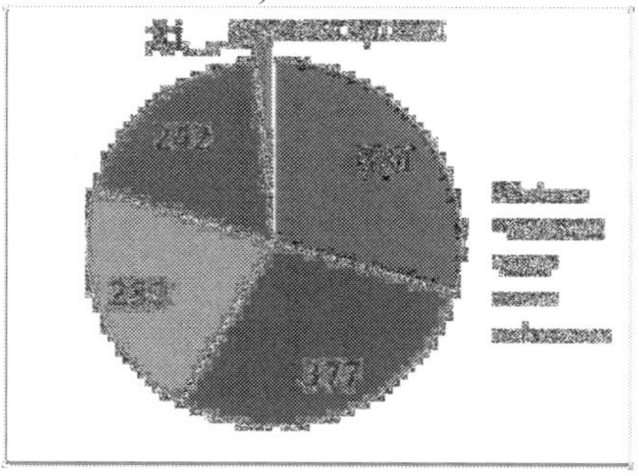

(only the pie slices are different colors instead of different shades of gray)

9. Create a line chart that shows the average GPA for each Level
10. Have the chart show the title "Average GPA by Level"
11. Add a second y axis that shows a column with the number of students in each level. (Hint, select the line on the graph and look under series options.)
12. Make the columns fatter and the lines thicker
13. Add a **gradient** color to the columns, using the color **blue** with the **rectangular** type in the direction of the **bottom right corner**.
14. Make the chart bigger so that it fills the box and make the legend appear at the bottom.
15. When finished, the chart should like exactly like this :

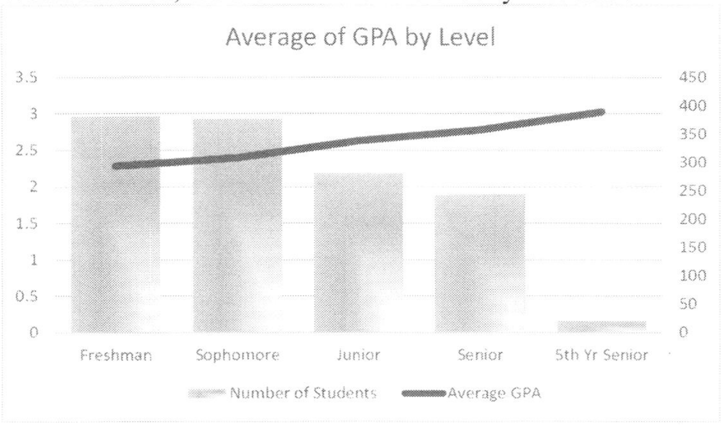


(only the columns are colored blue, not shades of gray)

16. In the Students sheet, in the I column next to the major numbers, and create a lookup so that instead of the major number, you see the major name. (The major names can be found on the sheet called **Majors**).
17. In the N column, lookup the assigned level by using the table found on the sheet called **Level**. (Except for fifth year senior, the responses should match the calculated levels in the O column).
18. In the CourseList spreadsheet, in the F column, write a function that will identify the course as a CORE course or not. The list of Core Courses is found on the sheet called **CoreCourses.**
19. Save the spreadsheet 10-1G_ReviewGrphVlkup_Stud_*UserID*.

## O. Protection

1. Open the file 10-1H_Protect_df.xlsx. Lock the cells with the white background and unlock the cells with the yellow background. Protect the whole sheet. Try to change one of the formulas. Change one of the numbers in the yellow cells. Did it work? Save the spreadsheet 10-1H_Protect_*UserID*.xlsx

## P. Create a Presentation

1. Start with a completely blank presentation.
2. Enter your name as the main title of the presentation. Use the subtitle: **All About Me.**
3. Using the outline data entry panel, enter information about yourself on the following topics (one slide for each topic): Family, Education, Hobbies, Friends, Future Career, Dreams.

4. For at least three of the slides, have at least four levels (with the title of the slide counting as the first level) of bulleted points.
5. Change the design or theme of the presentation to something different. Try a few so that you can see how each of the designs looks with your text. In the end, choose **Organic.**
6. Add a picture of yourself on the first slide.
7. Choose a relevant clip art and arrange so that it appears on every page (without pasting it on every page).
8. Change the transition options for the presentation. Use **Fall Over** with a duration of .5 seconds. Change the animation for the picture of yourself using **Wheel** so that it appears when you click the mouse with a clockwise wipe**.**
9. Save the file as 7-1C_AbtMePrs_*userID* .

## Q. *Modify an existing presentation*

1. Open the 8-1D_MstrDsgnPicAnim_df.pptx presentation file.
2. Change the design of the presentation to **Ion,** which has a dark background and light-colored text.
3. Replace the text Student's Name with your actual full name and the date with today's date. Add your course (BUS171) and section number (000) under the date. If the text gets smaller (because the box isn't big enough for 3 lines of text) make the box bigger. (Don't change the font.)
4. Open the file called 8-1D_FinPic.jpg in a graphics package. Reduce the size to just 15% of the original size. Save the revised file with the name 8-1D_FinPic _*userID*.jpg.
5. Place the revised version of the picture so that it appears on almost every slide in the presentation in the bottom right corner. (It may not appear on every single one of them – that's all right, but it should appear on most).
6. Change the bullets on all the slides so that the first bullet is an arrowhead (➢).
7. Add the following words to the slide entitled The Business Plan (cont'd) at the appropriate levels.

```
Administrative Plans
     Insurance
     Management and Human Resources Plan
Financial Plan
```

8. Move the Sources slide to just after the fifth slide (Title is **Many Resources Available**).
9. Delete the Questions slide.
10. Save the file as 8-1D_MstrDsgnPicAnim_*userID*.pptx.
11. Print a handout with nine slides per sheet in landscape mode using Pure Black and White. Close the file.

## 7-2. Self Study Test Questions:

1. What are the major categories of Software?
2. Name 2 or 3 examples of software and which subcategory they would fall under.
3. How has graphics software changed over the years?
4. Why is financial software so important to a business organization?
5. What is an ERP? Why is there an ERP Graveyard?
6. What is XBRL and why would the SEC demand that everyone use it?
7. What does CAD/CAM stand for? What is a BMS?
8. What type of user would be most attracted to Vertical Market Software?
9. What device was replaced by wordprocessing software?
10. What is a database? What is the main component of a database? What other components are included in a database?
11. Name several tasks that are easier to do in a database than in a spreadsheet.
12. Why would someone set up a form for data entry into a table?
13. What are at least 5 of the most common field types?
14. Why does it make sense to avoid spaces in field names?
15. What is SQL and why might it be confusing?
16. Why do we join tables?
17. What is a one-to-many relationship?
18. How can you tell which field is the linking field?
19. Why should you base your reports on queries?
20. Why shouldn't you put first and last name into the same field in a table structure?
21. What is concatenation? Give an example of a concatenated field.
22. Name all of the database programs you can think of. Which are legacy? Which are current? Order them from least expensive to most expensive. Order them from least robust to most robust.
23. What is Boolean Logic?
24. What is a What If Analysis? Describe what you need to set up a What If Analysis.
25. What is a Goal Seek? Describe what you need to set up a Goal Seek.
26. What is the first and most important task when creating a graph?
27. Why would you need non-contiguous selection on a graph?
28. What would happen if you include the totals and the details in the same graph?
29. What are the arguments for a payment function?
30. What are the arguments for a future value function?
31. What is important to remember about the Interest argument of most financial functions?
32. What is important to remember about the Present Value argument of most financial functions?
33. What function would you use to look up a value on a table?

34. What is the difference between the Hlookup and the Vlookup?
35. How do you know what number to put in for the index in a lookup function?
36. What is important to remember about sorting columns in a spreadsheet?
37. Before you choose the subtotal menu, what must be done to arrange the data first?
38. How do you protect your formulas and headings from being accidentally overwritten by someone else using your spreadsheet?

8

# Chapter 8: Process of Organization

Overview: In this chapter, readers will learn the following topics:

- ⇨ Role of Information Technology in the process of a business
- ⇨ Competitive Advantage of Information Technology in Business
- ⇨ Ethical Issues of Information Technology
- ⇨ The Essential Balance of Matching Processes to Technology

It's been a few chapters, so let me take a moment to reiterate the definition of *information technology*. Information technology is the people, processes, software, and hardware that make up the information flow in the operations of an organization.

We've discussed, in detail, the networking, hardware, and software aspects of information technology. Now we will focus on the process.

## Role of Information Technology in Business Process

Information technology plays four major roles within an organization: automating processes, streamlining processes, accelerating processes, and sales channel process expansion.

### *Automation*

When computers were first introduced into business, they filled a very specific niche role. Repetitive and high volume tasks were targeted for *automation* of a *business process*. A business process is a standard set of activities that accomplish a specific task. Automation generally meant replacing a manual process with an electronic transaction. For example, before computers, a retail store would have a filing cabinet containing papers upon which were written the results of the daily count of the cash register, as well as paper forms showing what items were ordered, and inventory lists showing which items were still on the shelves. To figure out what was sold, you counted the items on the shelves compared it to what had been ordered. Very time consuming and often not accurate.

With automation, the daily cash register count would be replaced with an electronic list of every item sold as the store clerk entered it. That automation drastically improved the accuracy of the inventory. Furthermore, if the items ordered, received, and shelved were also entered electronically, a comparison between what was in the store and what was sold (from the electronic register) would be even more accurate. The manager could easily determine what was sold and for how much, as well as what might be needed and whether it had already been ordered. Once the process had been fully automated, it became more accurate, took less time, and required fewer people.

During the early decades of information technology use, individual processes were automated individually. Often, each individual process utilized different software programs. The daily cash register would use a program (usually published by the cash register manufacturer). A separate company developed the purchase ordering program. And yet another company would write the inventory control program. At each step in the process, information would need to be re-entered, resulting in a great deal of duplication of effort. Or some processes would be automated, and other processes were still done manually. Automation of just some of the processes wasn't enough. To really get the benefits, the companies needed to streamline the process.

## Streamlining

The next role of information technology beyond automation is to utilize IT to streamline processes across different functions. When every process was manual, making the process go faster could only be accomplished by individuals within a process. For example, an organization might hire someone who could enter data faster and more accurately than other people, so the data entry took less time when that person was working.

Optimization of a single process could be achieved through automation, but doing so without considering its connection or role within the end-to-end process often led to sub-optimization. Sub-optimization could hurt the organization, causing problems and disconnects that made customers purchase from someone else.

Automation helped each individual process, but more dramatic improvements could be obtained by revising the interrelated processes from start to finish. Streamlining could achieve optimization of enterprise-wide processes that would lead to improvements in getting goods and services to the customers. *Re-engineering*[*] was the name introduced by Michael Hammer to the concept that processes needed to be reviewed, and re-engineered, across the entire enterprise. Business Process Engineering became a new discipline adopted by many large organizations. In order to accomplish this goal, many businesses adopted process improvement programs such as Total Quality Management (TQM), Continuous Improvement, and Six Sigma. Each of these business process improvement programs initiate teams of employees who focus on improving the overall business processes. These improvements often require changes to the way IT is implemented within the organization as well as changed to the individual processes that had been automated.

One function where IT was at the center stage of business process change was financial systems. Different departments or divisions of the organization (Sales, Accounting, Operations, Finance, Marketing, and Human Resources) had their own separate processes (seen in Table 19). Each was often responsible for their technology and independently made decisions about what to buy and implement. Businesses began to adopt ERP, Enterprise Resource Planning as they optimized enterprise wide processes. This allowed them to manage the entire supply chain instead of just one small part at a time. The term "supply chain management" was coined by Keith Oliver, a Booz Allen Hamilton executive in the early ‘80s, but it has grown to encompass an entire field made possible by ERP systems. ERP implementation would integrate, streamline, and optimize all the various separate processes.[†]

---

[*] Hammer, M. (1996). Beyond reengineering : how the process-centered organization is changing our work and our lives (Vol. 1st). New York: HarperBusiness

[†] More about ERP systems can be found in Financial section in Chapter 7: Application Software.

**Table 19. Individual Business Processes (handled by separate programs before ERP)**

| | | | |
|---|---|---|---|
| Accounts Payable | Customer Relationship Management (CRM) | Material Requirements | Return Merchandise Authorization |
| Accounts Receivable | Customer Service | Materials Management | Sales Forecaster |
| Bank Reconciliation | Demand Forecasting | Mobile Manager | Sales Forecasting |
| Bank Transactions | Distribution Management | Monitoring Metrics | Sales Opportunity Management |
| Bar Code Reader | Engineering | Monitoring Production | Sales Orders |
| Bill of Material | Estimating | Operations Management | Scheduling |
| Budgeting / Cash Flow | Financial Management | Payroll | Service Management |
| Business Intelligence Analysis | Fixed Asset Accounting | Planning | Shipping |
| Business Partners Management | General Ledger | Product Configuration | Shop Floor Data Collection |
| Capable to Promise Management | Human Resources | Production Management | Strategic Decision Making |
| Collaborative Workspace | Inventory Control | Progress Billing | Supply Chain Management |
| Competitive Analysis | Inventory Management | Project Accounting Software | Time Card Management |
| Configuration & Development | Job Costing | Purchase Order | Tracking Material Lots |
| Creating Routings | Job Ordering | Purchasing | Warehouse Management |
| Credit Card Processing | Logistics | Quality Management | Work Order Management |
| Credit Authorization | Magnetic Media Reporting | Quoting | Writing Reports |
| Customer Order Management | Manufacturing | Receiving | |
| | | Retail Management | |

In order to gain real competitive advantage in streamlining these functions, however, the organization needed to focus on alignment between the information technology and the business goals. This would lead to the next role of information technology and how well it aligns with the goals of the organization.

## Acceleration

Streamlining and optimizing all the processes within an organization is great - as long as the right processes are streamlined with the proper perspective. In order to effectively utilize information technology, a business needs to understand what they do well, what the major purpose for their existence is. Jim Collins[*] introduced the *Hedgehog Concept,,* which connotes the one most important thing that a business or organization does. Collins also noted that information technology can be a great competitive advantage because it acts as an *accelerator*. An accelerator is a force that speeds up a process and makes it stronger. But what Collins also pointed out was that technology could accelerate an organization into failure and bankruptcy as easily as it can speed up a technology for competitive advantage. When a business planned to utilize information technology, they had to implement it with an eye toward how it would impact the customer and employees. It wasn't enough to streamline the processes; the **right** processes needed to be streamlined. Too often, IT projects were implemented for the sake of implementing them, without any real clear understanding of how and why they would help the organization accomplish their goals. When a business invests in the proper IT that enhances and supports the goals of the business, it is said that *alignment* has occurred between the business and the IT departments. When a business invests heavily in IT without alignment, the organization suffers and often fails.

[*] Collins, J. C. (2001). Good To Great : Why Some Companies Make The Leap--And Others Don't (Vol. 1st). New York, NY: HarperBusiness.

Collins, J. C., & Porras, J. I. (1994). Built To Last : Successful Habits Of Visionary Companies (Vol. 1st). New York: HarperBusiness.

### Sales Channel Expansion

Until the proliferation of computers in people's homes and the ease of connection available with the Internet, IT was a "back end" function of most businesses. IT was utilized by the organization, but the customers themselves did not really deal with it directly. With the advent of the World Wide Web, however, businesses suddenly had another sales and marketing channel. Rather than enticing customers into their stores in fixed locations (referred to as *brick and mortar*), customers could review product information online in the comfort of their own homes, and order the products to be delivered.

Between 1997 and 2001 was the period often called the **Dot Com Bubble,** or the **Internet Boom.** During this period of time many businesses started utilizing the World Wide Web as a sales channel for the first time. A few high-profile technology companies went public, and a few early "wins" in the investment community started a frenzy of investment that ended when the bubble burst and the market crashed for technology-related businesses. Despite the roller-coaster ride of technology investment, however, there is no question that businesses needed to learn how the Internet and the World Wide Web fit into their own sales and marketing strategies. If the business decided to provide e-commerce (for their customers to purchase something online over the World Wide Web), then they had to re-engineer their processes and IT again in order to incorporate a public "face" where customers could order items. Sales is not the only IT issue; with the growth of social media, marketing also needs to consider additional marketing channels in order to get the attention of customers who are increasingly communicating and connecting online[*].

## Competitive Advantage of Information Technology in Business

Businesses that understand how to effectively utilize information technology can gain a competitive advantage over other businesses. Aligning IT with business objectives (i.e. the Hedgehog Concept) is important. One way to do that is to understand the difference between primary and support activities.

### Porter's Value Chain

Each business function can be described in terms of whether it is a *primary* value activity or a *supporting* value activity for the organization. Michael Porter, a university professor, introduced the concept of the *value chain*[†]. Primary activities included receiving and storing raw material, making the product or developing the service, delivering the product or service, marketing and selling the product or service, and providing customer service after the product or service had been sold. Support activities include the infrastructure of the business, human resource management, research and development (including IT) and procurement. Information technology impacts every level of the value chain, which in turn impacts competitive advantage strategies.

Competitive advantage strategies can be divided into Direct Competitive Advantages, and Indirect Competitive Advantages. Direct competitive advantages are: Price/Costs, Differentiation, or Focus as seen in Table 20.

***Price/Costs*** is attempting to provide goods or services for prices lower than the competition. Walmart uses its high-volume discounts to offer products for less than other retail stores who can't match the volume in order to get similar discounts from the suppliers.

---

[*] More details on the Dot Com Fever is discussed in the section on *Technology Revolutions* section in *Chapter 11: People - Foundations*

[†] Porter, M. E., & Millar, V. E. (1985). How information gives you competitive advantage. Harvard business review, 63(4), 149.

***Differentiation*** is convincing customers that the product or service is different than the competition. Uggs boots are known to be more expensive, more comfortable, and "cooler" to wear than similar looking boots.

***Focus*** is providing goods or services for a narrow market, such as a dedicated sock store or high end yachts. Starbucks is an example of Focus in that it focuses on coffee.

**Table 20. Direct and Indirect Competitive Advantage**

| Competitive Advantage | Type |
|---|---|
| Price/Costs | Direct |
| Differentiation | Direct |
| Focus | Direct |
| Barriers to Entry | Indirect |
| Lock In Buyer | Indirect |
| Alliance | Indirect |
| New Products/Enhancement | Indirect |

In addition to the three direct competitive strategies, a business can utilize technology to provide indirect competitive advantage: Barriers To Entry, Lock-in Buyers, Alliances, and New Products/Enhancements.

***Barriers to Entry*** is when a company provides a service or product that cannot be replicated, or where the cost to provide is very high. The cost for the infrastructure to build an airplane, for example, is very high, so Boeing has a competitive advantage to any other airplane manufacturer because they already have the infrastructure.

***Lock-in Buyers*** is similar, except that instead of preventing another organization from offering the product or service, the business finds a way to dissuade customers from purchasing the product or service from anyone else by making it extremely easy and/or inexpensive to obtain the product or service from the originator. Microsoft locked in buyers of their Windows operating systems by convincing computer manufacturers to offer Windows exclusively. Microsoft did that by giving an extremely deep discount on the operating system to the computer manufacturers if they provided Windows exclusively and automatically on every computer sold.

***Alliances*** can also be a competitive advantage; AT&T partnered with Apple when the iPhone first came out so that data plans were available only and exclusively through AT&T.

***New Products/Enhancements*** are also a competitive strategy. An iPhone can only be purchased through Apple and no one else, and Apple is well known for coming out with attractive new products or major enhancements to existing products.

## Productivity Advantage

Another major way to gain competitive advantage is to increase employee productivity with IT. When businesses are more productive with IT than a competitor, their costs are lower, which can translate either into a lower price to the customer, or higher profits to the business. Improved productivity, however, is not a given. To be more productive, employees need to understand the systems extremely well. Employees need to have optimized all the processes to work with the systems. They need to have learned all the little tricks and techniques for using the system effectively.

Employees sometimes take it upon themselves to become more knowledgeable and experienced in a particular technology. However, an organization can increase the chances by providing the proper training and create the proper environment to ensure superior productivity. People are unwilling to admit it when they don't know how to use IT effectively. This is human nature. But these obstacles can be overcome with the right training and environment

### *IT over Time*

Another factor in how productive an organization can be with IT is the amount of time the IT has been in use. The longer a process using IT has been implemented, the more productivity people will achieve with the process. Many times, businesses don't get the expected value from their IT simply because the process changes too often for people to accommodate. It takes people time to adjust to a new process. If the changes happen more quickly than people have time to adjust, they are constantly in that state of "learning but not knowing". That state decreases productivity. On the other hand, given enough time, people can adjust the process to fit. The people become more productive.

Finally, the more highly integrated and seamless the functionality of the system, the higher the productivity will be. Each disconnect or linkage break in a process causes potential problems and delay. Making sure all the IT is highly integrated increases the speed and reliability of the IT.

These combined ideas can be expressed as a non-numeric formula as shown in Figure 87. *Formula for productivity* .

Productivity = (People * Ability * Experience in IT) +
(the Process * length of Time process in practice) +
(the Network_Hardware_Software * Integration index)
or
P = (PE * A * E )+ (PR * T) + (NHS * I)

**Figure 87. Formula for productivity.**

It does not matter how much money a business spends on IT. It does matter how long the people, the process, and the network/hardware/software work with each other as they move toward the alignment with the Hedgehog Concept that provides the highest productivity possible. That is where true competitive advantage comes from.

## Ethical Issues of Information Technology Processes

The use of IT brings with it a large number of ethical issues. One of the reasons for the increased number of ethical issues is the feasibility of previously unfeasible unethical activities. To give an example, it has been possible since the invention of the photocopy machine to copy the pages of an entire book, thereby "stealing" the contents and violating the copyright. However; it was unfeasible. It would cost more to photocopy the pages (especially when you add in the time involved in setting up each page to be photocopied) than it would cost to purchase the book. When it became possible to post a file of an entire book on a website, however, the feasibility issue was completely different. Now it took just seconds for anyone to "steal" the contents of an entire book, violate the copyright, and sell or give away someone else's work. IT also makes it more difficult to get caught. Piracy of intellectual property (music, books, movies, pictures, software) became rampant due to IT making stealing feasible and easy.

Another ethical issue involves anti-trust concerns. Several companies in the IT industry have been hit with anti-trust and/or anti-competition suits. Many of the laws regarding intellectual property and competition were developed before IT became available. New laws had to be written, and old laws had to be re-interpreted, in order to deal with the new forces in the marketplace.

Privacy is a concern as well. The issue of junk mail has always been annoying, but at least previously there was a real cost to companies in the printing and the postage when they sent out junk mail, which provided some kind of limit. With email, there are no per piece costs; companies can (and do) send millions of emails to millions of people every day* whether they want them or not. Furthermore – how much privacy can we expect given our penchant for convenience? If a vendor makes purchasing their product easy and efficient by looking into your interests and providing a link to purchase something you've been researching on a search site, is that an invasion of privacy or a courtesy?

In the next section we will discuss the ethical issues involved for Individual Privacy, Anti-trust Regulations, Intellectual Property, and Piracy in more detail.

## Individual Privacy

Privacy has always been controversial. While we all want individual privacy for ourselves, we want transparency and knowledge of others - especially governments and organizations and criminals. Activity tracking and data mining are highly controversial privacy issues that are made even more complicated with new capabilities of technology. We don't want to enter our login and password each time we return to the same website, but we don't want any identifying information about us stored on their website or on our system. We want the ease of purchasing and obtaining information that comes from sharing our private financial data (i.e. credit card number, name, address, phone number, email), but we don't want that information used in a way that is not beneficial to us.

The problem is that beneficial is in the eye of the beholder. Most people allow websites to deposit some code (a *cookie)* on their computers so that they don't have to re-enter their login name and password each and every time they return to the same website. But some people may consider that an invasion of privacy. A company may well consider that they are doing people a favor when they send them information about an enhancement of a product they purchased or tell them about a new product. Some people, however, may consider it an invasion of their privacy to receive an unsolicited advertisement about a new product.

Even more insidious (according to some people) is the ability for data aggregators to combine data sources and cross-reference information in a way that obtains non-obvious relationship information (NORI). Does a health insurance company, for example, have the right to cross-check the different pharmaceuticals you might purchase in order to warn you about side effects or dangers of unintended drug combinations? Would it be held liable if it had that information and didn't warn you of the dangers? What about using that same database of information to let you know about a new drug on the market that might replace the drug you are already taking that was prescribed by your doctor? What about using the cookies on your computer to see which websites you are visiting to be able to send you information on other drugs that might be related to some disease you've been reading about? What about using that information to report to the authorities that you have taken out five different prescriptions from five different doctors for the same drug?

None of these issues are black and white. The main issues with privacy are accountability and control. Some people believe that the email address of anyone who downloads information or purchases a product is fair game unless they explicitly request **not** to receive such emails (*opt out*). Some people believe that companies should only be allowed to send emails to people who have specifically signed up to receive them (*opt in*).

Companies may be required by law to safeguard sensitive data they are storing from transactions or web profiles. It is obvious that ethically they should not be allowed to share that information with someone who might do harm by stealing someone's identity. But where does the

* Which results in the explosion of *spam* (unsolicited bulk email) discussed in the **Email Etiquette** section of *Chapter 6: Collaboration: World Wide Web and Other Applications.*

responsibility end? If security of that data is inadvertently breached, such as in the loss of a laptop computer with that data stored on the hard drive, does that mean the company must inform customers of that data loss? If someone steals the identities by using that information, is the company liable for the damage that is done to the finances and to reputation of the data owners?

What about their own use of the data? May they themselves use that data? May they share the data with trusted partners who provide related products and services? May they sell the data to a third party? These issues are still being wrestled in the courts. When you buy a product or service, most vendors assume that if you give them your email address, you are willing to accept emails from them.

The final issue regarding privacy of information transmitted over the Internet. The Internet itself has no protocol for privacy or security. Until recently when browsers started enforcing the use of https (secure hypertext transfer protocol using SSL (secure socket layer) encryption), every message, every web page, every email, every text, every chat was transmitted over the Internet in an open and readable protocol. Anyone with the proper equipment could see them all. To be truly secure, additional software and hardware is necessary, usually in the form of a VPN (virtual private network). *Encryption* means to transform the data so that it is unreadable without a *key*. When data is encrypted and then sent over the Internet, the information transmitted is hidden from prying eyes, generally. (Encryption keys can be broken, eventually).[*]

## Anti-trust Regulations

Business leaders are not allowed to get together and set price controls on products and services. Nor is any one business allowed to completely dominate any particular industry. These activities are barred by several anti-trust regulations (which make any conspiracy to restrict trade against the law) starting with the Sherman Anti-trust Act of 1890 up to the Clayton Act, (which was last amended in 1976). The Federal Trade Commission[†] publishes a list of anti-competitive actions that are forbidden:

- **Price Fixing**: an agreement among competitors that raises, lowers, or stabilizes prices or competitive terms.
- **Bid Rigging**: competitors agree in advance which firm will win the bid.
- **Market Division or Customer Allocation**: an agreement among competitors to assign sales territories or customers.
- **Group Boycotts:** an agreement among competitors not to do business with targeted individuals or businesses.
- **Other Agreements Among Competitors:** unreasonably restricts trade.

Anti-trust activity concerning IT corporations started early. In the eighties the US Department of Justice forced AT&T to break up into "Baby Bells" (independent organizations responsible for providing telephone services in various regions)[‡]. It tried to break up IBM as well, but IBM proactively broke itself up by selling software, hardware, networking, and services separately (whereas before they were sold bundled together). Microsoft was the target of a decades-long suit regarding their anti-competitive practices by the Department of Justice. In the end, they were forced to change some of their business practices, but for the most part they were not visibly hurt by it.

---

[*] More information on Internet protocols and security can be found in *Chapter 2: Networking*.

[†] Federal Trade Commission, Dealing with Competitors. http://www.ftc.gov/bc/antitrust/dealings_with_competitors.shtm. Retrieved October 15, 2020.

[‡] The Justice Department started pursuing AT&T in 1974, but it took more than ten years to conclude the monopoly case against AT&T.

At basic issue is the maintenance of fair competition. Some people consider fair competition an oxymoron because the concept behind competition is to gain an unfair advantage. Robert Bork[*] for example points out that to force everyone to "be fair" would destroy the motivation behind capitalism. He believes that market forces are the best regulating power, and that interference by the government destroys rather than enhances fair competition.

Furthermore, there is some question as to whether or not anti-trust laws are designed to benefit competing businesses or consumers. The reason to break up a monopoly, it is thought, is that a monopoly can charge unfair prices because there is nowhere else for customers to turn. But in the case of AT&T, for example, the monopoly enabled AT&T to subsidize higher cost local telephone service by using profits from the lucrative long distance services. When local and long distance services were unbundled, the cost for local service went up - by quite a lot - and became difficult for people in poorer areas to be able to afford basic telephone services. Hardly a benefit. Of course, in the long run, the break-up of AT&T spawned the development of many innovative IT products and services that would not have been developed under the yoke of a single large provider. In the end, therefore, customers were generally better served by breaking up the monopoly. But at the time, customers were disadvantaged by the breakup, so it is often difficult to tell which is the right path.

Until recently, for the past twenty years the Justice Department has been mostly hands-off, and has not accused any company of being a monopoly. But that may change – either because of the Justice Department, or because other companies may get tired of being bullied by tech giants Apple, Google, Microsoft, and Amazon.[†]

## Intellectual Property and Piracy

Copyright laws prevent people from stealing writings, photos, art, music, movies or software from those who originally created the works. The act of stealing intellectual property is called *pirating* or *piracy*. IT complicates the issue because it allows people to pirate other's intellectual property with ease.

In the seventies and eighties, music was sold on vinyl records that could not be easily reproduced. When tape cassettes and then compact disks became the prominent media upon which music was distributed and sold, the copying of the music to share became an annoying but minor problem. However; when websites such as Napster provided free downloads of millions of songs to millions of people (mostly adolescents, the major market for new music) it became more than just an annoying minor problem. If allowed to proliferate, such piracy would have destroyed the motivation for record companies to promote and fund musical groups. The record companies sued Napster, and the courts held that Napster was knowingly infringing on copyright by providing technology to enable members to share mp3 copies of copyrighted songs and shut Napster down. Napster filed for bankruptcy in 2001[‡].

The issue of intellectual property and the ease of publishing have, indeed, changed the dynamics of the music, book, and movie publishing industries. Previously the high cost of publishing meant that publishing companies controlled who got published and when. Publishing companies also had to invest significant dollars up front, and then took the biggest bite out of any profits from the music, books, or movies should they happen to be successful. (They also, of course, had the most to lose should the works prove unsuccessful, hence the importance of limiting who they backed to

---

[*] Bork, Robert H. (1993). The Antitrust Paradox (second edition). New York:Free Press. ISBN 0-02-904456-1.

[†] **When regulators fail to rein in Big Tech, some turn to antitrust litigation** by Reed Albergotti and Jay Greene. Posted August 21, 2020. https://www.washingtonpost.com/technology/2020/08/21/big-tech-lawsuits-antitrust/ Last accessed on October 15, 2020.

[‡] The website currently known as Napster.com, which is owned by Rhapsody, a legal music-subscription site, is not affiliated directly with the original Napster founders.

only the those with the highest chance of success.) A typical author contract, for example, provided between 6 and 10% of the book's **net** profits back to the author in royalties. Proven well-read authors might get 20%. But in any case, the publisher received the greatest rewards; 80-90% of the profits of any successful book. The same issue occurs in the music industry where the musicians get a very small portion of the income from a recording. Many musicians would like to simply share their recordings with their fans, cutting the record companies out of the supply chain.*

### Digital Rights Management

There is a great deal of difficulty, however, in trying to maintain control over the music once it has been released. Some companies have created DRM (Digital Rights Management) programs that allow them to control when and where a music file can be played. Sometimes this causes problems for the buyer who is unable to backup their purchase or transfer it to a different device or computer easily; something that they could easily do with tapes or CDs. For example, iTunes (now called Apple Music) was well known for restricting a buyer's ability to listen to music purchased on any other devices until 2009, when Apple finally gave in to user demands to provide DRM-free music on iTunes (though videos still use DRM). Before 2009 you could not, for example, transfer a song you bought from Apple to a non-Apple mp3 player to listen to it. Even now it is not easy to transfer music from one device to another.

The same is true for electronic books, which are becoming more and more common. Digital Rights Management systems for electronic books proliferate in order to allow publishers to control who can read an electronic book and when. There are several ebook formats that contain DRM available for books (EPUB, KF8, Mobipocket, PDF, and Topaz) that can restrict who can read the book and when. In general the concept of "owning" a piece of creative work is losing out to the concept of "leasing" the creative work for limited access.

At the present time, exactly how the rights of artists such as music creators, filmmakers, authors and graphic illustrators can be preserved in a digital world is still being worked out. There are several potential solutions including subscription plans, DRMs, and locking/unlocking devices.

### Safeguarding Software and Games

Piracy and rights management is also a problem for software. Mainframe and minicomputer software publishers have always only "leased" software to clients for a single machine. If the machine were to fail, all the software would be lost, and would need to be purchased again.

Before cloud software proliferated, personal computer software publishers sold the ability to use the software on **any** computer (for which it was designed). Unfortunately, unethical people started copying the program and selling or giving it away, cutting into the market of the software developers. Publishers use many different methods to control access including uncopyable "keys" built into CD or floppy disk media which had to remain in the disk drive while using the program, or required connections to Internet websites where ownership was authenticated. Often software publishers utilize long alphanumeric sequences called license keys, to be entered once when the program is installed. Often these keys are printed on the physical media (CD or DVD) or sent in email upon payment, which require their presence if the software must be reinstalled for some reason. Other times these keys are generated automatically through a website and are specific to the single machine upon which the software is installed. If the software needs to be reinstalled, a

* In a world where the cost of publishing is very low, the question arises as to how long the publishing companies will remain in control of the publishing schedule. Already "indie" movements (short for independent film makers or music publishers), and "self publishing" by book authors are making inroads into the market share of the major publishers. That, however, is a business strategy issue, not an ethical issue.

different key would need to be generated. All of these methods cause major problems for computer users who wish to back up software they already own so that it could be restored to another machine in the event that the computer fails. Some software companies, especially game publishers who sell to a very tech savvy customer base, have given up on trying to build copy protection into their product as it only adds an incentive to a large percentage of the customers to "break the code" and pirate the game as a protest against the concept of digital rights management.

In 1980 the United States Congress passed the Computer Software Copyright Act, making software copyrightable - apart from the actual language used to generate the code. In order words, it doesn't matter if the program was written in JAVA or Cobol, if the program functions the same, it would be considered a copyright infringement. The U. S. Supreme Court, in 1981, also held that software could be patentable if part of an original, novel invention. For a while businesses were getting patents for "look and feel" of their menus and operating systems as well as "business methods" such as the one-click purchasing. For a period of time in the nineties the patent office stopped taking applications for processes, citing the difficulty in separating out the process of the business method from the software. The courts are still arguing the issue and previously awarded patents for business processes may, or may not, be defendable.

In the past, large technology companies would sue each other for "Look and Feel" infringement. Lotus 123 successfully sued Borland for copying almost exactly, the menu structure for Quatro, their competing spreadsheet*. That led to a spate of IT companies suing each other for copying the "look and feel" of their systems, led by Apple Computers which has gone to court more than 350 times to restrict other companies from copying their physical and visual designs. For example, Apple Computer sued Samsung in April of 2011 for copying the look and feel of the iPhone for their new Galaxy phone. The fact that the original Macintosh look and feel was developed by Xerox before Apple was a company cause many to bemoan the seemingly hypocritical stance taken by Apple, but generally the courts have been siding with Apple.

Complicating the issue are Non-practicing Entities (NPE) patent suits. These are companies who are in the business of buying old-but-still-valid patents awarded years ago for business methods that are no longer non-obvious or inventive due to the speed of technology innovation. One example is a company called MPHJ who bought up a 2001 patent for scanning a document to email, a common practice today by all businesses. MPHJ started suing any company with more than 50 employees stating that the company had to lease the patent to scan a document to email, at $1000 per employee!† Large technology companies themselves are often hit with NPE suits – by the hundreds. The problem is so bad that several large companies (most notably Google, Apple, & Samsung) announced in 2014 that they were willing to set aside their own patent wars to work together with the Department of Justice to reform Patent Law that will make it more difficult for NPE suits to move forward. This idea was that it would cause a drastic change in the way technology companies deal with patents, and would allow the companies to spend their cash on new innovations instead of on lawyers to defend old ones. Whether or not the changes were successful is still an open question.

### *Difficulties with Copyrights and Patents*

The reforms to patent and copyright laws are needed because they are all outdated. The challenge with IT was that copyright laws were written to safeguard published work in a specific print medium, and patents were written to safeguard physical machinery and equipment from being copied and sold by someone else. When the works became digital, they no longer got "published"

---

* Shan, Linda (1994) The effects of the Lotus v. Borland case http://groups.csail.mit.edu/mac/classes/6.805/student-papers/fall94-papers/shan-copyright.txt, retrieved December 27, 2011

† Found on http://www.ibtimes.com/apple-samsung-google-stop-suing-each-other-start-focusing-patent-reform-1586677, written by Cameron Fuller on May 20, 2014. Accessed July 7, 2016.

in the traditional sense of the word. The intellectual property of software could not be "represented" by a visual description (which patents are based upon). The courts were in a quandary. Did the licensing agreement for the column previously written by an author and published by a newspaper mean that the newspaper could include the column in their archives without paying further royalties to the author? Should software be considered patentable or copyrightable? How do software developers safeguard their trade secrets?

Outside of the major companies' court cases, individuals are impacted by copyrights as well. Previously, there were exceptions to the copyright laws, for example, that allowed "fair use". For example, a teacher could copy and distribute a small section of an article or book for educational purposes without going to jail for copyright infringement. Newspapers could print excerpts of books and television stations could show excerpts of films without fear of suit. But does that mean a teacher could put a chapter of someone else's book on a website for students? Can a newspaper publish a digital copy of an artist painting on the website? Where does fair use end and copyright infringement begin? Napster tried to argue fair use when it provided a website where members could share music files with each other, but the courts clearly ruled that such an environment was not fair use.

The laws of the United States encourage creative works of art and patentable intellectual property to be safeguarded for a set period of time in order to encourage financial rewards for innovation. Copyrighted intellectual property is safeguarded for the lifetime of the originator plus 70 years, and patents last for 20 years from the date of the patent. Without safeguards for intellectual property, companies would not invest in research and development because they would not be able to recoup their investment. Without safeguards for creative works, authors, artists, musicians, and filmmakers could not make a living because anything they worked on could legally be sold by someone else.

### *Laws versus Ethics*

Listening to music, or watching a video, or playing a game which has been pirated is illegal and unethical. But often fair use is a personal ethics test rather than an issue for the courts. For example, is downloading a beautiful landscape from a website and using it as a background on your phone an unethical copyright violation or a fair use of intellectual property? The truth is that even if it were illegal, your personal use of the picture would most probably not be found by anyone, and even if found would probably not be prosecuted.

To determine whether the use is ethical (as opposed to legal) the original purpose or intent of the person who posted the picture on the website is important. If the landscape was a digital work of art that was meant to be sold to support the artist, it would be unethical to utilize it without reimbursing the creator. The same could be said for any creative work found on the web with an intention of being sold. However, some creative works posted on the web are there for purposes of free distribution. In general, downloading such creative work intended for free distribution and personal use is not unethical.

For the past few decades, there has been a movement of software developers to write programs and then release them for anyone else to use. This movement has been called the *Open Source* movement, and has gained the support of many major IT companies such as IBM. Open Source software is free, and is supported voluntarily by those who developed it. Most of the websites running on the Internet today utilize Linux, an open source operating system; Apache, an open source webserver; MySQL, an open source database; and PHP, an open source middleware program. This combination of free software is often called LAMP (Linux, Apache, MySQL, and PHP). Clearly, it is not unethical to use this free software and run a website; even when the website itself is a commercial venture designed to make money for you or your company.

However, if the creative work was posted without the permission of the original creator, or clearly posted identifying the cost to download and use, then doing so without fairly compensating the creator and/or publisher would be unethical. Additionally, using the creative work in a brochure or website designed to sell a product or service without compensating the original developers (or at least getting their permission) would be unethical. Benefiting financially by selling or using the creative work of another person without compensation is clearly a violation of trust and ethics even when the chances of being caught are small.

Now that we've discussed competitive advantage and IT ethics, let's move to the more essential information of HOW we align the IT with our organization and gain that competitive advantage (ethically).

## The Essential Balance: Matching the Process

Once we understand the process of an organization, it is important to be able to match the process to the other layers in the map (i.e. the network, hardware, software, and especially the people).

### *Assessing Process*

*Processes* are those procedures that people use to accomplish tasks. They include both manual and automatic parts of the task. Each process must have an input and output—perhaps several inputs and outputs. An input can be a telephone call, a note, a report, an e-mail, or a time/date. An output can be a telephone call, a note, a report, an e-mail, and so on.

The easiest way to identify a process is to start from the end and work backwards. Many processes are easily described. For example, the process to schedule a meeting is this: I receive a phone call or e-mail identifying when and where the meeting will be held; I check my schedule on my phone or on the computer to see if I'm free; if I'm free and have the interest, I enter the meeting location and purpose in my calendar at the meeting time. Some processes are more complicated, however. If I were a business process engineer trying to identify a process in an office that processes medical claims for an insurance company, the dialogue might sound something like this:

Business process engineer: **What does your office do?**

Worker: **We process medical claims for Euwish Insurance Company.**

Business process engineer: **And how does that work?**

Worker: **Well, the insured call us up to submit a claim. We process it.**

Business process engineer: **What do you mean by "process it"?**

Worker: **We investigate whether it's a valid claim and if so, we send a check to the vendor who provided the service for the insured.**

Business process engineer: **So the end result is a check to the insured?**

Worker: **Yes—but there's a lot more to it.**

Business process engineer: **Okay, but let's work backwards. How are the checks printed?**

Worker: **Suzy prints them every Friday on that printer over there. Then she puts them in the envelopes and drops them in the outgoing mail bin.**

Business process engineer: **How does Suzy know which checks to print?**

Worker: **After a claim is identified as valid, the adjuster enters the check information on this screen into this program on this computer.**

Business process engineer: **How does an adjuster know what to enter?**

Worker: **The name and address of the vendor are on the invoice, along with the amount charged.**

Business process engineer: **How does the adjuster get the invoice?**

Worker: **They come in the mail.**

Business process engineer: **Who sends them?**

Worker: **The vendor sends them.**

Business process engineer: **How does the vendor know to whom to send them?**

Worker: **The insured gives them the address to send them.**

Business process engineer: **How does the insured know the address to give to the vendor?**

Worker: **The insured calls us to find out the address and the claim number.**

Business process engineer: **Claim number?**

Worker: **Yes. The claim number needs to be on the envelope so we know which adjuster to give the mail to.**

Business process engineer: **So the insured calls you to submit a claim, and you give them a claim number?**

Worker: **Not right away. The insured has to be verified first.**

Business process engineer: **So the insured calls and gives you the information about the claim. Who takes that call?**

Worker: **One of the workers in the call center takes the call.**

Business process engineer: **And what do they do with the information they get on the call?**

Worker: **They fill out one of these forms with the name, address, and policy number of the insured.**

Business process engineer: **And what do they do with it?**

Worker: **They give it to Sally, who looks up the policy number in the computer and verifies that the person is actually insured with us. You'd be surprised at how many people call the wrong insurance company, or who don't know the policy number, or think they are insured for something they are not. It's a real problem. Sometimes Sally is backed up for days at a time verifying insurance policies.**

Business process engineer: **Okay, the insured calls and gives a name and address and policy number to the worker. The worker writes the information on this form and gives it to Sally, who looks it up to make sure the person is actually insured. Then what happens?**

Worker: **Sally gives the form to one of the adjusters.**

Business process engineer: **How does Sally know to which adjuster to give the form?**

This exchange, naturally, went on for another 10 minutes. But you get the picture—charting a process takes a fair amount of time and investigative work. It may be tedious, but for an information technology process to work effectively, every single tiny detail of the process must be known and documented. Note the frequency of the questions "How do they know which . . . " and "What do they do with . . . ." Once the entire process is known, it can be documented using flow-sheet boxes and arrows so that other people can see the process as well. It is important at each step in the process that these aspects of the step are written down: who, what, when, why, and how. It does not work, for example, to say "the mail is sorted." We must say, "The mailroom clerk sorts the mail and puts envelopes with people's names on them in their slot. Envelopes without names go to the receptionist."

But how do we begin analyzing this information? A good place to start is by matching the ability level with the technology used in the process.

### Matching Ability Levels Is Key

Let's look at a different example of where a mismatch between the experience and ability levels of the people, in combination with a poor process, ensured failure even with the best hardware and software.

At one company where I worked, the senior executives invested in an extremely powerful OLAP system to analyze the budgeted and actual expenses. OLAP stands for On-Line Analytical Processing, in this case a particular program called Hyperion. The software gives a company the ability to slice the numbers in a myriad of ways, providing a cube of data based on three (or four, or five) different factors at the same time. It is a highly sophisticated system. All the experts in the field of budgeting waxed poetic about the great ability of this software to quickly analyze and chart dozens of dimensions in a multidimensional cube.

The beauty of OLAP is its ability to keep drilling down into **live** data, seeing chart after chart in order to visualize the numbers and identify the issues. Unfortunately, the benefit of OLAP is lost if the senior leaders must describe to someone else what they want to see. Printing out a picture of the cube of data is worthless, and could more easily have been done with Excel or some other graphing program. The company would have been better off hiring a few more analysts dedicated to creating charts of the numbers and giving them to the senior leaders upon request. You can pay a harem of analysts for the cost of a single multimillion-dollar OLAP system.

***Powerful Software***

> As an analytical person who loves to play with financial numbers, I enthusiastically delved into this software and quickly learned how to get the most out of it. My superiors noticed how good I was with the new software and put me on the team responsible for implementing it. I got special training in how to use it.
>
> The power of the system was its capability to dynamically review live financial numbers and drill down into the data as needed. I got extremely good at analyzing the numbers and finding patterns and anomalies to investigate. But my expertise and facility with the software did not help my employer. Analyzing the numbers was not my job.
>
> The people who should have been doing the analysis, finding the patterns and identifying anomalies to investigate, were the senior leaders. The software was designed for them to play with the numbers and conduct the analysis themselves. The problem was that the ability and experience levels of the senior leaders were extremely low. (This was 1998. Many had not yet mastered email; they had their secretaries print out each email message, to which they dictated a reply that their secretaries then typed into the system.) They did not have the time to learn how to use this powerful tool, and they did not have the experience to pick it up as quickly as I had.

So the first problem was a clear mismatch between the ability and experience levels of the users, and the power of the software. But the next problem was even more influential in the difficulties in using the system.

Despite pointed requirements and strong pressure from me, managers across different departments failed to standardize terminology and agree on definitions prior to implementing the software. The result was the clichéd situation of GIGO (Garbage In, Garbage Out). To make the information available to the senior leaders, the underlying source data had to be hand massaged to clean it up sufficiently so that it reflected accurate data. That meant that the data couldn't be analyzed in real time, and instead had to be done in monthly batches. Since the beauty of an OLAP system is its ability to drill down into live data, the one thing it was designed to do could not be done.

The organization had spent millions implementing this new system, and still had inaccurate, old, and unintegrated data. This is a very common problem. Businesses often try to bolt a technology onto a poor process—when what they need to do is fix the process itself. In this case, fixing the definition problems at the source would have solved the problem of inaccurate and unintegrated data at a fraction of the cost of the new system. And then any graphing program could be used to easily illustrate the financial health of the different areas of the company.

So getting back to the process at the medical claims department of the insurance company; we first identify the process and match the levels. We need to ensure that the skills of the people (as defined in our assessment) and the technology properly match the needs of the resulting process. As noted earlier, problems are often identified when there is a mismatch between the skills of the people, the capabilities of the systems, or the flow of the process. In the OLAP project, for example, if we had done a detailed process flow prior to choosing and installing the system, it would have been obvious that the system could not replace the manually massaged numbers.

Can you tell where some of the difficulties are in the insurance process described in the previous section? Go back and reread the process to see if you can identify the problems before going on to the next section which will point a few of them out.

## Identifying the Issues

Let's dig into this case a little more. After investigating the features of the system, the business process engineer discovers that the system itself has a capability to print the letter with the claim number and the address. Additionally, the business process engineer finds out that the system contains a module that would allow quick lookups of people by name and policy number that can, within seconds, provide the information necessary to identify whether or not the person calling with a claim is actually the insured.

The business process engineer returns to the office to find out why the adjusters aren't using the system capabilities.

Business process engineer: **In further investigation, I found some capabilities of the system. I was wondering if you were aware of them.**

Worker: **What are they? You know they didn't give us very much training on the system, so there may be things that it can do that we weren't aware of.**

Business process engineer: **For one thing, there seems to be a module that provides the information about whether or not a person is insured pretty quickly—it's designed to be used by someone accepting the claim on the phone and immediately giving the claim number.**

Worker: **Oh yes—we knew about that. But we can't use it.**

Business process engineer: **Really?**

Worker: **The call center takes the call. They don't have this program.**

Business process engineer: **And they can't get it?**

Worker: **The call center people wanted to do it—they said that they are equipped to verify the claim and send out the letters, but our VP didn't feel comfortable that they would do it right. Besides—it would be too expensive. This program is $250 a seat. There are over 100 computers in the call center—it would cost us over $25,000 to purchase the program for the call center.**

Business process engineer: **I understand—that is a lot of money. I'm wondering, do you know how much time does it take for Sally to receive all the slips of paper, verify the accounts, assign the claim, and have the adjuster get the claim letter to the person?**

Worker: **Well it only takes a few seconds for Sally to verify the person once she gets the slip of paper. It takes a few more minutes for her to assign the account. If the adjuster is there, it can be done right away, but if not, however long it takes the adjuster to return to her desk. It could be a few days, even a few weeks, if the adjuster is really busy.**

Business process engineer: **What does the person do in the meantime?**

Worker: **What do you mean?**

Business process engineer: **Well I'm guessing that a person doesn't wait a few weeks to go to the hospital or see a doctor after an accident. If they don't have the claim number, what do they do in the meantime?**

Worker: **Oh I don't know. They must wait for the claim number.**

Business process engineer: **But what if they can't?**

Worker: **They should wait for the claim number, otherwise the vendor will use the wrong address and won't get paid.**

Business process engineer: **That must be a real problem.**

Worker: **You're telling me! It's insane—late bills, phone calls back and forth, angry vendors, angry customers. People are constantly giving the wrong address to the vendor, and then they get mad when the bill doesn't get paid!**

Business process engineer: **I can see that. Well—the other capability we discovered was the ability to send out a letter with the claim number and address—that can be done within the system, automatically.**

Worker: **Really? We didn't know that.**

Worker2: **I'm sorry to interrupt, but I knew that. I discovered it by accident one day. It saves me hours of time a week—that's why I'm able to process my quota every week—I don't have to spend so much time writing letters.**

Business process engineer: **That's great. How many people know about this?**

Worker2: **Oh—I think Janice knows, and Bill. I told Janice, and Bill found out on his own.**

**Business process engineer: Why doesn't everyone use it?**

Worker2: **Well, I told Mary, too, but she doesn't like to use it. So I stopped telling people about it. I told Sally about it, but she didn't want to force everyone to use it. For one thing, she was afraid our quotas would go up, and some people just aren't fast enough to keep up—even with the automatic letter printing.**

Business process engineer: **That's understandable. Thank you both for your help.**

What are the capabilities of the system in this scenario? What are the capabilities of the people? How is the process helping or hindering the people and the capabilities?

It's easy to see that there are a number of problems here*. The big lag time in getting the claim number and the address causes the insured to give the wrong address for billing to the vendors, who then get angry because they don't get paid. Then they bill the client, who gets angry after receiving a bill for something that was supposed to be paid for by the insurance company. The obstacle to fixing that problem is not technology—the system already has the capability to provide the claim number and address right away—on the phone in the initial call. The system also has the ability to send out that letter—automatically, right away. You could say that the problem is people—the VP of the area won't let the call center do that part of the function. But the VP believes that he is saving the company money ($25,000) and ensuring that the job is done right (which he believes only his department can do). So he has the best intentions and is dedicated to doing the right thing. You could say that the problem is the process—except that from the perspective of the workers, the process is working just fine.

Another problem is the fact that not all the adjusters are quick enough with the computer to understand and use the automatic letter feature. The adjusters are working hard, and are probably dedicated—but remember that ability curve. Not everyone has the ability to utilize technology without a lot of experience. And Sally knows that management will not permit the quotas to change—even temporarily—while people get used to the new technology. She is doing the best that she can for her people when she refuses to enforce use of the technology to accomplish the task.

### *Presenting the Solutions*

The difficulty is a mismatch between the process and the people and the technology. What is the answer? In this case, the business process engineer does find a way to improve the situation. Let's see:

Business process engineer: **I'd like to share some information with you, if you have the time right now.**

VP: **Sure. What have you got?**

Business process engineer: **Well, as you can see from this flow chart, the current process is working pretty well. Most people are able to achieve their quotas. Your overall customer satisfaction scores are pretty good.**

VP: **Well, of course!**

Business process engineer: **There are a few areas, however, that could use a little improvement.**

VP: **By all means, tell me what they are. We always want to improve if we can.**

Business process engineer: **Well, as you can tell from this flow chart, there can be a very long lag time between when people verbally submit their claims and the time they get the claim number and the address to which the vendors should submit bills.**

VP: **Yes, my people are complaining about that all the time.**

Business process engineer: **I've been told that you have investigated allowing the call center to provide the claim number to the people.**

VP: **I'm not convinced that the call center people can handle that. They simply don't have the right skills. What if they give the wrong claim number? Then when the adjusters send the letter, it will be a mismatch. Besides—it would cost $25,000 to give them the software—I don't have the budget for that.**

---

* I originally wrote this scenario based on the real-life insurance process I encountered after sustaining injuries from a November 2002 automobile accident. I figured that when I updated the book the process would have improved. I was so wrong. My husband got into a very bad automobile accident in 2017, and the exact same poorly-designed process was still in place!

Business process engineer: **Yes, I believe you are right. However, I wonder if . . . well, how well do you know the head of the complaints department?**

VP: **Jennifer? I know her very well. We have lunch together often. She's a really great person, but I wouldn't want her job. It's thankless! All those complaints.**

Business process engineer: **Do you think she would be willing to pay $25,000 out of her budget in order to decrease the number of complaints she gets?**

VP: **She sure would. If she could cut down on the complaints, she could cut down on overtime, which is extremely costly. She told me that last month she spent $45,000 on overtime alone!**

Business process engineer: **So perhaps she would be willing to use her budget to pay for the software?**

VP: **I'll bet she would. But I still don't trust the call center people to give the right claim number.**

**Business process engineer: What if the call center printed the letter, so the claim number they give over the phone and the claim number they send are always the same?**

VP: **I did discuss that with the call center manager. He said they don't know how to use a word processor and can't use the template we developed.**

Business process engineer: **But what if I told you that the system itself can automatically send the letter?**

VP: **I didn't know the system could do that!**

Business process engineer: **Yes—few people do. The first version of the software couldn't, but the one you are using does have that feature. I verified it—the automation would work—and the printers in the call center are already set up for using features like that—they have to send letters out for the life insurance department.**

VP: **Well, that does change things! I'll talk to Jennifer right away and then set up a meeting with the call center manager. This should be very helpful! Thanks.**

Business process engineer: **I didn't do anything. It was your people who pointed out the best way to do this. This has been a wonderful project to work on. I enjoyed it quite a bit. Thank you!**

As you can see, when everyone has the proper information, they generally make the right decisions. The key to ensuring that the process, people, and system match is to have full knowledge about all three of them.

- Understand the capabilities of the people (both ability and experience).
- Understand the capabilities of the system (all features and functions).
- Match the process to the people and the system.

It often helps to diagram the process visually, as in the example shown in Figure 88.

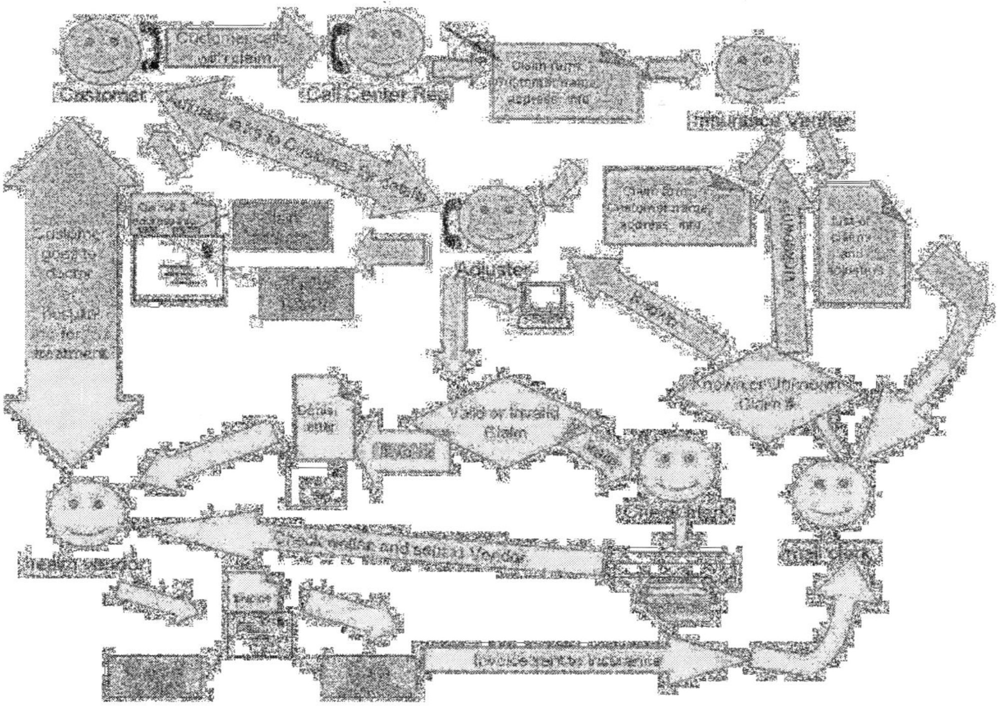


**Figure 88. Euwish Process**

Sometimes senior leaders purchase and implement new system after new system because they think the system is at fault. Or, they fire good people because they believe the people are at fault. In my years as a consultant, I have rarely found that either the people or the systems are at fault. In almost 99.9 percent of the cases, the problem is a mismatch between the system and the people.

That's not to say that systems don't have faults. Every system has faults. A company may invest in a new system that gets rid of a fault of the present system, only to discover that there are new faults that they must get around. Nor is it to say that people don't have faults. All people have faults. Again—you may fire someone to get rid of one of their faults only to hire someone who has a different fault.

The key to success is not to find a perfect system or perfect employees (neither of which exist in the real world). The key to success is to find a process that maximizes the strengths of the systems and people, and works around the weaknesses. Since process is the most flexible of the people-systems-process threesome, it makes sense to work hard at finding a process that matches the people and the systems before adjusting either the people or the system. To review a process, these guidelines are helpful:

- Review the process from end to end (i.e., not just one department, but from initial vendor to final customer).
- Accurately define the problem you are trying to solve.
- Brainstorm possible solutions.
- Envision the solutions to find alternative problems that might arise.
- Practice trial and error—try different solutions until one fits.

Often, finding the problem and solution is not the biggest stumbling block. The method that is used to identify the problem and communicate the solution is more important than the steps themselves. When investigating potential process-people-hardware-software mismatches, it is important to keep the following guidelines in mind:

- Find out why things are done as they are. "We've always done it that way" is not an answer. Find the real answer and assess whether the reason is still valid.

- Always maintain everyone's ego.
- Never look for, or lay, blame. Focus only actions and reasons behind them.

Go back to the business process engineer dialogue. What if the business process engineer had confronted the VP with what was found in a different way?

- Business process engineer: I've found a problem in the process.
- VP: Problem? What problem?
- Business process engineer: If the call center is given the software, they can validate that the person submitting the claim is insured and send out the letters.
- VP: We've already investigated that, and it won't work. It costs too much, and the call center people don't have the capability.
- Business process engineer: But I've checked with the manager and they have the capability. And the head of the complaints department is willing to pay for the software.
- VP: I said no. Now stop meddling. Pay attention to just my department—not what other people can do.

If the business process engineer starts out by saying there is a problem, the manager will get defensive. Despite the fact that the problem is identified and the solution known, a defensive manager will not allow a problem to exist and will not allow a solution to be implemented because it would impinge on his or her own ego—a common human trait. We want to avoid making people defensive or indicating in any way that they've done something wrong when trying to solve a sticky problem.

## *Making the Decision Involving Technology*

In addition to the process matching the people, the process has to match the system as well. Every technology, when implemented, will change processes. Some change them only a little; some change them a lot. But attempts to force the hardware and software to fit an existing manual process are like trying to build a house on quicksand. It may stay up long enough for pictures to be taken (and senior leaders to pat themselves on the back for a job well done), but probably sooner rather than later it will fall apart and sink.

There are several keys to maintaining the delicate balance that must exist between people, process, and systems. The first has to do with decision making, the second with communicating the vision, and the third with redirecting when we get off course.

Making good decisions is a talent that most successful leaders develop over time and is an essential part of the people-process-system balance. We can easily see why by looking at the steps of decision making.*

- Classifying the problem.
- Defining the problem.
- Specifying the answer to the problem.
- Deciding what is right rather than what is acceptable, in order to meet the boundary conditions.
- Building into the decision the action to carry it out.
- Testing the validity and effectiveness of the decision against the actual course of events.

It's that last one that usually is difficult and time consuming. All people make poor decisions initially, but they then, generally, make better and better ones as they go along. Executive leaders

* Drucker, Peter, Howard Raiffa, Alden M. Hayashi, Ralph L. Keeney, and John S. Hammond.2001. *Harvard Business Review on executive decision making.* Cambridge, MA: Harvard Business School Press.

often don't become successful until they are 40 or 50 years old. When they were in their twenties and thirties they made all the mistakes and have since learned from the poor decisions of the past.

Additionally, if we think about the actual stages of defining the problem—which usually include gathering information, delineating various options, identifying the advantages and disadvantages of each option—we can see that one of the most important factors regarding understanding the advantages and disadvantages is having all the knowledge necessary to define the options.

As described in an earlier section, both of these critical steps of normal decision making demonstrate problems when the decision is about technology. Technology hasn't been around long enough for most people to test the validity and effectiveness of their decisions against the actual course of events. Also, the people making the decisions haven't had time yet to gain the critical knowledge they need to properly define and assess the options. Many of the top decision makers in business today were educated long before technology became an influential factor. They may be knowledgeable in every business domain except technology.

Due to these problems, technology decisions (25 percent of the time according to my research) have been left up to the data-processing managers or the information systems experts. IT people, however, do not have the decision-making experience of today's leaders and due to their very nature often make very poor managers.

Understanding the roles involved in decision making is essential. That's more easily done if you keep these rules in mind.

- Don't abdicate.
- Remember that the vendor is not a decision maker.
- Ensure that both business and technology leaders agree on the goals.

### Don't Abdicate

As noted in the first chapter, information technology is involved in every major business decision these days. That scares some decision makers. I've known several CEOs who left IT decisions entirely up to CIOs and IT managers because they didn't feel qualified to decide. They hand wave the details. It's easier for the emperor to allow the wizard to decide, of course, as long as the wizard is a trustworthy person.

Any abdication, however, is a problem. If the wizard were good at emperor-type decisions, he or she would be emperor, not wizard. The key is this: Information about the technology should not drive the decision—information about the business should drive the decision.

The bottom line? Only the person who thoroughly understands the business has the depth of knowledge necessary to align the proper technology with the business. The business leader must be involved in the decision making for technology, but he must be knowledgeable enough to make the right decision.

### Remember That the Vendor Is Not a Decision Maker

Lack of confidence in their own technology knowledge, and the difficulties involved in obtaining the necessary information, often lead senior executives to relinquish the decision to the technology vendor. Abdicators often expect the vendor to tell them what can be accomplished with the new technology, how much it will cost, and how much money it will save the organization.

Entrepreneurs and business owners need to know the answers to these questions before talking to the vendors of any new technology. As noted in Chapter 2, a vendor will always be biased toward implementing technology, and it will always find financial and operational returns that seem to support its view. Despite what he or she may say, the vendor's job is to sell us technology, not to ensure that our business is healthy and growing for years to come.

As a buyer, we need to establish the parameters for the relationship, and we should clearly communicate that to the vendor.

## Ensure Both Business and Technology Leaders Agree on Goals

A couple is driving down the road. The husband is behind the wheel; the wife has a map open on her lap. It's obvious they've been driving for quite a while.

"Where's the next turn?" he inquires.

"Two miles ahead, on the right," she answers.

As they approach a quaint cabin, they both seem to perk up a little when she announces, "Finally! Stop here."

They get out of the car, stretch a bit, and look around. He looks puzzled and says, "Where is this? I thought we were going to the beach."

"Oh," she answers. "This is our cabin from last year. I thought we were going to the mountains."

Technology decisions depend upon the strategic direction of a business. As with our couple, sometimes the business leader has a completely different idea of where the business is going than the technology leader. The majority of the time, a decision is made either by the business leader (the CEO or the COO) or the technology leader (the CIO or IT manager), but rarely both. Out of the 584 organizations that participated in the surveys I conducted, 26 companies, less than 4 percent, said that both the CEO and the CIO had been among the technology decision makers (though less than 1 percent had the CEO and CIO as the only decision makers).

As can be seen in Figure 89, about half of the companies declared that technology decisions were made by the CEO-president-owner (50 percent). In 48 percent of the cases, 278 companies, the CEO was the sole technology decision maker. Only 18 percent (122 of the companies) included CIO/IT director/IT manager as part of the technology decision-making team. Ninety-one companies (16 percent) allowed the CIO to make technology decisions with no other input.

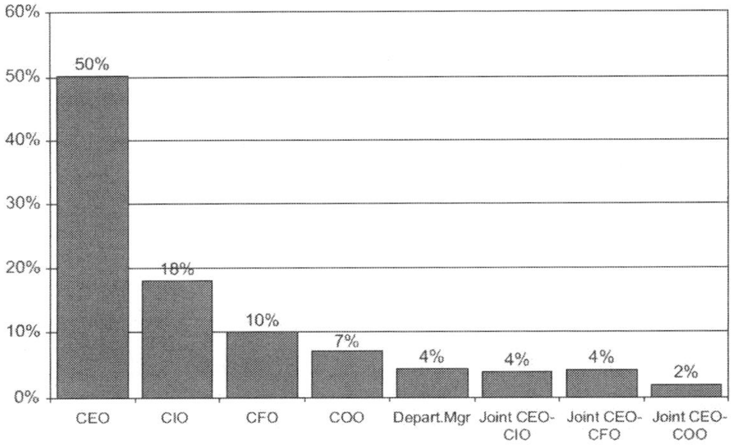


**Figure 89. Roles of Information Technology Decisionmakers**

Are you wondering which fared better in terms of new technology use—companies in which the CEO or the COO or the CFO made the technology decisions, or companies in which the CIO made the decisions? Neither. Companies where the CEO made the decisions did about the same as the CIO alone, which was about the same as any combination of decision makers.

However, when the CIO and the CEO got together to make the decision, the effective use of new technology such as the public Internet increased. It is easy to see in Figure 90. Only this one combination led to a majority of companies rated "Blossoming Internet Use" or "Internet Savvy" rather than "Bare Minimum Internet" or "Unversed in Internet" on overall Internet technology utilization.

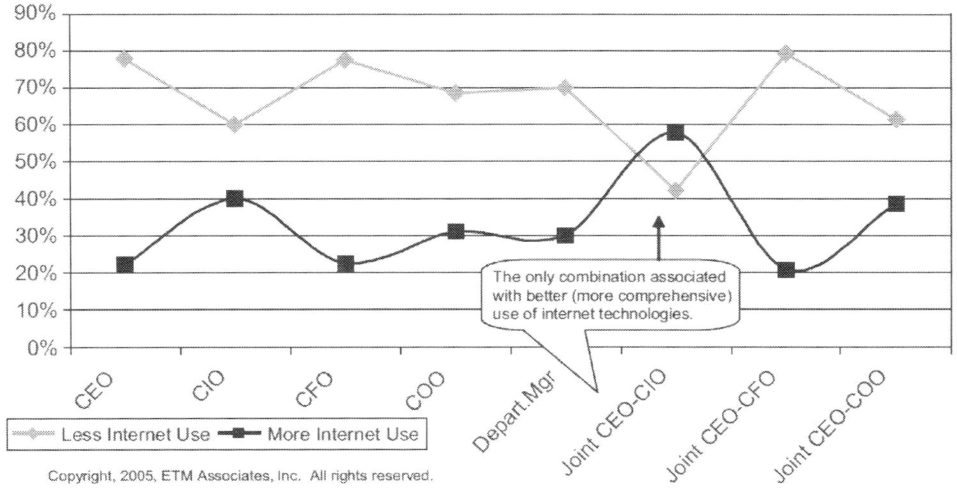


**Figure 90. Successful Internet Use by Role of Decisionmaker**

It appears, then, that when both CIOs and CEOs are included on the decision-making team, they improve the overall use of Internet technology. (You might also notice that the organizations that used Internet technologies least effectively were those who let the CEO and the CFO make the technology decisions.)

As already noted in the discussion of abdication, the answer is clear. CIOs should not be making technology decisions alone. Neither should the CEO, CFO, COO, or any other business leader (especially the CFO, as it can be clearly seen in the graph that including the CFO in the decision-making team always led to less Internet utilization). The best outcome appeared when both the business leader and the technology leader were involved in the decision-making process for technology projects, and when both agree on the final destination.

## Part of the Process: Communicating the Vision

Part of the process is **how** to communicate the vision of how technology fits into the overall process of the business. In a recent research study, I asked successful executives if they ever had to go back and do a project all over again. If they had, then I asked what they would have done differently. The number one most common answer was "spend more time communicating the vision."

There are several keys to communicating the vision:

- Start before the decision has been made.
- Keep the message simple and consistent.
- Design communication using multiple methods (visually, auditorily, kinesthetically, behaviorally).
- Reinforce the message frequently over a long period of time.
- Provide breadcrumbs in your message.

### Start Before the Decision Has Been Made

The worst time to start communicating a vision is after the decision has been made. No matter how participative we have tried to make the decision-making process, this will appear to everyone that the decision was made without their input. Human nature is such that everyone resents decisions that are made without their input. People are much more forgiving of a leader who goes against their advice than of a leader who has not even listened to their advice.

The easiest way to avoid this problem is to start the communication process before the decision has actually been made. We can plan personal contact with as many people as possible, asking them their opinion on the issue or problem we are trying to solve. This gives them a heads up that a decision is being made, and enables them to feel part of the decision-making process. It also begins the process of communicating a vision, because we can provide our vision as an option that we are considering. This process has the added advantage of getting input on the vision—which may turn out to be valuable. We may even modify our vision based upon the input. Often we think we know the perfect answer at first, but repeated presentations to interested parties are bound to improve the vision (or at the very least, how we describe the vision).

### Keep the Message Simple and Consistent

It is easy to think you are communicating well, only to find out that everyone misunderstood what you were trying to say. Keeping the messages simple, and expressing the *same* message over and over again helps with this problem. Additionally, remember to makes sure the message is appropriate and focused on the target audience. It's not what you need to say, it's what they need to understand.

Back in the days when I was an employee, I would always get a schizophrenic response to the assessment of my communication skills in my evaluations. On one hand I was told that I was a good communicator because I could write and speak well. My grammar is generally better than average (the product of tyrannical parents who constantly corrected me), and I'm highly animated and dynamic in front of an audience. On the other hand, I was told that I was a terrible communicator because my message was constantly being misunderstood by others. I was told that I was not clear in my communication, and that I didn't listen well.

It took me quite a while to figure out what my problem was. (Well, actually it took several expensive executive coaches hitting me over the head again and again with the lesson, but that's a story for another day). The problem was that I was too focused on my understanding of the issue or problem. Instead, I should have focused on the understanding of the issue or problem from the point of view of the person with whom I was communicating.

Furthermore, I was also a bit tyrannical when it came to accurately describing the circumstances, the problem, and the solution; I doggedly insisted on total accuracy in discussing technical issues. What I discovered was that complex and accurate descriptions often interfere with understanding, like the husband who keeps interrupting his wife when she is telling a good story. Although the techno-geek side of me always wants to give the whole story and teach everyone all the ins and outs of the problem and the solution, I learned through experience that I needed to give that up. I needed to listen to the people with whom I was communicating so that I knew what their current level of understanding was. Then I needed to figure out a way to communicate my message so that it matched their level of understanding. In order to see the vision, people have to understand it.

### Communicate in Multiple Ways

Rather than just hoping the vision gets communicated, it is important that we design the communication surrounding the vision using multiple methods: visual, auditory, kinesthetic, and behavioral.

The simple message can often be communicated with a short catchphrase, but the longer explanation behind the catchphrase needs to also be carefully crafted. A logo or picture can also help communicate the phrase. Concrete examples and physical equipment or office reorganization or furniture placement may be able to help with the communication. We don't need to hire a branding expert or spend a lot of money on graphic designers. We just need to choose the central theme to go with our vision. At every opportunity, we use the same theme to reinforce the vision.

Most important, behaviors need to change to accommodate the vision. Starting from the top and working its way down as well as starting from the bottom and working its way up—all people involved must catch themselves behaving the old way in order to purposefully change their own behavior to the new way to accommodate the new vision.

### *Measure through Performance Scorecard*

The easiest way to change behavior is to measure the behavior we wish to see. The concept of the performance scorecard provides a method to accomplish behavior change through measurement.*

All of the measurements must align. I worked for two different companies that claimed customer service was their differentiating factor. One, however, measured its customer service representatives in the quality of their interactions and the satisfaction of the customer. The other measured its customer service representatives on how short they could make their phone calls. Which do you think provided good customer service?

### *Reinforce the Message Frequently over a Long Period of Time*

Two of the most important keys to the vision are frequency and duration. Jim Collins talks about the flywheel and the doom loop. The doom loop is when an organization keeps announcing one wonderful new program after another, often with great fanfare, only to have enthusiasm and energy for the vision drop over time. Within a few years, no one will pay any attention to new visions because they know that within a year or two everyone will forget about it anyway.

Quality Management, Six Sigma, Process Reengineering—all are wonderful programs that can drastically improve performance of an organization. In a way, they all build upon each other. But if each one is tried, assessed, and then forgotten, no one benefits. Motivation can be sapped by too many improvement programs.

But if the same message is given, over and over and over again, year in, year out—people begin to get the message that the vision is real. It will not go away. Yes, it may get boring for the leaders to constantly quote the same statements and refer to the same pictures, but it is important to stay consistent.

At one of the companies I worked for, the phrase "Think Like a Customer" was printed over every single doorway, on every single computer screen. The dedication to the customer never flagged, never wavered. It was stated again and again. Year in. Year out. All of the performance scorecard measurements supported it. Customer success stories were told again, and again, and again. Each time, people (we weren't allowed to call them employees) got a little jolt of pride. We worked for a company that cared about the customer. The message was consistent and stable. Frequency and consistency worked.

### *Provide Breadcrumbs in Your Message*

In addition to the overarching vision, information should be communicated regarding where the organization came from, how it tried to accomplish the vision in the past, how it is accomplishing the vision now, and how it will accomplish the vision in the future. Like the breadcrumbs that Hansel and Gretel left to find their way back to where they had been, we must name the periods in the organization's past so that people know what the path has been. People within the organization should be able to track the different generations of technology and how they align with the vision. Constantly referring to both the past and the future enables everyone in the organization to maintain the balance even during periods of transition.

---

* Kaplan, Robert S., and David P. Norton. 1996. *The Balanced Scorecard: Translating Strategy Into Action.* Cambridge, MA: Harvard Business School Press, 1996.

# 8-1. Interactive

1. **Class Discussion**: Where do you think the courts should draw the line when it comes to "Look and Feel" suits? Look at some of the most recent court cases, and then look at some of the older court cases. How have the views changed?
2. **Class Discussion:** Do you think companies should use opt out, or opt in? How would you police email distribution lists? Review the incoming email in your own email account. How much of it was wanted? How much was unwanted? How much was unsolicited (remember to include wanted email). How much was solicited (specifically requested or a reply to your email.)
3. **Class Discussion**: Young people use texting more than email. How much of a role do you think spam plays in that choice? What happens when companies start sending texts to people's phones much the way they currently send emails to people's inboxes?
4. **Class Discussion:** Do you think IT is still the road to competitive advantage, or that it doesn't matter? Why or Why not?
5. **Classroom Activity**: Separate into teams. Each team should search and list any websites that seem to encourage piracy of intellectual property. Compare the types of websites found by each team.
6. **Independent Activity**: Using a variety of sources, find companies who have gained competitive advantage by using innovative new technologies in the past. List as many as you can, and identify the years when they made the most gains from the use of the new technologies. Now see how many companies you can find who failed after they adopted new innovative technologies. List as many as you can.
7. **Independent Activity**: How many times have you used, watched, looked at, or listened to something that was pirated in your life. Try to list at least 5. Did you feel at the time that you were doing something unethical? What justifications did you make? Do you think those justifications are still valid, or do you think now that you were unethically stealing someone's work?
8. **Independent Activity**: Think about a process within the company in which you work. Does the process run smoothly? Are there things that can be done to improve it? Are there disconnects that cause problems?
9. **Independent Activity**: Think about a situation when you were an employee. How did management communicate decisions about technology? Did you feel consulted? Did you agree with the decision? How did the decision turn out for the company?

# 8-2. Self Study Test Questions:

1. What are the four major roles of IT within a business?
2. What is business process improvement?
3. Describe a manual process, and then describe how it might be automated.
4. What are the advantages of automating a process? What are the disadvantages?
5. What is process re-engineering?
6. What does it mean when people call IT an accelerator?
7. What kinds of sales and marketing channels are now available because of the Internet?
8. What are the seven categories of competitive advantage through IT? Which are direct, and which are indirect?
9. What is the productivity formula?
10. What is intellectual property? How do people safeguard their intellectual property?
11. What is a copyright? What is a patent? What is the difference between them?
12. What is DRM?
13. How do most software companies protect their software from piracy?
14. What is the open source movement? List at least 3 open source programs.
15. How do you assess the ability and experience of the people?
16. Why do you identify the features and functions of the system?
17. How do you illustrate existing processes?
18. Why might there be mismatches between the people, the system, and the process?
19. How do you keep everyone's ego intact?
20. What is a good decision-making process?
21. Why shouldn't you abdicate the technology decisions to vendors?
22. Why is it important to communicate the vision and how each decision fits into the vision?
23. When should you start communicating about a decision regarding a change in process?
24. Why should you deep the message simple and consistent?
25. How many methods should you use when you design communication about process?
26. What is a performance scorecard?
27. Why should you reinforce the message frequently over a long period of time?
28. What is a breadcrumb, and why should you include it in your message?

9

# Chapter 9: Process of Managing Information Systems

Overview: In this chapter, readers will learn the following topics:

- ⇨ What Is Systems Analysis
- ⇨ How Does Systems Development differ from System Acquisition
- ⇨ What Is Outsourcing
- ⇨ Security Access and Controls
- ⇨ What Are the Different Types of Information Systems
  - Types of Systems
  - Types of Processing
  - Types of System Configurations
  - Types of Networking Configuration
- ⇨ Why Dealing with IT Vendors is So Difficult
- ⇨ How To Get The Most from IT Vendors

Part of process is the processes of the business itself. But another part of process is the process of managing the information technology of a business. It is this aspect that is covered in this chapter.

## Systems: Analyzing, Developing, Acquiring

*Systems analysis,* the art of designing a system, can take one of several approaches. In the early days of IT, corporations had no choice; they had to hire programmers and develop their own systems. Computers were purchased with storage, memory, and an operating system, but the concept of third party software had not yet been conceived. *Systems Development,* designing and programming their own software, was the standard practice of any business using a computer during the seventies and eighties.

Eventually, it was possible to purchase third party software. Large corporations continued to develop their own software both because they had a lot invested in their own systems, and because their processes and business methods were uniquely developed along with their own software development. For smaller corporations, however, and companies who had not yet started developing their own software, most of the in-house systems development surrounded customizing software purchased from a third party. Instead of system development, therefore, companies practiced *System Acquisition.*

Whether in-house software development or simply third party acquired software customization, it soon became apparent to business leaders that maintaining in-house expertise in programming and software development was very expensive. As geographic boundaries were eliminated because of Internet connectivity, it became feasible to hire another company (often a company who hired highly trained programmers from other countries willing to work for a lower salary) to actually do the programming, maintenance, and customization for the software rather than hiring the programmers in-house. Hiring a third party firm to do the programming is called *outsourcing.*

As Internet connectivity became more available, and bandwidth connections became faster and broader, there has been a movement toward what is called *cloud computing.* The concept behind cloud computing is that everything; the software applications, the data being stored; everything is managed and controlled by the company providing the service such as Google Drive, Dropbox, or Microsoft (who has gotten into the cloud market late, but with a vengeance, in order to fend off Google's head start).

Finally, one of the most important activities of IT people is security, access, and controls to ensure that data stays safe and restorable in the event of disaster.

This chapter will discuss these issues: system development, system acquisition, outsourcing, cloud computing and security.

## Systems Analysis and Design

The Systems Analyst is traditionally the title given to the person who designs the system. A systems analyst usually uses a *systems approach* which involves

1) recognizing a problem,
2) developing and evaluating alternative solutions,
3) selecting the best option,
4) designing the system to fulfill the option, and
5) implementing and evaluating the resulting system.

Another team member working with the system analyst may be a *Business Analyst.* A Business Analyst is someone who knows and understands the business processes undergoing automation, the IT environment, and the programming language. The Business Analyst may also be the liaison among the system analyst, the development team, and the senior leaders who are the funding decision makers, although often a *Project Manager* would take on that role. The final parts of the team include the *developers* (also called *programmers* or *system administrators* or *database administrators*) who actually create the code for the new system.

## Systems Development

For decades *system development,* writing their own programming for their own applications, was the only choice for businesses with computers. Highly refined process for planning, programming, and implementing the software have been developed over the years. These methods can be classified as either Waterfall or Iterative, though some combine features from both. The methods are System Development Life Cycle, Prototyping/Spiral, Rapid Application Development [which includes JAD (Joint Application Development) and DSDM (Dynamic Systems Development Model)], Rational Unified Process, or Agile Programming (which includes Extreme Programming and Scrum programming)*.

### Waterfall Method Systems Development Life Cycle

The most commonly used method by IT organizations for developing software from start to finish for mainframe transaction processing projects is called *Systems Development Life Cycle* (SDLC) . SDLC is a Waterfall method because it is linear; the results of each phase became the input for the next phase. The phases are known by various names and may be broken up in various spots, but they follow the general pattern shown in Figure 91: Planning and Feasibility, Requirements, Analysis, Design, Program/Code, Implement, Test, and Maintain.

* A good description of some of these methods, as well as a few more, can be found at http://www.itinfo.am/eng/software-development-methodologies/, published by the Association of Modern Technologies Professional, and accessed July 7, 2016.

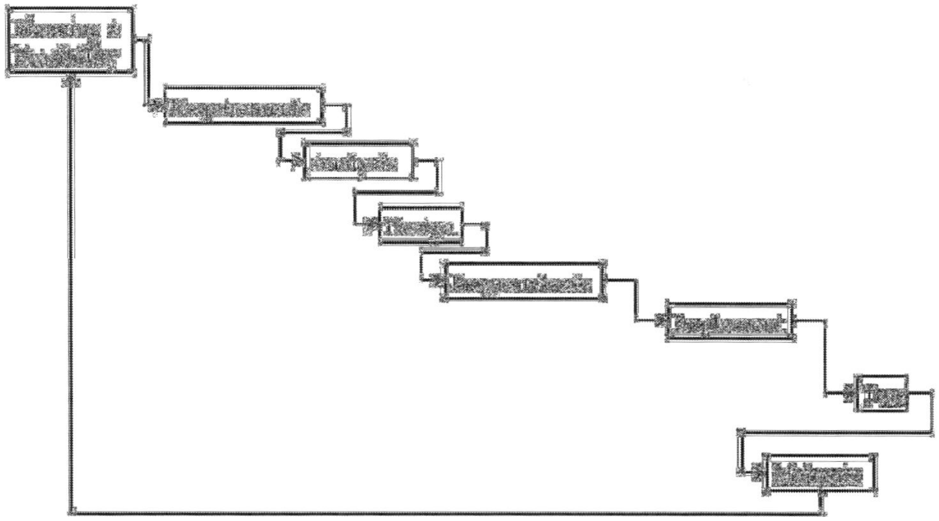

**Figure 91. Waterfall Method: Systems Development Life Cycle.**

During the Planning stage, feasibility studies are done (operation, economic, technical, organizational). During the Requirements phase the end users (employees) and (sometimes) customers are interviewed for their needs and wants. The new process is engineered, and the specific features and functions of the new system are documented (the business requirements). During the Analysis phase, the business requirements are translated into functional requirements (the design of the back-end database, the screen fields, the basic report parameters). The functional requirements are then used to design and develop the system in the next two phases. Once the coding is completed, the system is implemented, and then thoroughly tested. When the tests are completed, the system "goes live" and is used by the end users. IT personnel then go into maintenance mode, creating reports and making tweaks as requested by the end user.

The advantage of SDLC was control; each phase was rigorously controlled and documented. Generally formal sign-offs were necessary at each phase. The business requirements identified in the analysis stage were written by system designers, strictly programmed and implemented, and then rigorously tested against the original requirements. The disadvantage of SDLC was that business requirements were often impossible to identify in the analysis stage, or changed before the system was complete. End users typically don't really know how to communicate their needs, and often aren't aware of the variability of costs of different functions and features of a system. Furthermore, the cost of the system was front-loaded; by the time programming occurred in the implementation phase over sixty percent of the cost of the system was already expended, making it difficult to change at that point even if the original business requirements turned out to be unreasonable or insufficient. Due to the focus on documentation and signoff, SDLC projects were very long, usually lasting years. Many systems following the waterfall methodology were implemented only to discover the environment and needs had completely changed, making the system obsolete before it had even been used.

### *Iterative Methods*

The term *iterative* applies to all of the other system development methods because instead of linearly moving forward in a waterfall fashion, the programmers would constantly circle around back to the business requirement originators and review the system-in-progress so that changes could be made. Each would do so in a different way. In general, all of them emphasize constant communication between the end users and the developers; reviewing system prototypes, gaining feedback, and revising the system through dozens of iterations before doing the final coding, implementation, and testing. Usually, this results in a system that more closely meets the actual needs of the end users.

- **Spiral/Prototyping** – This method, called either prototyping or spiral methodology, combines some of the features of SDLC and more agile or rapid application development methods. The business requirement originators and the developers are in constant contact which is maintained through the project. Each iterative cycle around the spiral would contain four stages:
  1) Identify objectives, alternatives, and constraints;
  2) Evaluate alternatives and resolve issues;
  3) Develop deliverables;
  4) Plan next stage.

  A *deliverable* is any object, media, or material that is expected to be delivered, such as a report or the completed code for a feature. Generally the first prototype is developed using a graphical system (without worrying about scalability, speed, or volume issues), a pseudo-system is developed and shown to those who are writing the business requirements (often the end users themselves instead of system designers). Several iterations of prototypes may be reviewed before the final prototype is translated into a scalable system implemented on a larger scale.
- **Rapid Application Development (RAD)/ Joint Application Development (JAD)** – with speed as the utmost issue, programmers work directly with business requirements writers to create a working system, generally utilizing object-oriented programming or programming tools such as Rational Rose. RAD focuses on workshops or focus groups (called JAD sessions when using JAD), using prototypes and re-usable code, informal communications, and keeping to a schedule that lists features for future phases.
- **DSDM (Dynamic Systems Development Model)** – DSDM is both a RAD and an agile programming method. Its nine principles are: Active user involvement. Empowered teams that the authority to can make decisions. A focus on frequent delivery of products. Using fitness for business purpose as the essential criterion for acceptance of deliverables. Iterative and incremental development to ensure convergence on an accurate business solution. Reversible changes during development. Requirements that is baselined at a high level. Integrated testing throughout the life cycle. Collaboration and cooperation between all stakeholders.
- **Rational Unified Process (RUP) Methodology** – A method that recognizes the need for changes throughout the project. Elements are integrated progressively (a little at a time). Supports reducing functionality in order to meet a timeline, brings in testers and documenters earlier than other methods, and is designed to enable multiple iterations of the system.
- **Agile Programming/Extreme Programming/Scrum** – Agile methods are also attentive to speed but are more focused on the team doing the development than the specific method. Agile programming focuses on the needs and wants of the customer, and insists on constant two-way communication in the process. Also imbedded in this methodology is the concept that software is never "finished", but rather constantly undergoing development, getting closer and closer to the needs and wants of the user. Generally agile programmers work in a bullpen, with open desks, alongside testers, end users (or end user representatives), interface designers, and technical writers. These methods are very face-to-face oriented, and don't provide a lot of documentation. In extreme programming, the team goes a step further, having two programmers work together on the same system, so that they can double check each other's work. The focus is really on simplicity, trying to eliminate complexity whenever possible. Scrum is the most recent form of agile programming, and its focus is on self-managed teams, enabling programmers to have the kind of self-momentum so that they, and their scrum master, can overcome any obstacle quickly.

Any businessperson who will be involved in decision making regarding system development would be wise to become familiar with the different methodologies. Programmers are often partial to one particular method, and often will not work on projects that utilize alternative methods of development.

## System Acquisition

System development methods are focused on developing, rather than purchasing, systems. While many businesses still employ their own system developers and programmers, the trend is heading away from in-house IT programming. Instead, businesses purchase already-developed applications from third party vendors. (They are called third party because they are not the hardware vendors, which is the first party, or the operating system publishers, which is the second party.)

System Acquisition shares some of the same stages of the waterfall system development methodology. (Because the system has already been designed and programmed, none of the other system development methods can be used, although some may be applied during the customization of the software). The phases are Planning/Feasibility, Requirements, Analysis, Evaluate, Customize/Train, Implement, Assess, and Maintain. They can be seen in Figure 92.

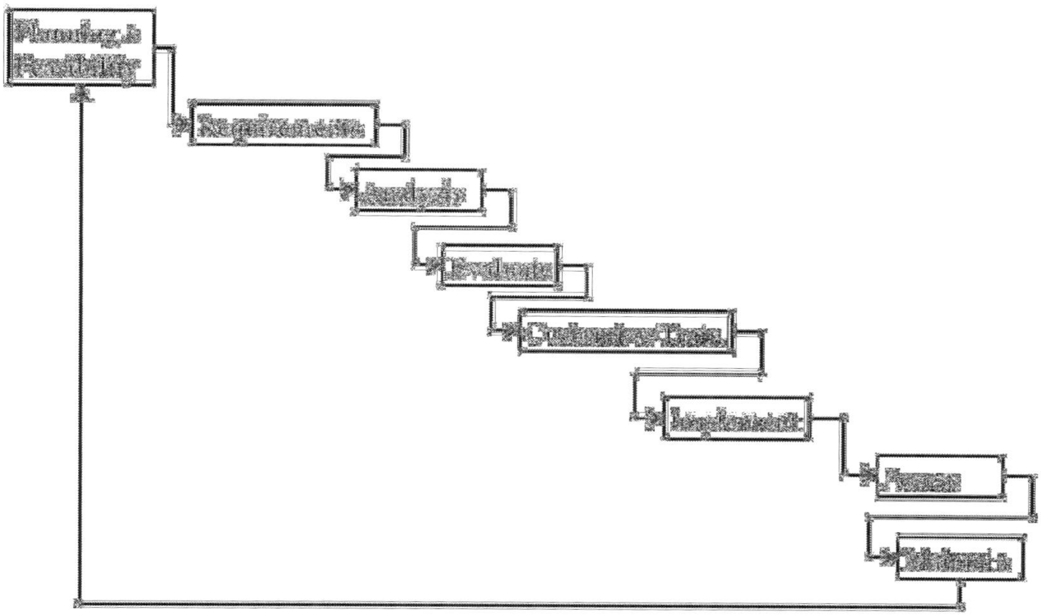

**Figure 92. System Acquisition Waterfall.**

Requirements remain at the business requirements phase, as there is no need to translate them to functional requirements. The Analysis is not an analysis of the coding needed, but rather an analysis of the business requirements and the IT environment. It is essential that the decision makers of the new system understand the hardware and operating systems (also known as the *platform*) that are in use within the organization. While there may be some necessity to bring in a new platform to meet the business requirements, it is not recommended. Each new platform that must be supported by the IT organization increases the costs to maintain the platform exponentially. Since each platform maintains a completely different upgrade, patch, and feature schedule as well as unique terminology and methods, an IT organization is usually required to have experts on-staff in each platform. Experts are expensive, so it makes sense to limit the number of platforms. (A *platform* is a combination of the client and the server hardware and operating system software.) In other words, if all of the current servers are manufactured by Dell and utilize the Windows Server 2016 operating system connected to clients' desktops with Windows 10 operating system, then it would make sense to look at applications that work in that environment. Bringing in an HP server requiring HPUX operating system that connects to Macintosh desktop clients would not be recommended. Not that there is anything wrong with the other platform, but if it doesn't match the existing environment it will increase the costs considerably. Of course, completely replacing the environment is a possibility, but that must also be investigated and the costs analyzed to see if the switch is worth it.

Once the requirements are well known, the applications that run within the existing platforms of the organization can be investigated and evaluated for how well they meet the state requirements. In order to complete this aspect of the process, it may be necessary to have some employees trained in some of the systems, and/or install them in a test lab in order to see whether or not they meet the requirements.

## Outsourcing

Rather than purchase a third party system, some companies opt to maintain their in-house developed systems (or have new systems developed), but they would rather not maintain programmers on staff. Hiring a company to do programming and maintenance is called *outsourcing*. Advocates of outsourcing claim that it improves financial planning, decreases costs, enables a company to focus on their core business, shortens implementation cycle time, reduces fixed costs, enhances security and provides high quality experts at decreased prices. Detractors of outsourcing claim that it often ends up with higher costs, lower quality, less competitive advantage and a loss of control, as well as the loss of dedicated and focused employees who understand the business better than outside programmers. There is also a cultural bias against outsourcing as it often entails moving tasks from high-paying jobs currently in the United States to low-paying jobs in other countries such as China, India, and Russia.

Despite the concerns, outsourcing continues to grow, and predictions are that cloud-based computing will only fuel the growth even more.

## Security Access and Controls

Some of the questions raised when speaking about cloud computing point out the importance of understanding all of the security and controls issues regarding digital data. Business people often don't think about these issues until they are audited, but it makes sense to consider security and control throughout the process of managing information systems. In addition to a financial audit each year, public companies are also required to undergo an information audit.

### *Physical Security*

Physical security is relatively easy to describe and see. A highly secure place is a bank vault. A locked safe is less secure than a bank vault, but more secure than simply a locked room or cabinet. In all cases, there is a combination code or a key (usually a physical key with a unique set of ridges) that opens the vault, safe, room, or cabinet so that the owner can retrieve whatever items they are safeguarding there. More sophisticated systems use biometrics (thumb prints, hand prints, iris scans or voice commands) to lock and unlock the space.

IT security begins with the physical; ensuring that important devices such as servers are secured behind locked doors, and media containing backups are secured in vaults, safes, locked rooms or locked cabinets.

Securing and controlling digital access is not quite so easy*. A device called a *firewall* is usually installed to prevent any unauthorized device from being able to connect to the network. Generally this is accomplished by checking the IP address of any incoming device against a list of authorized IP addresses. Unfortunately, knowledgeable digital intruders use *IP spoofing* to cause a device to look like it has the IP address of another device; one that would be authorized to connect to the network. The network itself is just the first level of access, however. At the next level is access to the computer server itself, and then access to the application on top of the server.

Making the application on the network secure from unauthorized people but still available to authorized people is usually accomplished through a username and password combination. In the early days of computing, people would often share usernames and passwords both because it was easier, and because the cost of the software increases with each names person using the software. But sharing usernames and passwords makes it that much easier for others to obtain them, which compromises the system. Most audit controls will require each individual person to have their own individual username and password.

Furthermore, strong passwords are becoming the norm. A strong password is longer than a weak password. A stronger password uses a combination of letters and numbers. An even stronger password also requires the use of a capital letter and/or a symbol. Furthermore, forcing users to change their password frequently also makes them more secure. Unfortunately, frequently changing passwords that require complex combinations are difficult to remember. Additionally, the proliferation of applications requires users to have usernames and passwords on each system, causing them to be forced to try and remember dozens of different username and password combinations. Users often write them down so they don't forget them, which then compromises their security.

The other problem with passwords is that they can be compromised in a number of ways. *Hacker* is the term used when someone accesses a system without the proper authorization. Sometimes hackers use *back doors*, which are undocumented ways of getting into a system without a username or password. Other times, hackers obtain a username and password by getting them from someone else.

*Social Engineering* is the term given to someone who manipulates people into giving them their username or password, usually by telling them a lie or tricking them into thinking they are helping someone. *Phishing* is similar, although generally done through email or a website. The email or website tricks someone into providing private information such as usernames, passwords, or financial information such as credit card data, often by presenting a screen that looks like a login form but that is actually fake.

For highly sensitive information, a combination of access methods may be used. In addition to the username and password, for example, the user may be required to insert a flash drive or special device (sometimes called a fob) which contains a constantly changing password or *token*. Alternatively, bio-metrics may be required as well as a password.

The IT department must also be concerned with the process by which usernames, passwords, and other security controls are assigned. It would hardly be helpful to set up an elaborate access control and then provide keys or set up users without a stringent check that the users are who they say they are and that they are authorized to access the system in which they are requesting access. Usually requests for usernames and passwords must be accompanied by specific documentation that proves the request is authorized.

* We did cover many of these items when discussing utilities software that deals with security, but it bears repeating here.

## Tracking

Another level of security is tracking what each user is doing on a system. For highly sensitive data, it is important to log (which means to record) any transactions or movements for each person for each screen. Although it may not be possible to catch a hacker at the moment the system is being hacked, it might be possible to identify what systems were compromised by reviewing logs of activities on the system. If the information within a system can be altered, the results can be devastating.

## Disaster and/or Business Continuity Planning

For someone responsible for business information systems, another responsibility usually taken on by the IT department is planning for what to do in the event of a disaster. This planning is usually called disaster recovery, or business recovery planning, or business continuity planning (BCP). BCP usually encompasses more than the systems and recovering data from backups. Since it is impossible to know what kind of disaster might occur, BCP provides general guidelines regarding who would be in charge, how people will communicate, and what the expectation should be. At its most basic level, BCP requires that a backup of the systems exist, and that it has been kept updated and stored off-site. BCP also usually considers backup to power sources, ensuring that there is an uninterruptible power source (UPS) available or generator that could power the equipment should electricity be cut off for some reason.

### Types of Disaster Recovery Sites

Disaster recovery sites can be categorized as cold, warm, and hot. A *hot site* is one that is up and running at all times, a complete copy of the working site including all the same hardware and software, with an up-to-the-minute backup of the data (synchronized) so that it could be switched on within seconds in the event of a disaster. A *warm site* may contain the appropriate hardware and software already installed, but the data is not kept up to date. Instead, there would be an expected lag time between the disaster and the alternative site coming up. Backups might be nightly, and therefore the system would be restored within a day or two of the original disaster.

A *cold site* is the least expensive, and generally doesn't even require that the equipment be purchased or the software installed. Often, a cold site is a plan, backed up by a monthly bill from a disaster recovery service, that the proper hardware and software would be made available in the event of a disaster. Usually a cold disaster recovery plan maintains the expectation that it would be a week or more before an alternative site would be up and running.

BCP and disaster recovery is very important to the future of any business, but senior business leaders often ignore it. It is expensive and time consuming to plan for a disaster that may never happen. The issue gained more focus after September 11, 2001 when terrorists attacked the World Trade Center in New York City. The World Trade Center housed the data centers for dozens of large financial companies that relied heavily on their IT. Though none of the companies considered the possibility of such a devastating loss, most BCP had adequate disaster recovery that they were not at a complete loss after the attack. But not all did, and some companies lost everything: their people, their servers, and their data.

The lesson is clear: all computers must have a backup, even if there is no large scale disaster. Hard drives crash. Systems lose files. Websites go down. All of these occur **frequently** even if you yourself never experienced it. Just because the system has not gone down in the last few years does not mean that it will not go down tomorrow. The question is not ***if*** the data gets lost, the question is ***when*** the data gets lost. Everyone, and every company, at some time or other, has lost information on a computer.

The only way to ensure digital information doesn't get completely lost is to have a backup that is personally controlled. There is no substitute for the peace of mind that comes with a full set of backup files stored on an external hard drive (preferably two sets - one on-site and the other off-site). In the old days, backups were done by copying the information to tapes or CDs, but external drives are so inexpensive now it makes no sense to spend time maintaining those old devices. An external hard drive or solid state drive, or an online backup service, is the most convenient and feasible. The most recent backup should be no older than the amount of information willing to be lost.

It is important to note that a cloud system is not a backup. If you delete a file on your cloud, it deletes on all connected devices automatically as soon as they connect. In order to make a backup of your cloud data, you must make a copy of the information on the local version of the files and put it somewhere else (i.e. in a different folder or drive).

#### *Types of Backup*

A backup can be *full, incremental,* or *differential.* A *full* backup is a complete backup of all files that can possibly be backed up. There are two kinds of full backups. One kind, an *image* backup, will back up the operating system files and installed programs so that the entire drive could be restored in the event of a crash. Typically creating an image backup is very time consuming, and the drive cannot be used while it is being created. Additionally, an image backup cannot be updated on a daily basis. (A large hard drive [anything over 500 gigabytes] may take more than 24 hour to create an image backup.) Usually most people create an image backup of their systems once a year or so. The other type of *full backup* simply backs up all the data files. This is the most common type of backup.

An *incremental* backup is a copy of all the files that have changed since the last time a backup was done. Incremental backups are cumulatively added to the full backup, so that if your system were to crash, you would first restore the full backup, and then apply each of the incremental backups (however many of them there are) in the order they were backed up.

A *differential* backup is similar to an incremental backup, except that instead of backing up just those files since the last time any backup was completed, the system backs up all the files that have changed since the last *full* backup was done. Because you are re-backing up all the files that have changed since the last full backup, you would only need to restore the full backup and then the *last* differential backup if the system were to crash. You would use a differential backup instead of an incremental backup if one or more of your files changed all the time (such as a database) which would make the incremental backups too large. A differential backup can overwrite the previous backup (which an incremental can't because it is only a backup *since* the last incremental).

Making sure all the data of all their employees is backed up frequently is the job of the IT department. Individuals, however, should always be sure that their own data is fully backed up at all times.

## Types of Business Information Systems

Information systems within a business can be classified in numerous ways, but the most popular is to separate out *Operations Support Systems* from *Management Support Systems.* There is a third classification, *Special Processing Systems* for those systems which could be classified as both Operations and Management or as neither one.

### *Operations Support*

Operations Support Systems include:

- **Transaction Processing Systems** - source systems that record an actual transaction such as a sale, a scheduled task, payroll records, etc.

- **Process Control Systems** - systems which monitor and control physical processes, such as the systems that control robots in a manufacturing plant, the system that monitors the temperature for the water heater, or the system that delivers the visual photograph of someone's heart during a coronary bypass operation.
- **Enterprise Collaboration Systems** - systems implemented by businesses to allow their employees, suppliers, and customers to work together better. This may include email, discussion boards, knowledge management systems, videoconferencing systems, and instant messaging capabilities.

Operations support systems are those systems used by all the employees, customers, and vendors of an organization. They contain primary source data: data that is entered directly through a scanner, a swipe, or a keyboard.

### *Management Support Systems*

Those in decision-making roles use Management Support Systems. They include:

- **Management Information Systems** - reports and compilations of the information stored in the operations support systems.
- **Decision Support Systems** - interactive systems that enable managers to get answers to questions and help them make decisions regarding the organization.
- **Executive Information Systems** - decision support systems that are tailored for those who are at the very top of the organizational chart, the senior leaders of an organization.

Management Support Systems are only used by a subset of the organization. They usually only contain secondary sources of data, that is, data which is compiled, aggregated, and sorted but which come primarily from the operational systems.

### *Special Processing Systems*

Often a business has a specialized type of system that could fall under either of the previous categories. These systems might be known as expert systems, knowledge management systems, strategic information systems, or functional business systems, and they are specialized for a particular business in some way.

Whether the system is an operations system, or a management information system, or a special processing system, the way data is processed will be one of two types: batch or real time.

## Types of IT Processing

We can think of IT Processing in two different ways; the processing of the transactions themselves, or the type of process used within the network.

### *Batch or OLTP*

In the old days, customer data was stored on tapes that had to be manually taken off the rack and mounted (or loaded) onto the machines. Every transaction was done in a *batch* – all the transactions of the day were gathered and processed or "run" at one time. When data storage started using hard drives instead of tapes, the programs did not change much, but instead of manually waiting for a tape to load, the program would automate the running of the batch from the information stored on the hard drive. Typically, the programmer would set the process of running the batch in the middle of the night when usage of the computer for entering transactions was lower.

There are many processes that are still batch processes. Banks, for example, reconcile checks and deposits in the middle of the night – which often means that you can't access the money you deposited today until it "clears" tomorrow. Payroll is always a batch process because you can't get paid for each minute of working as you are working; you must wait to turn in a timesheet or for two weeks to have passed.

These days, however, it is possible to have immediate, up-to-the-minute processing for many types of transactions. This is known as *OLTP (On Line Transaction Processing)* or *RTR (Real Time Reporting).* Airline tickets, for example, are posted the moment they are purchased. When you register for college classes using an online registration system, the moment you commit to a course, the seat for the course is reserved for you*. If you order a book from Amazon.com, the payment is posted immediately, and the online available inventory is modified immediately. If you book a concert ticket using an online program, the ticket is reserved for you the moment you enter your credit card information to pay for it. None of these OLTP transactions had to wait for a batch to finish processing.

## Telecommunications and Network Configuration Processes

In the early days, the entire network was completely owned and controlled by the company. All the connection points were *point to point* leased lines (usually leased from the phone company and configured to the specifications identified by the *telecommunications* expert in the company). *Telecommunications* was a highly complex skill requiring detailed understanding of the mainframe computers, the networking protocols (usually SNA, Systems Network Architecture, the only networking protocol used by IBM mainframe computers). The networking protocols were *circuits*†. When a leased data line establishes a circuit, two devices (usually a terminal and a computer, but also perhaps two computers directly) could send messages to each other back and forth in long streams of data of variable length. Only those two devices could communicate with each other (hence the *point to point* connection.)

Paying for a specific leased line between every single terminal and every computer within a business is an expensive proposition. Data line costs are based on distance and size. The further away the points and the more data transferred (the *data rate)*, the higher the cost for the line. A business might order a 128k data line which transfers 128 kilobits per second, (plenty for text-based terminals of the time), or, for really intense needs, a T1 line (which transfers 1.544 megabits per second). A business wanting to establish a T1 connection to a terminal or computer in Japan from the United States, for example, might spend $30,000 dollars a month ($360,000 a year) for a single connection. Telecommunications expenses took a big bite from the budget, and only the largest companies could afford far-flung offices.

When packet-based communication was developed in the next decade, it changed the telecommunication requirements and lowered the data line cost drastically. Instead of a 100% fully dedicated leased line between any two devices, a business could have multiple devices *share* the leased line, giving rise to the *frame relay.* Telecommunications became even more important, because divvying up the purchased leased data line into *virtual circuits* and establishing the frame relay grid was a complex skill. While the per-line costs decreased, the reliance on telecommunications in general increased, so the number of lines kept increasing. The IT budget kept getting bigger and bigger. As computers became less text and more graphics, T1 lines were no longer enough; T3 lines (the equivalent 28 T1 lines, or 44.7 Megabits per second) and more became the norm.

* And for those of you who are college students now, you have no idea how good you have it regarding registration. Thirty-five years ago, a student had to stand in line (sometimes for hours) in order to pick up a computer punch card for the class they wanted – *for each individual class.* If you got to the front of the line and the class you wanted was full, you had figure out what course you could take, and then wait in another line to pick up the punch card for that one. Registration was an all-day event.

† More technical details on the difference between circuit based communications and packet-based communications can be found in Packet Switching vs. Circuits in Chapter 2: Networking.

Packet-based communication was also the foundation for networks to communicate with other networks, internetworking. This (as you know, of course) gave rise to the Internet, the single world-wide network that is shared by everyone.

The impact on business was astounding. In addition to being able to utilize the *World Wide Web (*which was the system of *hyperlinks* that ran on top of the Internet making it very easy to use), a business could also connect to any other computer anywhere in the world **without paying for a leased line.** This economic issue was a major influence in sparking the revolution in business caused by the Internet. Furthermore, geographical region limitations disappeared, and businesses could sell products and services anywhere in the world without the previously high costs associated with expansion.

All of these changes impacted business strategy. Companies who understood how to take advantage of new information technologies gained a huge competitive advantage. However, there were obstacles as well. Most laws had been forged prior to the introduction of information technology, which caused some issues regarding privacy and control of intellectual property (many of which are still being wrestled today).

## The Process of Dealing with IT Vendors

Why is purchasing technology so difficult? Good question. It is very difficult to establish good relationships with information technology vendors for several reasons:

- Immaturity.
- Obfuscation.
- Speed of change.

Let's look at each of these in turn.

### *Immaturity*

The patent for ENIAC, the first real computer, was filed on June 26, 1947. Since then, we have seen technology swell into a world-changing revolution over the past 50 years. With the exception of science fiction readers and writers, the world of today would be unrecognizable to the typical adult of that period. Fifty years, however, is not a very long time. As a result, many of the difficulties we have with purchasing new technology arise out of the immaturity of the entire field.

There are several issues that are influenced by the immaturity of the field. IT is a relatively new domain of knowledge without the background of other business domains. Being fairly new to the world makes IT difficult to understand. The typical leaders in today's corporations were more likely to grow up racing in soap box* derbies than texting with their friends. And if you ask a typical teen today what a soap box derby is, he will look at you as if you were losing your marbles.

Most people who rise to the position of decision maker in a company have learned from school or from experience quite a bit about several different business domains: sales, financials, customer service, logistics, and operations. There are thousands of years of history upon which to rely for most business decisions, and hundreds of business schools willing to teach them. The fields are well documented, and the terms are well-defined. Business leaders either learned the optimum way of doing things within various areas themselves, or they found people who could be trusted to

* If you are too young to know what a soap box derby is, I'll explain. In the depths of the depression, soap and fruit would be delivered in large crates to local stores, who would then allow local children to add wheels to those crates and race them in downhill streets that parents would block off from traffic. By the time most baby boomers were children, the soap boxes had graduated to kits and sophisticated systems that enabled children to travel via gravity alone up to 30 miles per hour, and soap box derbies became a common pastime for children all over the country.

provide them with the right information to make the right decisions within those fields, or they educated themselves from the easily accessible sources of knowledge on those domains.

Information technology, however, is something new. There is not much of a track record, and no history to speak of. Documentation leaves much to be desired. The egos of supposedly self-assured decision makers prevent them from asking too many questions or revealing their own ignorance—especially when talking with vendors or consultants, or even their own technology employees.

Business leaders probably perceive that vendors and consultants treat them to arrogant chatter in an impenetrable language. Alternatively, vendors might stick to vague business concepts that don't convey the critical information necessary for optimum decision making and end up making the decision themselves. Either way, the business leaders lose.

Another problem is that the infrastructure is not yet in place for smooth technology communications. This is not without precedence. In the early days of the automobile, people had to pave their own roads to connect with the public streets. Even the public streets weren't all that great. Stops for animal crossings were frequent, as were bumpy dirt-covered roads. There were no end-to-end highways ensuring a smooth fast ride from one state to the next.

Contrary to popular belief, the public Internet is still just barely out of its infancy. It's beyond the covered-wagon-trail stage, but still full of back roads covered in dirt. It will become a superhighway within the next decade. There are many toll roads, and no agreement on how big the roads should be or how much to charge for them. Incompatible routes, frequent outages, and no guarantees are the rules of the day.

Today's Internet consists of more private networks than public interstate highways. Currently, we provide our own road (a private Internet) up to a certain point, and then we get on the open street (the public Internet). A typical problem is a big truck hogging up both sides of a little side road, preventing anyone else from passing (someone in your neighborhood hogging the bandwidth to download a movie or video). Another problem is when larger organizations try to speed up their own traffic by mixing private Internets with the public Internet. Private Internets are like railroads in the middle of a city. While they speed up the traffic for those in the train; they slow down all the traffic on the public roads around them. Like the railroad, the crossings between the public and private will eventually disappear until there are only a few left in small towns. At that point we will be able to count on a paved, highly stable infrastructure. Until then, using the public Internet will be a bit perilous.

Another major problem is that IT components aren't purchased already assembled or standardized. In the early days of automobile purchasing, buyers put up with an extremely difficult purchasing process. Many of them built their own cars and had their own mechanics on staff. Just because you could drive one car didn't mean you could drive another since, at the time, there were no industry standards in place.

How easy would it be today to purchase a car if you had to go to one place for the chassis, another for the tires, another for the engine, and another for the steering wheel? Imagine learning to drive if one car had a round steering wheel, while another vendor used a stick to steer.

The automobile industry matured, however, so that now you go to one vendor for all the parts of the car—already assembled and tested together. You can find mechanics in many places and you only need them a few times a year, so you needn't hire one full time. You know that the steering wheel will be round, and that the brake is a pedal you push on the left while the gas is a pedal you push on the right. You can count on the fact that the car will work as it was designed to work on the interstate highways that stretch out waiting to take us where you want to go.

When it comes to technology, we've actually devolved. Strangely enough, in the early days of computing, purchasing a system from a single vendor was the norm. But the breakup of monopolies like IBM and the popularity of Microsoft operating systems and personal computing software put an end to that approach. Think about how we purchase IT these days: We buy the hardware from one vendor, the operating system from another, and the application software from yet another vendor. Then we work with different vendors to configure the network: one for Internet service, one for the routers, one for the switches, one for the wiring in our building, and a different one for the cards that connect functions within our computers. We fully expect that certain parts will not work as designed, and will need to be replaced several times before they are correct. We hire programmers and developers to create or customize the software so that they work with our processes, which were designed by a totally different vendor or consultant. And there are only tiny, crowded public connections between our system and the outside world, so we still end up building our own highways in some cases. And that requires a whole host of other vendors.

The immaturity of the entire industry means constant shakeouts. Companies that once dominated the industry, like DEC or Commodore or Gateway are dead and companies that were started out of a college dorm twenty five years ago (like Dell) or after a fun dorm party just fifteen years ago like YouTube have inhaled market share with each breath. Thousands of IT companies have started and ended in a very short period of time.

***Matter of Timing***

Even a lowly school teacher without any formal training who had just a smidgen of talent with computers and no sales or business experience at all was able to build a sizeable technology consulting firm because businesses were willing to pay big bucks to gain the productivity technology promised. Even more amazing, within a short time that same lowly teacher (me) sold the whole business to another technology company for big bucks, who then sold it to another for bigger bucks, who then sold it to a public company for really big bucks, who shortly thereafter went out of business due to the dot-com crash.

When acquisitions and mergers happen in more mature industries, the customer isn't impacted much. You can go to any grocery store today and buy the same Cashmere Boutique soap and the same Palmolive dish detergent that you could in 1928 before the merger of Colgate and Palmolive. But when mergers happen in a technology industry, the impact on customers is substantial.

A Fortune 500 construction conglomerate, one of my clients, purchased a highly touted computer system produced by a company called Data General in 1999. All the industry magazines at the time published glowing reports on how wonderful this new technology was. And, indeed, it had far surpassed any other system I'd seen in terms of reliability, speed, and flexibility.

But a few months after my client bought the Data General system, another company, EMC, a fast-growing vendor of storage devices, purchased Data General because of a new storage device it had developed. Within a very short time, all other vendors stopped supporting Data General hardware. It was very difficult to find technicians who understood it. New programs wouldn't run on it. Parts could not be found for it. My client was forced to purchase an entirely new system within three years.

The problem is not limited to hardware. A few years later, this company undertook a multiyear study to figure out the exact right financial system to purchase. After identifying all the requirements and evaluating all the systems, it finally chose one that met all its needs: the JD Edwards Real One system. Negotiations ensued, prices were established, and pilots were successfully run. Within a few weeks after the company signed on the dotted line, however, PeopleSoft purchased JD Edwards. Within a year, Oracle successfully took over PeopleSoft. JD Edwards Financial software is at risk of being orphaned, which means no one would be updating the code after a certain date. If the system is orphaned, the company would probably be forced to change again in a few years. Because technology requires a continual level of support from the vendor, mergers and acquisitions almost always bode harshly for the customer.

Even in stable technology corporations that hire hardworking, honest people, there is a risk. No IT vendor can be trusted to be around for the long term. One difficulty is holding on to really brilliant

IT gurus. This problem plagues IT vendors of all sizes. Customers just start getting comfortable with their vendors and, poof, the core guru of the service decides to find a better pond.

Also, because good technologists are hard to find, they tend to leave employment when they don't feel fully appreciated, which their lack of people skills causes to happen a lot. Just because a vendor project starts with superior technologists doesn't mean they will be there at the end.

Stock bubbles don't help. The irrationally exuberant dot-com feast of the 1990s had technology-related companies shooting for the stars, and sometimes getting there. Kids straight out of high school were able to get $90,000 a year programming for the Web.

***Personal Experience***

Think I'm exaggerating? When I was the CIO for a short time at Millstar, an e-commerce software development company, $90,000 was the **standard starting salary** for Web developers who knew how to program using a middleware program called **Cold Fusion.** Few had graduated from college, and all just happened to hit the job market with the right skills at the right time to be vaulted to guru status immediately. Venture capital was plentiful and everyone got jolly playing with the Web toys. New companies were popping up left and right. No one worried about profits or business process. It was a heady time – for a few years.

And then reality set in.

Information technology vendors did not do well after the dot com fever crashed and the Internet bubble burst. The sticky air of desperation still hovered over technology salespeople—those that survived—for years and years after the bubble burst. They jumped at every opportunity, and drooled over every marketing lead. The legions of unemployed programmers, project managers, and systems analysts finally thinned as people left the field and joined so-called normal industries. Salaries came back down to earth. But only until the next round of bubble, burst, recession, and growth.

## Obfuscation

Obfuscation, or the art of appearing to state something clearly while actually hiding the meaning, leads to several issues. For one, IT is fraught with hype. This is caused by the fact that IT seems to have amazing capabilities, but it also has limitations that make no sense to most people. Hype is fueled by these unrealistic expectations, and obfuscation can be used to write service level agreements that seem to meet expectations but actually don't.

***Hype hurts everyone***

How does hype hurt? I once worked for a CEO who loved to read industry magazines. However, he had no technology background. Every Monday morning he would come into my office and share some gem he found over the weekend—a carefully highlighted article or advertisement on some hardware or software. The truth is, he rarely even understood what the software or hardware product or service was, much less whether it could help us. The CEO was a trusting soul. He would read all those wonderful, glowing phrases and thought they really meant something. I finally stopped trying to educate him on the issue and just said "thanks" and filed the ad in the round filing system.

I wasn't alone. At a Society for Information Management meeting I attended, a CIO humorously proposed a motion that we lobby Congress to ban technology ad–filled magazines on airplanes for just this reason.

To demonstrate how easy it is to make a product sound like a winner, here are some actual ads from magazines (with the product names disguised). In the following examples of advertisements that use hype, see if you can tell what the products and services really do.

- What if you had someone to watch your back? Someone to integrate your back-end logistics with deadly efficiency. Someone to manage your warehousing, delivery and financial settlements. Someone to protect your entire e-commerce operation. What if one company could handle your entire back-end operation? That's the idea behind Product A. Built around an open and extensible transaction network, our integrated approach enables you to outsource any or all of the logistics behind your buy button. A partner to plan it, implement it, and stand by it. In short, someone to make sure it pays off.

- Customer Care is a Five Way Street . . . A bundled "point solution" that handles one or two communication channels is no longer the answer. Product B, a complete and integrated Internet Relationship Management system, can arm your business with self-service and full-service customer care solutions over any communication channel.
- Introducing Product C. You enjoy suspense in a good mystery novel, not in a working relationship! To successfully manage all the complexities and risks involved in any business you need collaboration. With someone who brings all the experience, tools, and understanding to bear, in order to help you anticipate and deal with any unforeseen problem. Someone who will share with you knowledge, practices, risks. And results. Discover the Collaborative Business Experience, discover Product C, a partner you can count on day after day. A partner who is committed to helping you achieve faster, better, and more sustainable results. And puts it on paper.

If you have a good technology map, you can identify these products and services from the nuggets of fact among the mountains of hype in these ads. But to most people, these ads might sound like they solve all your problems. They make it seem as if technology is easy, and that if you just make the right choice, your problems will go away.

Vendors do not hype because they are bad people. Vendors hype because the market forces them to do so. If Vendor A tells a buyer something can't be done, and the buyer goes to Vendor B who tells the buyer that it can be done even though in the end it can't, the buyer is the loser, not Vendor B. If a buyer is looking for a particular feature that is actually useless, and Vendor A tells the buyer the truth and Vendor B tells the buyer about a product with that feature and sells it, the buyer loses again. Furthermore, because business leaders' egos will not allow them to admit they don't understand what a product or service does, it is easy for the vendors to use meaningless phrases as if the buyers understand, when in reality they don't.

Hype is also the source of *vaporware*—a term used to describe software with all the features we want and need at an affordable price advertised, sometimes heavily, by a technology vendor. The only problem is that the software doesn't exist yet. Someone has put together the requirements and specifications, but the programming to actually make it happen hasn't been completed yet, and may never be. Technology vendors such as Microsoft frequently use vaporware to scare off rivals from developing competing products.

Years ago the term *marketecture* began to be used to describe marketing methods for nebulous products like .NET (pronounced "dot-net") or SOA (for Service Oriented Architecture) . Not quite content to simply advertise software products that don't exist, IT vendors moved on to describe an entire computer networking architecture that didn't yet exist. In descriptions these new architectures sound wonderful: secure, easy, convenient, cheap, connecting to the world seamlessly through the desktop. But in real life, they rarely live up to the promise. As of this writing, .NET is completely dead – having barely gotten off the ground. And SOA never became a real product – just a description for good IT business practices. Descriptions of remarkably integrated systems abound, whereas actual examples of integration are few and far between.

I've always thought that the technology industry was **uniquely** filled with hype. In 2007 I carried out a research study to confirm this fact, and instead discovered that there is nothing unique about IT products and services. All advertising is filled with hype. But we don't have as much of a problem with other hyped products. I finally decided that the difference must be that everyone recognizes, without thinking, the hype surrounding normal products. Do we really believe that if a young man wears a certain aftershave the girls will have trouble keeping their hands off him? Do we really think that if we start using this brand of color copier, our customers will flock to us because of our wonderful color brochures, no matter what is inside? Of course not. And yet, when a vendor tells us that this piece of network management software will "allow us to relax on the beach while the system heals itself" we believe it. When a technology vendor tells us that installing this CRM program will "increase our sales," we believe it. Because IT is not as well understood, it is more difficult to separate actual capability from hype.

One of the major problems is that IT seems to have some amazing capabilities. When you think about what computer systems have accomplished, it is sometimes hard to believe. How fast can they compute formulas? Billions of computations per second! Huge complex formulas—in nanoseconds. They work 24 hours a day, seven days a week, 365 days a year. Tirelessly. They remember word for word trillions of documents and can recall a single requested phrase within seconds. These are things even genius human beings have a hard time doing. Information technology systems can do all of these things quickly and easily. They have achieved amazing results.

Given the unbelievable capabilities of computer systems, it is sometimes difficult to understand how and why they can't do certain things. In the late 1980s, expert systems, artificial intelligence, and natural language processing were all the rage. Scientists and engineers were convinced that within a very short time frame we would have computers that could understand simple spoken commands just as they do on *Star Trek*. This misconception is now laughable.

People who have worked with computers for a while understand intuitively what computers can, and cannot, do. They understand that computers can follow simple step-by-step logical instructions. But they cannot handle directions that are incomplete or ambiguous in any way.

Which is why they can't handle natural language yet, not even in the written form. A two year old has better language capability than the most sophisticated computer in the world.

***Difficulties with Human Knowledge***

For over fifty years, scientists have been trying to get computers to be as intelligent at humans. Television and movies often portray computers as thinking and talking, providing analysis and the "right answer". This was a common scene in science fiction even as early as the original Star Trek where James T. Kirk would ask the computer for its analysis of complex sets of data, which it dutifully reported verbally in a few seconds of blinking lights and beeps and boops. There are several problems with this scenario, making it most unlikely ever to be real (at least not in our lifetime).

Computers can follow simple step-by-step logical instructions. But they cannot handle directions that are incomplete or ambiguous in any way. Computer don't "get" humans. Take something as simple as happy or sad. Despite millions of dollars spent on studying what seemed to be a very simple problem, computers still cannot recognize human emotions from subtle body language. Humans, even children, do this all the time without even thinking about it.

***Instructional Process Model***

Before being convinced by my dissertation committee that I wanted to build an expert system (because at the time, that was where the government money was), I had worked for nearly two years on a different topic; identifying the characteristics of a good teacher so that those characteristics could be reproduced in a computer aided learning system. After all that time, hundreds of articles and thousands of hours thinking about it, I finally came to the conclusion (which, in truth, hit me like a ton of bricks in 1988 as I was walking down my stairwell one day) that no matter how much information I got or how hard I tried, a computer couldn't come close to mimicking a good teacher. The reason was that good teachers utilized the *look on the face* of the student to determine what topic to teach next. If the student looked confused, the good teacher would step back and teach a component topic of the current topic. If the student clearly understood the lesson, the good teacher would move forward to the next lesson, perhaps even skipping a few if it was obvious the students already understood it [again, simply from the look on the face of the student and the speed of understanding of previous lessons].

If a computer can't discern the emotions or state of mind of the student by the look on their face, there was no way that a computer-aided-lesson would properly match the level of understanding of the student. That's why computer tutorials are usually useless; they either go way too slow for the students who already understands most of the topic, or way too fast for the student who doesn't. In the final analysis, good teachers use a *characteristic model* of the student while the computer aided instruction could only use a *buggy model* (i.e., wait for the student to make a mistake and then re-present the missing concept again even though repeating the same lesson over and over does not enable the lesson to be learned.) I came to this conclusion over thirty years ago, and I'm still waiting for a computer that can interpret the emotions behind the facial expression. It still doesn't exist.

***Natural Language Processing***

Computers still can't quite handle natural language, though SIRI and Alexa and all the natural language processors that SIRI spawned make a really good effort. But the truth is that a four year old has better language capability in a larger variety of contexts than the most sophisticated computer in the world – still, today, in 2021[*].

***Identifying the Context of Business Data***

Business data must have context. Take the following inventory list:

- 4 tires
- 2 windshield wiper blades
- 5 quarts of oil
- 1 oil filter
- 4 petunias
- 6 spark plugs
- 2 headlight lamps

Anyone reading will immediately recognize six of the items as normal automobile maintenance supplies. A 10-year-old child could probably tell you the fifth item doesn't belong. But how would a computer know? It wouldn't. There is no way to program a computer so that it would immediately recognize that flowers don't belong with automobile accessories without explicitly telling it so.

[*] Faller, A. (2020, Feb. 7, 2020). "What is Natural Language Processing and Why is it Hard?" Retrieved March 28, 2021, 2021, from https://www.colorado.edu/earthlab/2020/02/07/what-natural-language-processing-and-why-it-hard.

No computer, no matter how powerful, can identify the *relationship* between two items meaningfully without being explicitly told. Computers can store billions of facts and figures—but not in context. Computers can sort and organize information, but they cannot turn that information into knowledge. Computers cannot analyze. Computers cannot make judgment calls. If you cannot clearly and unambiguously describe the rules that are accurate 100 percent of the time, the computer cannot follow the rules. That means they cannot follow common rules of thumb. There still are no computers that can predict the weather as well as a human being can (though they have gotten a lot better in recent years and can now confirm that it is raining this minute if it is indeed raining this minute. Weather predicting supercomputers are independently accurate for about a six hour window. Beyond that, they need humans to clarify the models).

The volume and quality of relationships surrounding nouns and verbs that can be stored by the human brain is exorbitant. The brain is much more capable in this regard than a computer. A human brain has over 100 billion neurons. If you include the glial cells that support and nurture each neuron, you have millions times that. In comparison, there are only a little more than 50 billion computers on the public Internet. Even if you factor in each integrated circuit chip, there is still more processing power in the brain of a six-month-old baby than there is in all the computers of the world combined.

***Human Context needed***

Why is this so hard for a computer? Well, the easiest way for me to explain is to share the sentence provided by my artificial intelligence professor at Lehigh University when I was working on my doctorate. Dr. Hill asked us to consider the following sentence:

> *The boy hit the other boy with the girl with the blond hair with the hammer.*

Picture it. What vision does this sentence conjure in your mind?

Logically this sentence makes absolutely no sense. It says that one boy picked up the girl and hit the other boy with her. We know that's not what happened and we know why.

But to a human being, this sentence makes perfect sense. But how do we *know!?* How do you know the hammer is in the hand of the first boy and was the instrument used to hit the other boy? How do you know why the boy hit the boy? How do you know the blond hair is on the girl and not the boy?

We know that the boy was jealous of the other boy because he was with the girl with blond hair. We know this because we read the sentence in the context of the human condition where girls with blond hair and boys might mix with jealousy and end up with violence perpetrated by hammers. There is no computer in the world that could figure out the meaning of the sentence without the human context.

**Figure 93. The boy hit the boy with the girl with the hammer.**

***Dirty Data and GIGO***

One of the tasks that computers simply cannot do is clean up dirty data. This task has got to be one of the most difficult, laborious, and thankless jobs in all the world. (Well, okay, maybe cleaning up latrines beats it, but not by much.) One would think that given its importance, all data stored by computers by now would be clean and pristine—but they are not. Most of the information stored is contaminated in some way: duplicate records, inaccurate entries, incomplete information, and so on. Dirty data is the scourge of every computer system: Garbage In, Garbage Out.

Though still a problem, we have gotten better at cleaning up dirty data because we started recognizing how important it was to put the data in exactly the same way (i.e. using picklists instead of having people type information in). Plus the volume of data tends to overcome the difficulty of dirty data when analyzing and aggregating. If you have 1000 records to analyze, having 200 of them be duplicates or incorrect would nullify any conclusions you might draw. But if you have 100,000,000,000 records, the fact that 20% of them are wrong means that you still have 80,000,000,000 valid records upon which to base your conclusions. Going back to the dirty inventory; one can still predict purchase trends even if the granularity is not consistent.

Or not. Which is one of the mucky little secrets of information technology data; despite the volume, it is still possible to come up with the exact wrong answer. If your goal was to see whether the red candles or the blue candles were going to sell better next year (based on existing sales), and the red candles might be under the names magenta, scarlet and maroon while the blue candles are all called blue, you might come to the conclusion that blue will sell better because it has a higher number but that's only because the red candles were split among different colors. That's why it is up to human beings to develop the *art* (and yes, it is an art, not a science), of interpreting the meaning and context behind the myriad of data in order to discern real patterns and form generalizable conclusions.

Since computers cannot clean up dirty data, only a human being can do it. Expecting the computer to do so is unrealistic, although a common misconception. Often decision makers think that a new technology implementation will magically solve the problems caused by dirty data. I have seen highly intelligent leaders choose, purchase, and implement a new system—while at the same time claiming that there isn't time to clean up the data from the old system. Somehow they think the computer itself will do the cleanup. Months later they are shocked (shocked!) to find out that the new system wasn't any better than the old system, since the problem wasn't the system, but the dirty data all along! What they obviously didn't realize is that they could have had all the benefits touted by the vendor simply by cleaning up their data and saving themselves a bundle by continuing to use the perfectly adequate existing system.

***Unrealistic Expectations***

Whole businesses have been built on unrealistic expectations. I once had a client, Jan, who tried to get me to identify a system that would allow her to provide transcription services to doctors and lawyers without using people typing into word processors. She had seen a demonstration at a conference of Dragon's Naturally Speaking system. She actually saw people who talked into a computer while the system typed their words into the document. She figured she could set up a telephone system and have doctors and lawyers call up on the phone and dictate into it. On the other end would be a computer that would transcribe their words. What is so scary is that she'd already gone to a bank and gotten funding for this new business, and so she was just looking to choose the hardware and software to do it! That was twenty years ago, and still, twenty years later, there is no system that would allow doctors and lawyers to speak into them and have their words flawlessly transcribed.

The devil is in the details. Yes, voice recognition really works—in a very limited way. A system can recognize a myriad of voices but only a few select words (such as numbers or a limited list of known phrases). Or a system can recognize many different words, but only if a single person spends dozens of hours training it specifically to his or her voice. Additionally, the system is only about 80 percent effective on a good day—and having a human being going back and cleaning up the 20 percent that is wrong takes more time than if you had simply hired someone to do the typing to begin with. Truth be told – sometimes businesses today are wasting money on voice recognition systems to replace operators and receptionists because they think it is cheaper. But the truth is, often people are cheaper and more accurate – not to mention much more customer friendly!

Vendors often take advantage of the tendency toward unrealistic expectations. Why else would enterprise network management systems sell at all? Buyers of these six-or-seven-figure systems expect the system to warn them before a network device fails. Occasionally, when the situation is just right and people who know the network very well have tuned the configuration for months and months, the system will provide the right people with the right warning at the right time to avoid a problem before it happens. More often, the system will let them know that the network has gone down after the fact (big whup!). But if the buyer asks the vendor, "Will this system warn me before the network goes down?" The vendor will answer, truthfully, "yes."

Let's look at an analogy. We want to get to Africa, but we're a bit ignorant about exactly where that is. We go to a dealership selling Mercedes Benz vehicles. We ask the dealer, "Can this car get me to Africa?" The dealer says "Yes! Absolutely!" Obviously, a Mercedes can get us to Africa as well as any other car can get us to Africa. We buy the car and drive toward Africa. But we have to stop suddenly when we run into the Atlantic Ocean.

The Mercedes Benz dealer hasn't lied; the dealer just took advantage of the fact that when we heard "yes," we assumed that the Mercedes was *all we needed* to get to Africa. Because the dealer knows that no car can get us to Africa, the dealer is not worried we will buy from someone else. Besides, we still need a car to get to the airport or shipping port, don't we? Is it the dealer's fault we thought a car could get us all the way to Africa?

***The Story of E-CommerceLinks.net and Unrealistic Expectations***

I have been a victim of unrealistic expectations quite a few times—even after I thought I knew better. In the year 2001, I was deciding whether or not to join a new start-up. The high-powered references of the CEO checked out. But when he told me that he wanted to develop and sell live video customer service across the public Internet, I initially told him flat out he was nuts. The infrastructure did not yet exist, I said. High-quality live video could not be done over the public Internet, I said. I explained that the public Internet was not stable enough and did not have enough bandwidth for the kind of traffic video generated. It would be like trying to take a train through the intertwining streets of your neighborhood.

The CEO insisted that it could be done and was already being done. He flew me to the West Coast to meet with people from the telecommunications company, and they demonstrated high-quality video over the Internet (they said). Perhaps the technology had improved more than I had realized, I thought. After all, I'd been a bank VP for seven years and out of the networking game. When the telecommunications company promised to provide circuits to any of our customers for $300 a month, no matter where they were in the country, I crunched the numbers and I realized that at that price, the CEO was right—the service could be cost effective. I joined up and became CTO of Commercelinks.net. I invested quite a lot of time and money on the venture, led the team to develop the top-end live video customer technology that would utilize the Internet protocols, and applied for the preliminary patents. We were all ready to go.

Our unrealistic expectations came falling down on us just as the same thing was happening to everyone else involved in dot-coms. The telecommunications company reneged on the $300 per month connection promise. During my investigation into why, I got one of the technologists to admit that the demonstration they showed us was sort of rigged. They had not demonstrated the technology over the public Internet, but rather over a private Internet (their own private network using Internet protocols, something that's easy for a telephone company to do since they own the public lines). The cost to provide the kind of quality we required far outstripped the $300 per month quote unless every customer we worked with was already a customer of that particular telecommunications company.

There are plenty of technologies that will work fine over a private Internet, but not work well at all over the public Internet. On a private Internet, the network traffic pattern can be controlled from end to end. On the public Internet, no one controls the traffic, jams are frequent, and there are no guarantees. The difference is like traveling to China via a charter plane or taking an unreliable, slow boat (not even a cruise liner) with 400 other people squeezed into tiny cubicles all around you. With private lines (i.e., the charter jet) live video customer service is a viable alternative. But at $30,000 a month, it is not feasible.

Failing to differentiate between the public Internet and a private Internet is still a common misunderstanding among technology businesses everywhere today.

Unrealistic expectations can really do damage when naïve buyers presume information technology vendors and consultants are giving them unbiased information. In the early 1990s, in my role as a consultant for the company that bought my first business, I was often paid to do a study to identify the best option for systems management or network management for customers. It was always understood by the customer that my time would be spent identifying which of the many options **offered by my company** would best meet their needs. Referring them to a competitor's products or services was not an option. Another choice that wasn't on the list of options was "None of the above". There would always be a technological solution to their request because we were in the business of selling technology. Although I always made this clear to my clients, I am aware that many consultants do not.

Let's again look to the analogy. Upon arriving at the lot of the Mercedes-Benz dealer, would you expect the salesperson to answer that the Cadillac dealer down the street was a better choice when asked "Which car is the best for me?" Of course not! However, in the immature technology world, it may take weeks of interviewing and learning current processes in order to determine which service or system would work for the business—and someone has to pay for that time. Vendors cannot afford to spend time to make recommendations without charging for them—but that doesn't mean that they are looking at all options—usually just the ones that will make them money.

It is possible, but rare, to find unbiased consultants whose income doesn't depend upon vendors' projects. However, even independent consultants who evaluate several systems usually get a referral fee from the vendor of the winning bid. Even third-party consultants like Gartner Group indirectly make money on technology solutions. There are very few unbiased reliable sources to whom a decision maker can turn when they want to know whether or not to do something.

The final place where obfuscation raises its ugly head is a popular solution touted by many consultants and vendors: the service level agreement (SLA). Putting in place an SLA was supposed to solve typical problems between buyers and sellers of technology. The idea sounds convincing. An SLA puts in writing exactly what the seller is going to do and what the consequences are if the promises aren't delivered. Huge outsourcing contracts where big technology vendors took over all the technology needs of Big Business use SLAs.

Here's the problem: Because technology is immature, both the vendors and the buyers are in the same bind. Neither can understand the challenges they face as they try to implement new technology. IT projects are not like construction projects. The construction industry has thousands of years of history to let them know what works and what doesn't. IT projects are more like promises to build a bridge for a price without knowing how wide the river is or whether the banks are sand or rock. No one can estimate the time involved, or the costs, or even the long-term risks in most IT projects, although we keep pretending we can.

An example: One of my clients had a five-year contract with a company to manage the company-wide help desk for 8,000 employees. The agreement was several hundred pages long, with financial penalties for every month that the quality standards were not met. The quality standards? That every help desk call had a response within 24 hours, and that every new PC was deployed within three days of delivery. The agreement also required that 10 help desk people be available round the clock, 24 hours a day. Sounds reasonable, doesn't it? Just what we want, right?

In this case, the help desk vendor usually met a 10-minute response time instead of a 24-hour response time. Every time people called the help desk, they received an immediate response saying that someone was looking into the problem. The vendor also deployed every new PC within two days of delivery instead of three, and they averaged 12 help desk people every day. The client should love them, right?

Wrong. Within months, employees stopped calling the help desk because, despite the immediate e-mail response, they were rarely actually helped. The tech support people staffing the phones seemed to be high school dropouts who didn't know very much about the systems implemented at the customer's site. And that quick response time for new PCs was really helpful—until the employee tried to actually use the PC and discovered that the mouse drivers had never been installed and the wrong printer driver had been. It usually took four or five calls to the help desk to actually fix all the problems so that the PC was usable. I was called in to help document the hundreds of costly problems so that the lawyers had a case to nullify the contract without paying the huge penalty for early termination.

Armed with my documentation, my client finally cancelled the agreement after two years and said good riddance to the vendor. Unfortunately, they didn't learn the lesson. Instead of taking my advice, they chose to replace the vendor with another one, getting more lawyers involved in negotiating yet another multi-year deal with another SLA. They wouldn't admit that the attempt to codify technical support was a losers' game.

The nature of IT people can also contribute to the failure of SLAs. When there is a shortage, and there usually is, vendors put the technologists with people skills on the sales end of the process, and put the more heavy duty, and often more technically knowledgeable, engineers on the implementation side. For buyers, what sounded great when described by the sales engineer turns out to be not so great when implemented by a different, more recalcitrant, technical engineer. This problem is worse when the sales engineer is both good with people and good at technology. Technically proficient sales engineers describe a great system that could be great if they were the ones implementing it, but because they spend all their time in sales, they don't really have any hand in making it happen and the technical engineer may not be as good as they are, or simply unwilling to go the extra mile.

The bottom line is that most SLAs are worthless. SLAs either have to be written so vaguely that they actually mean nothing (and it comes down to whether you can trust the vendor—in which case you can save yourself the lawyers' fees), or they are so specific that they are unworkable in real life, or obsolete before the signature dries on the page. It is better to spend more time finding reliable and trustworthy vendors than to spend time on lawyers fees to try and write a bullet-proof agreement that ends up with hundreds of holes.

## Speed Of Change

> Two suited business colleagues walking into their office together: "I knew we shouldn't have gone to lunch. Now our technology is obsolete!"

You don't have to be a marketing guru to recognize that in recent years the product lifespan for technology has gotten shorter and shorter. In the 1970s and 1980s, technology was expected to last for 25 or 30 years. Nowadays, a company is lucky to have the same technology for five years in a row.

### Subscription Services, Not Products

Of course operating systems these days are more like a subscription services than a one-time installed products. In other words, Windows and Android and iOS are all on a continuous update cycle so that the product only lasts for a few weeks, until the next update, at which time our systems might get more secure and have earlier problems fixed, or (unfortunately frequently the case) become less secure and break our previously working system. In either case, we can no longer measure the life of our operating system in years, because now the life cycle is weeks, and we are paying for a subscription rather than a product. (This is actually a throwback to mainframe days, when all software was leased and not sold.)

Vendors might argue that customers demand new features, that a public outcry for improvement has shortened the life cycle. However, I would argue that more often than not customers would prefer simply to have software that works, and works well, 100 percent of the time. Continuous changes often means continuous problems.

I would also argue that most new features of products are unnecessary, unneeded, and unwanted. Unfortunately, it is very difficult for any individual company or consumer to fight the trend and stick with the same hardware or software for long. Vendors typically stop supporting declining products and share horror stories about how much it will cost if there are problems later down the road. Vendors like to lead customers to believe that they would be up a creek without a paddle without upgrades, and that they shouldn't take that chance. Many technologists echo this sentiment. It's impossible to fight anymore.

The short product cycle would be bad enough if the products all fit together modularly, but they don't. There is no technology independent of the technology to which it is connected. So when one component changes, all the other components must change as well.

One of my CEO clients, Kim, wanted to install an optical wireless mouse. Kim ordered the mouse —just like anyone would order a new knob for the credenza cabinet of a desk. Kim didn't want to bother the technical people when it arrived, so early one morning Kim sat down and followed the directions to plug it in.

Fourteen hours later Kim called me to come and fix the system since nothing seemed to get the new mouse to work, and now the old mouse wouldn't work either. The system was stuck. Of course, being a CEO, Kim preferred to pay big bucks for me to come and fix the problem rather than let the technical employees know about the new mouse. Kim said "I don't need their 'sneers'."

As soon as I started working on it, what happened became obvious. It is a simple example of what happens when you want to make one tiny change in any working computer system.

In order for a wireless mouse to work, there have to be connections in five places:

- The mouse hardware must be able to talk to the interface software.
- The interface software of the mouse must talk over the wireless connection of the mouse to the wireless receiver.
- The wireless receiver of the computer hardware must talk to the USB connection.
- The USB connection must talk to the interface software of the mouse on the computer.
- The interface software of the mouse must then communicate with the operating system.

Since the mouse, the hardware, and the operating system were developed and sold by three different, unrelated manufacturers during different generations, the chances were astronomically small that all the details would have been worked out between the three vendors on every peripheral. In this case, the breakdown apparently occurred at the wireless connection of the computer hardware, which wasn't talking with the interface software of the mouse. Bottom line: To install an optical wireless mouse, Kim would have to install an entirely new system.

### Speed of change impacts ROI Analysis

The speed of change in information technology does not necessarily mean that other fields such as finance and accounting have kept up. I used to think that the rules of capitalization were an accountant's exercise—highly boring, pretty useless in the real world, and certainly not of any interest to the manager of a technology project. You wouldn't find me under green eyeshades and cuff protectors poring over columns of numbers to determine the depreciation of an item, no sir!

But then a strange thing happened. As a VP for a Fortune 500 financial services firm, I had to create a spreadsheet template that would calculate the return on investment (ROI) to be used by the budget analysts when requesting funds for projects. So in 1996, there I was, poring over columns of formulas in a spreadsheet (sans eyeshades and cuff protectors) grunting my surprise at what seemed to me a new accounting discovery: The rules of capitalization could make the difference between a positive ROI and a negative ROI*.

We're not talking pocket change, either. Using a different option for capitalization rules on one of the projects I analyzed, for example, switched the payback period from eight months to eight years.

I became very concerned about the consequences of this finding. I knew the top decision makers in the company didn't appear to know—and certainly didn't appear to care—*which* rules of capitalization were used in assessing the value of the project. I'd probably drop from shock if I ever heard the CEO of a $4 billion company ask, "Did you capitalize or expense the programming on that project?" At the same time, my experience showed how centrally important the question is to information technology projects.

Here's how it works. If we pay a flat amount for the hardware and software, an amount that includes the consultants to customize and install the system and train the employees, we are allowed to capitalize the entire purchase price. In other words, if we spent $15 million, only $1 million had to be charged to expenses that first year. Only the depreciation, 15 years for an enterprise-wide system, affected the current year's profit.

On the other hand, if we used hardware we already had, and had our employees or even outside consultants, but not as part of the purchase price, write the software, implement the system, and train the employees, we had to expense the entire project. While the project itself would probably be cheaper (say, perhaps, only $10 million instead of $15 million), the entire amount had to be

* Yes, I know. If you have an accounting background this should seem obvious – but I didn't. I just assumed that the rules of capitalization wouldn't make that much difference in the return on investment analysis.

charged to expenses, and therefore would impact the profit in a major way. Instead of $1 million cost, we faced a $10 million cost even though we just saved the company $5 million!

Keep in mind that we are talking about two different ways to accomplish the same goal. Nothing in the hardware, software, programming, training, or project implementation was different except how it was **billed**. The operational outcome for these two choices would be exactly the same, but one would show a positive ROI, and the other would show a negative ROI.

The accountants I talked with didn't seem concerned about this issue. They declared that the rules of capitalization had been well established, were used consistently, and made perfect sense. But I could see a mile-wide gap in their knowledge about technology projects that led to their indifference: **The exact same system could be described as either off the shelf and configured or customized to the client's needs, or developed by programmers.** The difference was only semantics, not an inherent difference in the technology itself.

This is why technology vendors sell customized programming and development as if it were an off-the- shelf product. Most ERP systems, for example, are actually custom development projects although they are talked about as if they are packages of software.

***Shipment Idiocy***

When I worked for a system integrator, we carefully shrink-wrapped and shipped boxes containing nothing more than a piece of paper with a license number on it. The software (worth $25,000 in 1989, valued at $57,693 today) was actually being delivered dynamically by our programmers, but the customer needed to **receive a shipment of the product**, so we sent them the license number (without which the system wouldn't work). I now see that the capitalization rules were probably to blame for this bit of idiocy as well.

Outdated rules of capitalization is just one of the problems with generally accepted accounting principles that causes the ROI calculation to lead entrepreneurs and business owners to make the exact wrong decision on technology projects. ROI calculations have difficulty accounting for concepts such as ongoing maintenance, residual value, infrastructure separation, and the risk of unproven capabilities. Projects using new technology are much riskier than projects using stable technology. Projects requiring new infrastructure are more expensive (but sometimes have higher residual value) than projects using existing infrastructure. Project using common technologies will have much lower ongoing maintenance costs than projects using less common technologies. Until ROI algorithms have been modified to take into account these issues so that we can separate a technology project from a building (where none of these concepts apply), a positive ROI calculation will lead to approval for the projects most likely to fail, and rejection for projects most likely to succeed.

When I discovered this in 1996, I lobbied hard to get a text field included in the ROI results stating whether the technology in the project proposed was obsolete, stable, new, or leading edge. I couldn't imagine a senior leader making an informed decision without that basic information. Unfortunately, although I was successful in getting the information into the presentation, I was not successful in communicating its importance. The senior leaders continued to make decisions without looking at that information.

### Getting Technical Details

These problems are further complicated by the inability of buyers of new technologies to get accurate technical documentation. In the 1980s and 1990s, vendors would make available technical descriptions of their products in documents called white papers. A white paper usually described exactly what the new technologies were under the covers: how they worked, how they differed from previous technologies, and so on. They were written in technical jargon, and were meant for IT people, but at least they were available.

When the speed of change accelerated drastically, due, in part, to the public Internet, and as leaders with less IT knowledge started making technology decisions, the marketing department took over. White papers lost their technical edge. Instead, they evolved into the same type of marketing fluff that one might find in a product brochure. Even if buyers are relatively knowledgeable in the world of IT, they cannot get access to the underlying structure of new technologies in order to assess their stability and "fit" into an organization's existing technologies without purchasing and installing it to see.

### *Getting the Most From Vendors*

By now you must be thinking, "Whoa! Those are a lot of obstacles. How can anyone ever get around them? Given the problems, how can we ever know if technology will help us reach our goals? How can we ever get a good partnership with hardware and software vendors?"

There are solutions to each and every one of these obstacles. It is possible to have good relationships between buyers and sellers of technology. Step one is to recognize the difficulties involved in each one. Step two is to neutralize the problem with knowledge, attitude, and behavior. And that's just what we'll cover in the next section.

## The Inside Secrets of Software and Hardware Vendors

It is quite possible to mitigate all the challenges and come out with great decisions about information technology and good relationships with vendors regardless of the obstacles. All we need to do is keep the following rules in mind:

- Become technically literate.
- Recognize the immaturity.
- Define your own standards.
- Maximize periods of stability, minimize periods of instability.
- Cultivate long-term vendor partnerships.
- Ignore hype.
- Get off the bleeding edge.

Let's look at each of these in detail.

### *Becoming Technically Literate*

> Please grant me the knowledge to understand the technology I do;
> the courage to ask about the technology I don't;
> and the wisdom to know the difference.*

Just as the first step on the road to recovery for an alcoholic is recognizing the problem, the first step on the road to better technology decision making is recognizing the problem: having the necessary knowledge, and understanding it, is a prerequisite to make a good decision. Since we know that the gap between those who understand IT and those who don't is costing us money, we are faced with the difficult task of learning what we need to know. There is no alternative.

There are several ways to learn so we can recognize the best deal when we see it. First, we start, as suggested in Chapter 1, by creating an internal technology map that places, in relationship to each other, all the current technologies our organization is using. Second, we gather additional information from IT vendors, making sure that we understand which layer of the technology map each one is on. Third, we need to find people in whom we can place our trust who understand IT. We can't abdicate our decisions to them; we must spend the necessary time to pull from them their knowledge so that we can make the decisions ourselves.

---

* Just in case you didn't catch it, this is a parody of "The Serenity Prayer."

As discussed in the first chapter, an accurate technology map of the different technologies used within the business is essential. Developing this map may initially require us to tamp down our egos a bit. First we have to admit what we don't know. Then we have to ask technology vendors and staff to explain every buzzword, clarify every glossy phrase. We may have to put up with an occasional rolled eye and impatient sigh. The result, however, often means the difference between a good decision and a bad decision. The temporary condition of not understanding will shortly be replaced with easily keeping up with the ever-changing technology landscape. Eventually we will understand the technology better than the salespeople, which is a good position to be in.

***Missed Opportunity of Senior Leaders to Gain Value from Systems***

It always behooves senior leaders spending money on IT to understand it. When I was working as an enterprise network management consultant, I would facilitate a series of concept of operation meetings with the business leaders of the company during the first few weeks of the project. I can tell you from experience that the business leaders often did not understand their role. Even the rare occasion when we did get the right people in the room (the CEO, COO, CFO, and CIO), they were not patient enough to become knowledgeable enough so they could see how their decisions connected to the concept of operations or the software implementation.

Despite my attempts to explain, they were limited by their own misunderstanding. They thought that we were simply installing a piece of software like Microsoft Word or Excel, and didn't understand the decisions they were making. Normally, I ended up making many of the concept of operation decisions based upon my best guess as to what they needed from our discussions. (In other words, they abdicated to me.) But the value of the system to their business would have increased dramatically if they'd simply understood how they could align the way the software worked **with** their business. That understanding would also enable them to make adjustments in the future as the business changed. As it was, I suspect that after the engineers implemented the concept of operations requirements that I defined, the senior leaders never thought about the system again. They missed an incredible opportunity to achieve commensurate value for the hundreds of thousands they had paid for the system.

Knowing as much or more than salespeople do can help us evaluate the information they give us. Remember that IT salespeople often don't have a very deep understanding of IT, either; they defer to IT engineers. Often it is very difficult to get technical information from IT vendors. If we can get an IT vendor to provide us with detailed technical information without purchasing, we should do so. If not, one way to get the information (albeit a bit sneaky) is to ask for detailed proposals from many different vendors as if you are ready to purchase. A well-done proposal can teach us many things about the hardware, software, and services—facts you cannot get in any other way. (Apologies to all technology vendors for this advice. You can avoid spending hours on proposals that won't necessarily lead to commitment by providing information in technical white papers devoid of marketing gibberish.)

Sometimes vendors are willing to do a pilot project to resolve any technical issues before the buyer commits to purchasing the technology. If it costs absolutely nothing, go for it. Be aware, however, that many vendors use doing a pilot as a ploy. Free trials are never free. In many cases, the actual hardware and software are extremely cheap, and the real cost for the system is in the customization and development, for which you are charged even in the pilot. Be careful. Make sure that your commitment of time and money is low, and that you won't end up with a big bill if the pilot is successful. Make sure you have an out.

Another way to get information is to ask vendors for references and talk with other users of the products and services. When discussing it, however, keep in mind the differences between their businesses and yours. One size does not fit all.

Additionally, it is up to us to be sure, before we ask a question, that we are asking the right vendor the question. It doesn't make any sense to ask the person filling the potholes which car is the most reliable, and the car dealer won't be able to tell us which type of material makes the best roads. So why would we ask an application vendor which server to purchase, or rely upon the advice of the networking salesperson about which financial system to purchase?

Consider the following question from a Bank Vice President to a Sales Engineer about enterprise networking software:

> VP: Can this software check the entry of the clerk so that the transfer always goes through?
> Sales Engineer: Absolutely—yes. Every time a packet gets sent an acknowledgment is returned so that the sending computer knows if it was successful. The transfer always goes through.

The VP was on the **application** layer and the Sales Engineer was on the **network** layer. So while the engineer answered truthfully (like that Mercedes-Benz dealer), the Sales Engineer was not answering the question the VP asked because the network has nothing to do with the clerk's interface or the database entry. If we have a technology map we would know better than to ask a **network** engineer a question applicable only to the **application** program.*

When we hire consultants, we want to sop up all their knowledge like a sponge rather than allow them to keep their expertise to themselves. Another way to ensure that money paid to a consultant is not wasted is to pay them for more permanent knowledge in the form of documentation and formal training. Penny-pinching managers often skimp on documentation hours, and technology staff who hate to document are only too happy to comply with a lack of emphasis on the documentation. But documentation often can save multiples of its cost by providing the exact right information necessary at a critical decision-making time.

***Importance of Documentation***

> One of the projects I managed as VP for a Fortune 500 financial firm required reconnecting our new systems with all three credit reporting agencies. Three years earlier, a consultant had done the original connection for our existing systems. This time, the entire project was put on hold for days because none of the people on the team knew what needed to be done to update this connection or whom to contact at the credit reporting agencies to make the change. The team estimated the time needed to reconstruct the missing information: two weeks. This would cost the project well over $300,000 in lost opportunity and team time. This would have been an unexpected expense we couldn't afford on a project already behind.
>
> In the midst of this crisis, however, almost by accident I happened to be talking with someone who had previously worked in the department. He told me exactly where to find the documentation the previous consultant left, along with the names and phone numbers of the people to contact at the credit reporting agencies. Within 24 hours the connection was done. I whispered my thanks to the heavens for the unknown manager who had insisted on that documentation. (Now if they had only insisted on making sure the next manager knew where the documentation was so there wouldn't have been any delay!)

## *Recognizing the Immaturity*

The first step to mitigating the immaturity of the information technology industry is to accept that it is not going to respond like a mature field. I like to watch for signs that it is maturing—such as that you can now get a computer, operating system, monitor, storage drives, and printer from the same manufacturer, (although there are still problems in getting them all to play nicely together). However, we shouldn't be surprised when IT falls down once in a while and fails. IT is still at the youth stage rather than the mature adult stage, and we should expect mistakes.

It is because of the immaturity that we can't expect the same from technology vendors as we do from car dealerships or appliance stores. We must remember that, more often than not, the vendor is in new territory and doesn't have the answers any more than we do.

We also need to recognize which technologies are easily purchased commodities and which are not. We've come a long way over the years toward not needing consultants for every little

* This was an actual conversation I overheard when training people on Enterprise Network Management systems. I pulled the VP aside and explained the difference so that there was no mystification on this project, but I'm sure that conversation happened hundreds of times.

technology decision. Think about word processing. Deciding which wordprocessing system to purchase used to be a huge project.

***The Wordprocessing Decision Project***

Almost 40 years ago, I was teaching for a local school district, and it had hired a consultant's two-person firm to do a six-month study that would determine the clerical staff's word processing needs. The outcome was a Wang word processing system costing tens of thousands of dollars, which was used in the district for the next 10 years. Word processing was a complex topic back then. There were typewriters with storage capabilities. There were huge systems like Wang that could handle 20 or 30 word processors (secretaries) at the same time. There were PC-based systems like WordStar and MultiMate. The features available in these different solutions were leapfrogging each other, with capabilities changing month to month. Most people had experience only with a simple electric typewriter. They were confused by all the choices. They needed an expert consultant to do a study and indicate the best choice. Purchasing a word processor was an expensive multi-month proposition.

Can you imagine spending six months today to identify whether you want to use Microsoft Word or Google Docs? Of course not! Today we might give the decision five minutes' consideration before choosing Microsoft Word. Word processing decisions no longer require the expertise of a consultant.

Twenty years ago, Internet technologies were at the same level as those old word-processing systems were in 1984. During the Internet bubble, businesses hired e-strategy consultants to help them get on-line, and often spent millions of dollars doing so. Today, the public Internet is cheap and ubiquitous. Today, a company can get on the Web within hours. There are hundreds of public Internet sites that can register domain names, provide us with a Web site template, and take us through an easy-to-use wizard to build Web pages.*

At the height of the Internet boom, getting a simple Web site on the public Internet might have cost hundreds of thousands of dollars. It is hard to believe that what was once an expensive undertaking can now be had for pennies a day (or free). But this is a standard path for technology.

It is not easy to figure out exactly when the expensive new technology becomes cheap old technology. Think about the consultants and staff people at the school district that implemented the Wang word processing system. Five years after it had been installed, were they going to share with the school district decision makers how cheap WordPerfect had become†? The employees who ran the system weren't about to let their bosses know that the whole thing could be replaced for a fraction of the annual maintenance cost. The staff who had struggled to figure out how to use the difficult-to-learn system weren't about to change even if the new system was easier. The school district continued to pay whopping fees for five more years after the cheaper alternative was available because it didn't know any better.

* One such site is Go Daddy. I used their webhosting services from 2001 to 2021. Started in 1997, Go Daddy became the largest domain registrar on the Internet. At one time, 1 in every 20 websites was hosted on GoDaddy. Sadly, their service has gone downhill since the founder left, and I would not recommend them to anyone. GoDaddy Review: Can You Trust Their Shared Hosting? By Josep posted on Oct. 8, 2020. https://www.websitetooltester.com/en/hosting/godaddy-reviews/ Last accessed October 16, 2020.

† Microsoft Word was not yet available. WordPerfect was the top wordprocessor in the early eighties.

In every technology's life is a tipping point, (a phrase made famous by Malcolm Gladwell*) The *tipping point* is the moment when a previously expensive and customized solution becomes a cheap-and-easy commodity. We can tell when that point occurs in a product life cycle by watching for the following signs:

- The majority of businesses already use it successfully.
- Dozens (or hundreds) of smaller software firms have consolidated into two or three large competing publishers.
- We do not see articles or success stories about the technology in industry magazines anymore.
- Salespeople stop calling on us to sell the technology.
- Consultants don't recommend the technology anymore (because its use is a **given**).

In most cases, **we want to wait until technology moves from innovative product to commodity** before investing in it. When we purchase commodity items, we should not need the help of a consultant.

When faced with highly complex and unintegrated technologies, however, entrepreneurs and business owners do need to rely upon unbiased consultants to help them make the right decision. As noted in the previous section, the first step is to recognize the difference between the easy commodity stuff and the newer innovative stuff. Business Analytics, Data Science, Internet of Things, Blockchain, Robotic Process Automation (RPA), Virtual Reality, Edge Computing, Intelligent apps —there are still plenty of technologies for which there is no other choice but to hire a consultant†.

These, too, will become commodities within five years of their tipping point, but for now, if we need them, we should hire a consultant to ensure that we understand what we are getting, how much it should cost, and how to implement it. Of course, when we do that we want to pay very close attention to ensure that there are no behind-the-scenes partnerships or deals that would bias the consultant. If we look, we can find consultants whose income doesn't depend upon vendors' projects and who don't get a referral fee from vendors. It makes sense, if we can, to only work with consultants who will not profit from the choices we make (if we can find them). Most important of all, we want to make sure that "none of the above right now" is one of the options considered. We don't want to spend money on a new technology that we don't need.

## Buying in Generations

Because of the application's and hardware's reliance on the OS, people purchasing computers are best off buying them in "generations", trying to, as much as possible, purchase the hardware, OS, and application software that are all at peak popularity about the same time. If, for example, the hardware was manufacturer before the OS existed, then chances are it isn't going to be able to utilize some of the advanced features of the OS. (The same is true for the OS – it's best to purchase software applications that have been revised specifically for the OS rather than trying to get application software to run on an OS that was developed after the application was developed.)

---

* Gladwell, Malcolm. 2002. The Tipping Point: How Little Things Can Make a Big Difference. Back Bay Books, Boston, MA.

† To emphasize the speed of this transition, when I revised this book in 2006, the list of technologies which still required the use of a consultant was security, streaming video, voice-over-data lines, databases over the public Internet, wireless implementations, radio-frequency ID codes, and customized handheld applications. Not a one of them would require a consultant now, in 2021.

***When is the best time to buy a computer?***

I used to say that the best time to purchase a personal computer is about 2 years after the major release of a new OS. Earlier than that, and the hardware manufacturers and application publishers did not have enough time to upgrade their components for the OS. But this advice fell apart when the software publishers start automatically updating their OSs rather than release a new version and applications because leased versions that constantly changed instead of packages that are bought and installed. The Windows 10 of today is far away from the Windows 10 of 2015 when it was first released.

Then I said the best time to purchase a personal computer is about 2 years after a major release of new Internet security protocols. Because security requires base-level changes in the computer processes, older hardware and OSs had a hard time adjusting to the new protocols. But that advice also falls apart since new security protocols are being released all the time in the never-ending quest to stay ahead of the hackers. So now I say the best time to purchase a new personal computer is when your old one can no longer run the apps you want it to run, or recognize the peripherals you need it to recognize.

## Defining Company-Wide It Standards

Technology leaders are known for lauding the cost savings of standards. It has been shown again and again that simplifying the technical environment can save a company a great deal of money in the future. But how standards help organizations save money is still a mystery to many.

First, let's differentiate what we mean by the word *standards*. In a later chapter, we will talk about the term standard in conjunction with *standards bodies*. That type of standard is **not** what we mean here. Nor do we mean implementing a similar-looking desktop on all the PCs (e.g., the same screen saver and the same menu choices on all PCs).

***When is a Standard Not a Standard?***

One time at a workshop I heard a manager brag about how all company's computers were completely standardized because they locked down all the desktops with the same screen and application programs. Yet later I found out that their technical environment included DB2, Oracle, Sybase, SQL, and Access (all different databases) as well as Solaris, HPUX, Linux, Windows NT, Windows 2000, and AIX (all different operating systems). I wouldn't exactly call that standardized.

In this case, standards refers to the list of hardware and software, including models and versions, in use in the organizations. Within a business, it is important that we define what is current, what is stable, and what is future. *Current* is the brand and version of each component we are purchasing today. *Stable* is the component we are no longer purchasing, nor replacing, but just leaving in place to work effectively. *Future* is a component that may be selling currently, but which we are not purchasing yet—though we probably plan to switch at some point in the future. We might even have a few systems that are *Obsolete* but which we are actively working to replace. Once we've defined obsolete, current, stable, and future, we have documented what we have and what we want, as well as what we are not ready for yet. All of this is considered part of the technology map.

The secret for saving money is not to push. We don't get rid of what we have just because it is no longer current. But the next time a purchase decision comes up, we simply look in the current column and purchase that. We don't purchase the future components yet, even if available, because we have not yet decided to move to that new technology. Perhaps we haven't fully tested it in our environment. Perhaps the ROI does not yet show a positive value. For whatever reason, we are sticking with what we currently buy until there is a valid reason for switching to future.

The only hardware and software we push to replace with current is anything not listed as stable (i.e. obsolete). The longer our current technology remains stable, the better value the system provides the business. Our standards chart, part of our technology map, is the tool we use to implement those guidelines. The best value is obtained when the current, stable, and future are all the same, which is only possible in already mature technologies. This is the goal for all levels, but not under our control.

For each multiple in each column, the cost to support it increases exponentially because we need technical support people in multiple domains. To pay for technical expertise in DB2, Oracle, MySQL, SQL, Access, HPUX, Linux, Windows Server 2019, AIX and i, we need cadres of people, at least a dozen. If we have 1,000 employees, the cost per employee would be $600 per year just for ongoing support. If all 1,000 employees were supported by simply Oracle, HPUX, our ongoing technology support costs would drop down to $100 per employee.

There is no such thing as the best standard for all businesses—each will be unique and can only be determined after carefully reviewing the existing environment. Undoubtedly, the first time the standards are documented, there will be multiple answers in the Stable and Current columns. But over time, as the technology matures, we will have fewer and fewer standards and save more and more money.

The goal, then, is to slowly move toward simplification. We cannot push it because existing applications don't always port to other environments. Eventually, however, our environment will mature. We will have only one answer in each category, and we will save scads of money on maintenance and support.

## *Maximizing Periods Of Stability, Minimizing Periods Of Instability*

We can't do anything about the fact that the IT industry itself is unstable and constantly changing. But we don't have to give in to that method ourselves. One of the best ways to ensure smooth computing is to hold on to our existing information technology configuration without making changes—neither people nor process nor network/hardware/software— as long as possible.

Holding on to existing configurations without changes is not easy. One problem could be that the vendor goes out of business or gets acquired. When considering a new technology, we'll want to know where the market currently resides in this cycle of technology innovation, capitalism, and financing in order to prevent this problem. We want to stick with established profitable companies—companies with financial stability that aren't dependent upon a constant influx of venture capital dollars even when the market for some new technology is just beginning. Additionally, funding often depends upon the amount of buzz surrounding the new technology. Since the amount of buzz often depends upon hype rather than actual revolutionary change, we need to be careful of products in the growth phase of a new market cycle. So we want to stay away from whatever everybody is writing about and talking about.

### ***Product Life Cycle***

We want to pay attention to the life cycle. Within the overall market cycle, each component of technology has a life cycle. In general, we can say every product goes through four stages from birth to death. The phases of a product life cycle are the following:

1. Development
    - Introduction of idea
    - Developing, testing (alpha and beta), feasibility
2. Growth
    - Commercial introduction (piloting)
    - Commercial acceptance
    - Expansion
3. Maturity
    - Stability (longest phase for successful products)
    - Decline
4. Obsolescence
    - Gasping
    - Completely dead

During the development and growth stages, expect quite a bit of hype about the product. After maturity, the hype will fall away, leaving only the actual capabilities of the product. After maturity, the vendor will stop issuing press releases and the version number will not change much. After a while, the vendor won't even help with problems of older versions and will just tell you to upgrade before they even attempt a fix. The vendor will eventually charge you money to upgrade (to encourage people to upgrade earlier in the new product life cycle or to keep up with automated updates). Finally, at the gasping stage, the vendor will not even give you credit for having purchased earlier versions. When almost all customers have given up using the product, it is completely dead.

The following guidelines will help us to recognize when in the product cycle to purchase and how to fight the pressure to constantly change your IT:

- Buy in the expansion phase.
- Follow the crowd.
- Buy only in generations.
- Don't switch lightly.

### *Buy in the Expansion Phase*

The best time to purchase technology is near the end of the expansion period, but before maturity has set in. While the terminology below may seem to fit software best, it also applies to hardware (operating systems or firmware releases) and process methodologies (TQM, Six Sigma, Re-engineering, etc.).

It is not easy to identify when a product is in expansion. We cannot rely upon the vendor of the technology. A vendor will always tell us that a product is already mature before it is, or that the technology is obsolete before it actually is. Remember, the shorter the product life cycle, the more money vendors make.

I've already discussed one of my clients who purchased a highly touted Data General system in 1999. I noted that all the industry magazines at the time spouted in glowing terms how wonderful this new technology was. And, indeed, it had far surpassed any other system I'd seen in terms of reliability, speed, and flexibility. But a few months after my client bought the system, EMC purchased Data General. Within a very short time, all other vendors stopped supporting the hardware. It was very difficult to find administrators who understood it. New programs wouldn't run on it. Parts could not be found for it. My client was forced to purchase an entirely new system within three years.

If my client had waited until the expansion phase to purchase the hardware, the system wouldn't even have been in the running because it never made it to that phase. They bought it at the commercial acceptance phase, when a relatively small number of customers had installed it and the industry was beginning to notice how good it was (a prime moment for buyout in a venture-capital-based company). My client simply bought it too early.

Unless we are already an expert in the technology, we are faced with perusing existing clues to find out which phase a technology product is in. Here are some general signs to recognize that a product has matured:

- There are many companies similar to ours who are already using it.
- New releases and updates have slowed.
- Upgrades are not actually available yet, though they may have been announced. (Remember also that replacement software and hardware might be vaporware, announced but never produced.)

We can find out this information by searching the vendors' sites on the World Wide Web and by asking a few questions of the account manager (i.e. salesperson):

- How many companies are already utilizing this system?
- How often do new releases or updates happen?
- Is there a technical community support site? How many questions are answered? How many are left unanswered or not answered satisfactorily?
- What comes after this version? Is there another major revision on the drawing board?
- How financially stable is the vendor? Is it capitalized with venture capital?

We must be careful not to share with vendors what our hoped-for answers are prior to the questioning or they will tell us just what we want to hear. The answers should be as close as possible to the following optimum answers:

- About half of our competitors and other companies similar to ours are already using it.
- Last year, there were five minor updates (for example, three in the first quarter and two in the second quarter). This year, there might be one in the first quarter, but no more are expected.
- The developers are working on the next major version—expected to be released more than a year from now.
- There is a technical community support site, and over eighty percent of the questions are satisfactorily answered. None are not answered.
- The vendor companies are profitable and have not spent beyond their means recently in advertising or non-direct capital investments.

Here are the danger-point, too-soon, still-in-early-growth stage answers:

- Less than 10 percent of our competitors or similar companies have installed the technology.
- The vendors released five minor updates within the last quarter and expect another one soon.
- The vendor is putting all of its efforts into improving the existing product.
- The vendor has plenty of cash, and it is backed by some successful venture capital companies.
- The vendor has been written up recently in some industry magazines.
- The vendor expects to be profitable soon.

Here are the danger-point, too-late, almost-obsolete answers:

- Almost all of our competitors or similar companies installed the technology years ago.
- Many of our competitors or similar companies are beginning to uninstall the technology for newer types of systems.
- The products are very stable. There have been no major upgrades within the past 18 months.
- The next major versions are due to hit the market next quarter.
- The companies have been profitably in business for more than 30 years.

- The technical community site has many unanswered questions.
- They recently began a new expansion with the release of major revisions of their products—the first in the last 10 years.

Why is it so important to purchase at the expansion time of the product life cycle? There are several reasons:

- Purchasing after bugs have been fixed decreases time we spend being frustrated by products that don't work as promised.
- Purchasing after market acceptance ensures that there are plenty of other people who purchased it—which increases the chances for a long life.
- Purchasing before maturity ensures a stable product for the longest time.
- Purchasing from financially secure companies ensures that we don't get surprised by an unexpectedly short product life often caused by acquisitions.

### Following the Crowd

For several decades there was an oft-stated truism: *No one ever got fired for buying IBM.* This is not just a truism, it is sound advice*. Basically, buy what everyone else is buying at the time.

In an immature industry, a technology with a huge installed base is as close as we can get to a guarantee that the technology will not become obsolete in a matter of months. When we are using the same hardware and software as most people, the vendor cannot afford to simply drop support. They are forced, by the sheer volume of past customers, to continue to invest in the future of the product. Nothing will prevent the obsolescence of the product, but the life cycle can be stretched significantly when there is a huge installed base.

In technology purchasing, if we bet on what is popular, we bet wisely.

Technology history is replete with examples of better technologies disappearing because the installed base of the inferior technology was large enough to make switching unfeasible. Consider Microsoft Windows 3.1 versus IBM OS/2. OS/2 was a superior operating system in every way to Microsoft Windows (as well as all the other choices at the time including Mac, DOS, and Unix). But the better mousetrap does not necessarily bring market share. Due to the sheer number of installations, applications were developed for Windows and not for the other operating systems. The sheer number and variety of applications have made Windows the most common desktop operating system in the world. If we had purchased OS/2 in 1992, we received the last release of it in 1994. If we had purchased Windows 3.1 in 1992, we would have used it until 1998 or 1999. And we would still be using a version of Windows today. Some of those applications we purchased in 1992 would still be running on Windows.

Another example discussed is Ethernet versus Token Ring. Token Ring was faster, more controlled, could handle many more computers, and was more reliable than Ethernet. Despite being of superior quality and based on an industry standard, Token Ring simply never had the installed base to compete. If we invested in Token Ring in 1985, we would have had to replace it around 1992 or suffer spiraling costs for network hardware. If we invested in Ethernet in 1985, we would still be using the same network today. Ethernet was less reliable, but cheaper, and before you could blink an eye, there were 1000 Ethernet networks for every Token Ring network. The volume of sales caused the price to drop even further until there was a huge differential in the price. Ethernet's installed base was the competitive advantage.

Vendors know that buyers want to buy what everyone else is buying. So you can't trust the vendor when it comes to the numbers or users. Many will not reveal what their actual installed base is, or they exaggerate how many companies are using their technology. Some will only publish the

* Well, except for the IBM part since IBM is past its prime and no longer the one company everyone buys from. But for the first forty years of the technology world, when you spoke about technology you spoke about IBM. I would now say that Microsoft is starting to show its teeth, except that the web alternatives for software are not yet common enough. But when Microsoft switches to web-only, its competitors may surpass it since they have more experience in that space.

number of packages out there, not the number of packages installed. Since many applications come prepackaged with certain hardware, the delivered number is never equal to the number of systems that are using the software. Others will use faulty sampling to provide incorrect or misleading data.

I noticed this problem years ago when I signed up for a webcast that promised to reveal which infrastructure standards were commonly in use. I wanted to know how much had changed since I last looked at this sort of information, so I registered to watch it.

The Webcast was sponsored by an industry publication and presented by a well-reputed research company, so I had expected unbiased statistics on what businesses today were using in their networks. I was disappointed. The presenter's statistics made it look as if the majority of networks were using Gigabit Ethernet, which at the time was a very fast, and very new, version of the Ethernet. I knew that was ridiculous. I knew from experience that among the billions of networking devices out there in the world at the time, Gigabit Ethernet was barely a blip—about on the ratio of Rolls Royce vehicles to common cars. I also knew that you couldn't use Gigabit Ethernet with the standard ethernet port that came on a computer; you had to buy a special external card. So no way was what the vendor was saying true.

It took me a while to figure out how the presenter got his numbers (proving Mark Twain's comments about "liars, damn liars, and statistics"[*]). The presenter had taken statistics from a vendor who specialized in Gigabit Ethernet devices and extrapolated those numbers as if the average network reflected only what was being sold by that vendor at that time. That would be like looking at sales of off-road motorcycles in April of 2020 (at the height of the pandemic when sales went through the roof) and stating that 70% of all motorcycles were off-road![†]

Imagine someone without any real experience or knowledge attending this webcast and thinking, "Oh no! Our network is only slow Ethernet and our competitors' networks are faster—we'd better get moving on those RFPs!" Not good.

The lesson is this: We can't depend upon vendors (and sometimes not even supposedly unbiased research firms) to tell us the real installed base for any particular technology. Instead, we should rely upon our own eyes and ears. We should ask other businesspeople around us what technologies they are using and plan accordingly.

### *Buy Only in Generations*

We generally want to buy technology in bunches. If we are smart, we buy new hardware, software, peripherals, and connectivity methods all at the same time. We want to answer the questions regarding phases presented earlier for all the interrelated components of technology. If the answer from the software application vendor is optimum (i.e., expansion phase), but the answer from the hardware vendor is subpar (i.e., growth phase or obsolescence phase), we have to weigh the value of each piece of the technology before committing. The technology chain is only as strong as its weakest link.

We will never find a perfect solution with optimum everything. The objective is to strive for as close to optimum for as many of the layers of technology as we can. If we follow this simple advice, we can maximize the technology life cycle for our own company in real time—extending the normal three to five years into 10 to 15 years. The key is to get IT running smoothly, and then ***leave it alone***.

---

[*] Prompting me to write an article about the problem. Rhoads, C. J. (2005) "Lies, Damn Lies, and Statistics." Unitofl.

[†] In the midst of the COVID-19 Pandemic when sales of most industries were in the toilet, off-road motorcycle sales jumped 35% due, undoubtedly, to the fact that off-road biking was one of the activities still allowed with the pandemic raging.

### Don't Switch Lightly

Think about the whole product cycle. It often takes three to five years for a new component to be fully debugged and stable. And it takes even longer for the rest of the interrelated hardware and software to adjust to a new technology so that we can upgrade our systems to the new generation.

We want to be very careful when switching to new technology. If we are constantly changing one thing or another, we are constantly experiencing the productivity dip of new technology. The opportunity for peak productivity is never reached. This commonly happens under three circumstances:

- When companies try to change just one component of a technology at a time.
- When they choose technology either too early or too late in the product life cycle.
- When they try to implement every new major upgrade for a component.

How long we can keep mature technology going depends solely upon the success of the expansion phase and the number of other customers who are still using the technology. The more customers who use the technology, the longer the maturity phase and the slower the decline phase. With the right combination, it can be years and years, perhaps even as long as 15 or so. There should be an element of "If it ain't broke, don't fix it." This attitude can be very profitable under the right circumstances.

One requirement for following through on this plan is to ignore the pressure to upgrade put on us by vendors and techies. If we listened to them, we might get very nervous using technology after the vendor announces that it will not be supported anymore.

But what, really, is support? The vendor certainly does not help us if our systems go down. They aren't concerned with error messages or problems we have getting their software to work. To a vendor, support means sending out releases that by rights should be fully tested before they are installed anyway. As long as we are not changing our configuration (i.e., installing new software or hardware), our systems will continue to chug along without problems (barring a hardware failure, of course), so it is economically beneficial to stretch out the life cycle as long as we can. When we replace it, we replace it all with a completely different generation of technology.

In brief, for most companies and for most technologies, maximizing stability and minimizing instability will enable employees to gain peak productivity before the next generation of technology is forced upon them.

## Long-Term Vendor Partnerships

There are several guidelines to keep in mind as you work on creating long-term vendor partnerships. First, you should stop haggling. You don't want to buy solely on price. You want to avoid cheap technology, because it may end up being the worse value. As explained earlier, the only exception is when you are purposefully keeping an older system going with cheap parts until it either fails completely or our planned next generation of IT is being installed.

Though many a purchasing manager would like to deny it, there is a difference between toilet paper and computer systems. Trying to get the best deal on information technology will almost always cost you more than you bargained for.

I'm not saying that you can't save money. But there is an art to spending bottom dollar on technology. Haggling over the component prices of each item is not the way. Neither is using hard-line negotiating tactics with software or hardware vendors. A purchasing manager may feel good about saving $1,000 on a $10,000 invoice because he or she doesn't see the $3,000 cost that follows up on the savings.

***When is a Savings Not a Savings?***

I once talked to a CEO who was bragging about getting a 30 percent discount on software licenses—only to discover that he had negotiated for named seat licensing instead of concurrent seat licensing because he hadn't realized that there was a difference. A named-seat license means that you pay for each individual named person who uses the software. Concurrent-seat licensing is when you pay for a certain number of people to use the software at the same time. For this reporting software, the company had about 30 people who needed to use it, but every day only five different people needed to use it at the same time. What would have cost them $3,218 for concurrent seat licensing was now going to cost them, with his remarkable discount for named seats, $14,850!

Someday, the industry will mature. Someday, we won't have to read the fine print in every software licensing agreement. Someday, it won't make a difference if we purchase the software from Vendor A or Vendor B. Someday, we will have more power to negotiate a better deal.

But today is not someday. Hard-line negotiating tactics will only force the good vendors to flee the table in search of buyers who understand the costs and are willing to pay them. For every dollar you get the vendors to cut, they will find a way to cut services that will cost you twice as much to mitigate.

In 1999 the Data General system was an expensive system costing well over $200,000. In 2001 the same system could be purchased for less than $50,000 because EMC had purchased Data General. But if you purchased the cheaper system in 2001, the cost of keeping that technology going would skyrocket within a few years. If you added the $70,000 a year that it would cost to find someone who understood the system, or the $20,000 it would cost to replace a single burned-out card, you can see that even at a quarter of the price, you are not getting any bargains. The cheaper price was great for those who had already purchased the system and wanted to keep them going as long as possible (and therefore needed extra parts), but being cheap should not be the reason for the purchase.

Another important factor to which we should pay attention is financial stability of the vendor. Fast increasing growth in sales may be a danger sign. Venture capitalists love successful new companies, but they tend to pull out at the peak when the returns are largest, and to hell with the ongoing technology support needs of the customers. Therefore, look beyond growth and success. Buy from the vendors who have the wherewithal to go for the long haul.

There are thousands of services and products that were, at one time, the top of the heap, but are now broken and bleeding or simply gone. Once again I use as an example Data General being bought by EMC at its peak, leading to Data General's chief product becoming obsolete within a year. If anyone had looked at the balance sheet of Data General at the time, it would have been obvious that the company was going to be sold, and soon. DEC, Netscape, Ashton-Tate's, Foxbase, Compaq, Iomega, Alta Vista, StumbleUpon —all these companies had great technology that fell by the wayside after they were acquired.

Companies funded by venture capital are not necessarily bad. But an important prerequisite to financial stability is a healthy amount of cash and financial control by those who are interested in building the company rather than obtaining a financial return. We might want to avoid vendors who have had several rounds of funding without the assets to continue on in the foreseeable future.

Even more important than financial stability is integrity. In a world where the CEO of a charity steals the money for himself, where CEOs of major corporations lie about sales and go to jail because they stole money from their shareholders, it is difficult to believe that there are honest people in the world, much less at the top of a technology vendor. I'll admit that I've seen more of the former type than the latter. But people of integrity do exist. There are companies that try to put the needs of their clients and customers above their own needs whenever possible. To paraphrase the words of Warren Buffett, regarding the characteristics he looks for in his people: *Integrity, intelligence, and energy. Hire someone without the first, and the other two will kill you.*

We want to look for vendors who know what they are doing, who have the intelligence to understand our goals, and who have the energy to accomplish them. But none of that means anything without integrity. As I noted earlier, because of the immaturity of the industry and the number of unknown issues that typically arise in every technology project, even formally drawn up agreements such as service level agreements (SLAs) can't safeguard your interests. With vendors of integrity, we don't need SLAs.

***Bigger Vendors are not always Better***

When I had another client in the same position as the failed help desk contract discussed earlier in this chapter, instead of one huge contract outsourcing their help desk, I suggested that they have smaller contracts with different local fix-it shops—each near one of their remote office locations. They had staff people on-site for the main office. This is how we arranged it: The local fix-it shops got paid a fixed amount each month that all problems reported were successfully solved. It was a great rate when the problems were few and far between, and when they could be fixed within minutes on the phone. It was a horrible rate (for them) when the problems took multiple visits or hours to solve. Only the caller at my client company could "close" the ticket as successfully solved. No long-term contract. No paperwork.

Within a few months, the fix-it shop staffs had gotten to know the customer systems like the back of their collective hands. They spent quite a bit of time in the early months figuring out how to prevent future problems so that in the later months they could get paid without having to spend much time. They focused all their energy on making every person who called a happy customer, usually by fixing reported problems within minutes. In only one case did a vendor have to be replaced because it grew too big and began to ignore calls. The client used this method of partnering for years and is still going strong with it.

Instead of spending months negotiating a multimillion dollar deal with legions of lawyers, it makes much more sense to put our efforts into developing long-term partnerships with knowledgeable vendors of integrity. We need to know who will be responding in times of need and how experienced and knowledgeable they are. We want to pay enough so that our vendors stay in business and don't seek alternative sources of income. Our dollars are better spent investing in vendors with integrity rather than constantly negotiating and searching for a better deal.

## Ignoring Hype

As discussed in earlier, hype is so rampant that we've come to expect it. Technology vendors usually tell us what we want to hear instead of the truth. Knowing this, I am always surprised when people actually believe what a vendor tells them, but it happens frequently. It's easy to recognize hype once we know to look for it. To see it, we need to understand market cycles and product cycles. That will help us recognize marketecture and vaporware, and differentiate between what's possible and what's not. Finally, we can use a Hypometer (my own little invention) to measure the amount of hype.

One key to recognize hype is to understand the market cycle, which is intertwined with the nature of technology innovations and capitalist financing. Whenever a new technology hits the market, there is a predictable flurry of new companies exploiting and selling the technology. This is a period of fast-track growth with much public adulation, followed by a shark-fest of mergers and acquisitions. These finally unmask the two or three winners who will dominate the market for another decade.

To see the impact of hype, I use an imaginary machine I call a Hypometer to dissect ads and claims so I can distinguish the hype from the real information. I took one of the ads from the hype section in a previous chapter and put it through the Hypometer so that we could see the fact from the fiction.

*What if you had someone to watch your back?*

Pure hype. No one is going to watch your back, and no software or hardware can do it either.

*Someone to integrate your back-end logistics with deadly efficiency.*
There's no such thing as deadly efficiency. But there is a factual clue here—this company has something to do with integration. They have no product of their own, but they integrate other products.
*Someone to manage your warehousing, delivery, and financial settlements.*
Some examples of the products or services of their clients, obviously, but it says nothing about the integrator, except that it doesn't specialize.
*Someone to protect your entire e-commerce operation.*
Again—a dream. No one will protect your operation but you. They also seem to specialize in e-commerce rather than internal technologies.
*What if one company could handle your entire back-end operation?*
They meant to say that their company can handle all your software integration needs—they don't want you using their competitors also. But this phrase feeds the dream that you can just hire someone else and all your problems will go away. Hype.
*That's the idea behind Product A. Built around an open and extensible transaction network, our integrated approach enables you to outsource any or all of the logistics behind your buy button.*
An "open and extensible transaction network" could mean anything. I don't think it means software. I think they're referring to their network of vendors with whom they've worked on projects. This means that they give you the opportunity to pay them for the approach and to pay an outsourcer or vendor for the actual work.
*A partner to plan it, implement it, and stand by it.*
Here's a real clue: They don't program, they plan. Sounds like a company that does project management for integration projects using other vendors' software and hardware. The "stand by it" is hype—I've yet to see a vendor provide a full money-back guarantee if the client is not satisfied for any reason.
*In short, someone to make sure it pays off.*
Here's the question for the contract reviewer: Are they willing to return our money if one of the vendors doesn't live up to its promise? In writing? I'll believe it when I see it.

Well, the Hypometer gives this a 12.5 on a scale of 100 (12 factual words out of 96 altogether).

Hype is fueled by unrealistic expectations about what IT can do. As noted earlier, we use the terms *vaporware* and *marketecture* to describe solutions for which there is a deep need, but for which there are no shrink-wrapped products. Sometimes vaporware and marketecture are announced by technology vendors to dissuade other vendors from attempting to develop a solution to fulfill a gap between available products and the needs of business. Other times vaporware and marketecture are simply ways for vendors to describe what could be in the future.

Strictly speaking, many blockchain systems would actually fall under the definition of marketecture or vaporware. Most blockchain systems aren't actually commercial products at all, but rather a defined way for information to be stored. This is one of the reasons there have been some spectacular failures amidst the smattering of successes for blockchain systems; the quality is entirely dependent upon the skills and talents of the developer and the proposed application rather than the software itself.

How does vaporware and marketecture hurt us? It doesn't—as long as we recognize what it is and what it isn't.

***Project Management Vaporware Story***

I remember one time in particular when I was at odds with my boss because he wanted me to implement a project management software. After an initial analysis, I suspected that the software was vaporware. My boss didn't agree. He waved away my concerns, saying that the consultant recommending the software was highly regarded.

The system my boss was pitching was an application from a small firm in Scotland. They had implemented this so-called package at exactly one location, a bank. When I asked them to send us a CD so I could install it, they had to send the developer from Scotland, on our dime, to install it. I watched as he wrote hundreds of lines in programming code in order to customize the software. When he first installed it, all of the screen forms and reports still had the name of the Scottish bank on them—obviously hard-coded from the previous client.

When I talked to my boss again about my misgivings, he again minimized my concerns. He claimed that when the consultant came to do the best-practice training, I would see the value of the application. Furthermore, I was expressly forbidden from sharing my misgivings with the CIO, the final decision maker.

The highly regarded consultant came to do the best-practice training, and I tried my best to keep an open mind. However, after the first half-day session, the project managers came just short of booing him out of the room due to his contemptuous treatment of their current project-management methods. All refused to attend another session. Privately I was still lobbying to drop the entire project which I had concluded was a bad project with a capital B. My boss ignored my advice, and instead mandated the use of the software by all project managers in the company.

I still remember that decisive meeting when my boss and I presented "our" views to the CIO using the materials provided by the vendor. Publicly I was still doing my boss's bidding and supporting the software, though privately I harbored the tender hope that the CIO was sensitive enough to read between the slides. "Please," I silently directed him, "ask these questions":

- How many other companies have implemented this system?
- What do potential business users think of the software?
- How will this system help the business meet the needs of the customer?
- What do the project managers who have met him think of the consultant recommending this software?

The result? My ability to predict a poor decision outstripped my ability at telepathy. The CIO asked no such questions. My hope had been dashed against the rocks of my boss's enthusiasm*.

Once we understand how prevalent vaporware and marketecture are, it is easy to recognize one or more of the signs of software not-yet-ready for commercial use status:

- Described only in terms of benefits and features.
- Doesn't fit into technology maps of existing hardware and software layers.
- Difficult to get pricing breakouts. One price for a nebulous list of items, or many prices for modules of unexplained roles.
- Tiny installed base (i.e., only a few companies have actually implemented it).
- Software has no installation modules or runs entirely on the cloud with no access to customization modules.
- Hard-coded variables instead of soft-coded variables.
- Comes with a requirement for customization.
- Difficult to answer the question, "Exactly what does it do?"
- Coupled with "methods" and "training" and "best practices."

**Hard-coded** means that instead of using a variable that can be changed, the programming code contains the actual contents. When developing a program for one company, it is quicker to hard code everything. When multiple companies will be using a program such as in a shrink-wrapped commercially produced software package, soft-coded variables are set up for each customer to fill in. The cost differential between writing a program for one company and writing a program for many companies to use is dramatic. One way to tell if a particular program is off-the-shelf or custom programming is to ask to see the screen where the user configures the variables. Custom programming won't have such a screen, instead relying upon the programmer to code the variables.

The bottom line is: Vendors do both themselves and their customers a favor when they try, as much as possible, to eliminate wishful thinking and generic benefit lists. The entire industry and all the customers of information technology would benefit from a lot less hype.

* I'll bet you can guess who got the blame when the project publicly failed. Indeed, my boss blamed the whole fiasco on me. He subsequently tried to get me to quit, and when I wouldn't quit he tried to fire me. But the company fired him first, so I was saved for a short time. Of course, the whole company was doing poorly so it was only a matter of time before 2000 people were laid off – me among them. But at least I got a nice package when I got laid off.

## Getting Off the Bleeding Edge

Haven't you heard since the dawn of the first microchip that you don't want to be left behind? That you want to be competitive? That you need the newest to be the best? Isn't the leading edge where you should be?

In a word, **no**. We use the term *bleeding edge* instead of leading edge for a reason. Even though some vendors will proclaim this advice as scandalous, you should be behind in technology, not ahead (with the noted exception when dealing with your Hedgehog Concept as explained in a later chapter). Many business leaders discovered this truth when they surfed dangerously out in front of the wave during the Internet bubble. They found themselves dashed against the rocks and sands of reality when the bubble burst.

Paddling behind the wave where it is quiet and the surfboard bobs gently may be boring, but it is much more profitable. For the most part we are better off watching other companies do all the surfing, falling, and wiping out. For most technology, we should wait at least a year, sometimes two, after any new technology has hit the market before purchasing.

First, if you buy too early in the growth stage, other interconnected technology products will not work with the new technology (as discussed in the buy in generations section). Second, vendors always release products before all the bugs have been worked out. Early users can count on spending many frustrating hours with software and hardware that does not work as promised (as discussed in the recognizing immaturity section). Third, many products never get past the growth phase (as discussed in the buy in expansion phase).

In brief, except for some very specific situations, and in the research labs of our top technology gurus, our operational technology should be older than everyone else's technology.

## Big Data and Analytics

Speaking of bleeding edge: let's talk Big Data and Analytics. Recently the term "analytics" has been popping up everywhere. Colleges and Universities are offering degrees and certificates in it. Employers, especially, looking for anyone who has experience in analytics. Courses and workshops are abundant in analytics. But what exactly is analytics?

We'll tell you what analytics is not. It is not new. Analytics have been around for centuries. Two of Aristotle's works on deductive reasoning written about 300 BCE is purported to have been called ***Prior Analytics*** and ***Posterior Analytics***[*]**.** The term was also used in 1590 and defined as a *method of logical analysis*[†]. Other definitions include:

- a process in which a computer examines information using mathematical methods in order to find useful patterns.
- a system for performing analytics on received data.
- the science of logical analysis.

---

[*] In case you didn't know, Aristotle lived around 350 BCE (Before the Common Era) and wrote over 200 books and treatises, though not a single one survives to this day.

[†] (2021). "analytics." Oxford English Dictionary Retrieved March 7, 2021, 2021, from https://www.lexico.com/definition/analytics.

(2021). "analytics." Oxford English Dictionary Retrieved March 7, 2021, 2021, from https://www.lexico.com/definition/analytics.

(2021). "analytics." Techopedia.com Retrieved March 7, 2021, 2021, from https://www.techopedia.com/definition/30296/analytics.

(2021). "analytics." Dictionary.com Retrieved March 7, 2021, 2021, from https://www.dictionary.com/browse/analytics.

(2021). "analytics." Cambridge Dictionary Retrieved March 7, 2021, 2021, from https://dictionary.cambridge.org/us/dictionary/english/analytics.

(2021). "analytics." Merriam-Webster Dictionary Retrieved March 7, 2021, 2021, from https://www.merriam-webster.com/dictionary/analytic.

- the analysis of data, typically large sets of business data, by the use of mathematics, statistics, and computer software.
- the patterns and other meaningful information gathered from the analysis of data.
- the scientific process of discovering and communicating the meaningful patterns which can be found in data.
- concerned with turning raw data into insight for making better decisions.
- relies on the application of statistics, computer programming, and operations research in order to quantify and gain insight to the meanings of data.
- Information resulting from the systematic analysis of data or statistics.
- The systematic computational analysis of data or statistics.

You will note that the term can apply to three aspects:

1. the domain (i.e. the science of …),
2. the process (i.e. the method of …),
3. the results (i.e. the information resulting from …).

The sentence *In the science of analytics, analytics produces analytics* is a perfectly valid and grammatically correct sentence*.

***Why the term Analytics?***

I personally suspect that the term analytics is used in such a widespread way because people were turned off from the term statistics, which is really the synonym of the word analytics (and is equally applied to the three different aspects of the term: domain, process, and results). When people think of statistics, they tend to think of an extremely difficult college course they were forced to take, of which they remember little except that they never thought they would ever actually need to USE that knowledge. For generations statistics/analytics was the domain of isolated academics in their ivory towers who were good in math because of the extensive mathematical equations necessary in order to get a decent answer from a set of data. Often, when I talk to people who have been doing analytics for years, when I mention that they are doing statistics, they have no idea (or actually deny it) because they never realized that they could do statistics without complex formulas and terminology. Statistics is actually the science of seeing patterns in data and estimating the chances that the reality represented by the data is true or false. So is analytics.

In any case, one of the reasons analytics is talked about so much in this day and age is because data† is so much more accessible. In years gone by, the only way to get data was to laboriously set up surveys and polls and experiments. Generally the data was painstakingly and manually entered into a computer system. Then once you had the data, you conducted (or hired someone to conduct) the appropriate statistics. This analysis would answer whether your hypothesis was likely to be true or false. The results also provided you with an appropriately couched answer to the statistical question you originally asked.

Few people knew the ins and outs of normalizing and categorizing the data so that it could reveal real information. College-level statistics classes notwithstanding, few understand what is meant by all the statistical terms such as Type I or Type II error, variance, standard deviation, correlations, significance, etc.. Outside of statisticians and pollsters, people don't usually

---

* Similar to another of my favorite grammatically correct sentences: *Programmers program programs.*

† For years I have argued with editors about whether terms like data and analytics and social media are singular or plural. The style guides will call them plural, forcing odd-sounding phrases like *the data are,* or *social media say … .* Throughout this book these terms are treated as singular **because they ARE singular**. Data **is** one thing, just as a ream of paper is one thing despite there being 500 pieces of paper in a ream. Analytics **is** one thing, and it produces one thing (analytics). Social media **is** one thing, despite the billions of records or hundreds of different sites which are components of it.

understand the underlying meaning of research conclusions. The news media would often report conclusions in a misleading way (and social media even moreso).

***Social Media, News Media, Vendors, even Researchers often Lie***

I've seen misleading conclusions reported, time and time again, even in academic research when the researcher didn't really understand the underlying statistics. I have read "the result wasn't statistically significance, but it was positive". That's a lie. No, it wasn't positive. If it wasn't statistically significant, it means that **no conclusions** about positive or negative can be stated. It means we still don't know the answer because the sample wasn't big enough to make a conclusion based upon the results received.

## Why Do We Need Business Analytics?

Business analytics is essential to improvement processes. Business analytics helps businesses optimize their performances. Implementing analytics into the key performance measures of a business can help reduce costs by identifying more efficient ways of doing business.

A company can also use data analytics to make better business decisions and help analyze customer trends and satisfaction, which can lead to new and improved products and services.

### Big Data

These days, data abounds. Tons of it. The term "big data" has been applied to the huge amounts of data in a variety of structures that are available to those who know how to dive in and get it. While computers have been around for more than 50 years at this point, it took human beings that long to figure out how to more easily store and "code" data so that it could be accessible to those who wished to conduct statistics on it.

Big data can be large or huge, structured or unstructured, discrete or continuous. Some people characterize it by the 3 V's: Volume, Variety, and Velocity. These three vectors refer to the amount of data (volume), the different types of data (variety), and the speed at which the data can be processed (velocity). Also remember that big data can have multiple sources; the real power of big data is being able to combine information from multiple sources.

### Benefits and Pitfalls of Big Data

As noted, one of the benefits of big data is that there is a much larger variety of types of data available in a timely manner. Often insights and patterns can be discerned simply by sampling and looking at big data sets. For academics, utilizing big data enables them to "find significance" much more easily. Since there is a bias against publishing research unless if finds significance, that can make it easier to publish and share findings.

Regardless of whether data is big data or not, the old GIGO adage about garbage-in, garbage-out applies. Unfortunately, much of the data that is available is dirty – i.e., it has been entered or gathered in inconsistent ways which make the underlying data suspect.

For example, many people review the so-called "analytics" on their website and believe that the "hits" to their website is equivalent to the number of people who have seen their site. Nothing could be further from the truth. A "hit" might be a human being looking at the site, but it might just as easily be another computer looking at the site. (Programs roaming on the world wide web called "bots" go to the websites for a variety of reasons, and they count as "hits"). While the analytics attempts to differentiate between human and bot, it rarely does so accurately. Furthermore, each time a single human being clicks on something or scrolls somewhere on the site, it is counted as another hit. Most web hosting systems do not bother to explain this (because everyone would like to believe the myth that hundreds or thousands of people are coming to their website every day).

Another example is inventory systems that use inconsistent descriptions or different levels of granularity. Any operational manager responsible for inventory control can describe the myriad of difficulties in keeping track of inventory. There are many common problems, mostly surrounding the difficulty of granularity.

Granularity refers to the level at which a unique item exists. Imagine that we own a mom-and-pop knickknack store. We sell decorative candles. We can order a skid of candles from any one of three different vendors. We take the skid and break it apart into three levels of inventory: cases, boxes, and individual candles. The candles come in either small or large, and either red, blue, green or white. For a small knickknack store none of this is any problem. When we sell the items, we know that the individual small candles are $2 (regardless of color) and the individual large candles are $4. A box of 12 small candles is $20, and a box of large candles is $40, a case of small is $240, and so on. When we want to reorder, we just take a look at the inventory that is left. If all the red candles are gone, we break open another box of candles and convert the 1 box into 12 individual candles and/or put them on the sheet to reorder.

This process is easy for a human, but for a blind and dumb computer system it is very difficult. It is very difficult to properly figure out what items are in stock, how much to sell each candle for, which vendor to purchase from, and which items need to be reordered. Most inventory systems in use today don't work at that level of granularity, choosing to ignore color, for example, when identifying the SKU (stock keeping unit), resulting in only 18 unique identifying numbers instead of 72. These systems rely upon a human being who can easily recognize which color needs to be ordered. Unfortunately, that wouldn't work in an online system because (oddly enough) people don't want to be surprised by what color candle they ordered. Some online systems get around this by adding an "attribute" field and expecting the buyer to type in the color, or choose the color from the list. This, then, gets into a huge programming issue because each product would need a different attribute list. This is just one example of the complexity involved when trying to provide an online purchasing experience and clean up an inventory. And inventory that is not clean leads to sales records that are not accurate or reliable for many purposes.

***Big Data and Conspiracy Theories***

The ease with which researchers can find significance is also one of the downfalls of big data. Often patterns are perceived *which are not real.* In the television show *Star Trek Voyager* in the late 90's, there was a character named Seven of Nine who was a former Borg with a lot of technology installed in her brain. In one episode, she had downloaded too much data, and it made her thinking process go haywire. She started making connections and concocting conspiracy theories, one right after another. She started to think, and act on, crazy ideas. Similarly, big data can lead us down the rabbit hole into seeing patterns that aren't there and making conclusions that are patently false. The computer cannot detect the difference between coincidence and real connection; it is up to thinking human beings to disqualify items that are not likely given our understanding of human dynamics.

## Where Does Big Data Come From?

There are numerous sources of big data. Among them are:

- OLTP (On Line Transaction Processing)
- ERP (Enterprise Resource Planning)
- SCM (Supply Chain Management)
- CRM (Customer Relationship Management)
- Social Media (sites such as Twitter, Facebook, Instagram, Snapchat, Pinterest, YouTube, Yelp, etc.)
- Search Engines (sites such a Google, Yahoo, Bing, Baidu)

The growth of these systems means that what formerly were off-line human-to-human transactions (such as going down to the corner grocery store and buying candles) has now become one or more records of data in a computer system somewhere. The drastic drop in the price of storage and the increased speed of data access have both contributed to the ability for businesses to represent all of the different aspects of the supply chain from idea, design, source, manufacture, inventory, order, sell, pay for, delivery to after-sale followup activities.

***Speed of Technology***

Back in 1999, I worked for a large credit card firm just as big data was becoming, well, big. I was responsible for the project management of the VLDBS (Very Large Data Base) which was to store all of the credit card transactions for all of our customers so that the data could be deep-mined in huge data warehouses and resold to other companies who could then market new products and services to people based upon what they'd already purchased. Back then, this idea was very new, and the senior leaders of the firm thought that this was just the thing to take the company into the next phase of growth. Back then, however, storing and analyzing all that data was also relatively new, and it was expensive. We had hired a high-priced expert who had implemented 2 other VLDBS of 10 terabytes of data. We also purchased special technology large enough to handle up to 30 terabytes of data. The project had a budget of about 30 million dollars.

Alas, the project was never finished. Due, in part, to the firm missing its financial targets and its subsequent announcement of the plan to sell people's data, the bottom fell out of the entire banking industry and most banks lost 30% of their stock price in one day. A year later the government swooped in with privacy concerns and the senior leaders of the company were demoted or fired. (By that time, I had already gotten laid off with a nice package, so it wasn't a big deal to me personally.)

Fast forward twenty years. As I noted in the chapter on hardware, in 2020 I purchased, for myself and my own tiny company, a 10 terabyte hard drive. It cost less than $200. It was added to the slew of other multi-terabyte hard drives that I own, so that altogether I have about 75 terabytes of data sitting in my home office (though most of it is redundant backups).

## Tools of the Trade

Just because we need human beings to cognitively and creatively conduct the analysis, doesn't mean we don't need computers. We do. New technology, both hardware and software, has recently transmuted the world of analytics from the ivory towers of mathematicians and statisticians to the real world, dealing with real data, for any typically intelligent manager. The ability to review thousands, millions, or billions of records graphically and pictorially has changed the game. Now we can easily see patterns without math.

There are several software tools that help us do that. I am hesitant to list exactly what software is in the category of analytics because they change so rapidly that by the time you read this, many will be different. To demonstrate, here is a list of business intelligence software published by Gartner Group, a consulting firm that rates various types of business software in what they call a "Magic Quadrant"*. I reviewed the 2011 and the 2021 Magic Quadrant list in Table 21.

| Name/Publisher of Software | From which list |
|---|---|
| IBM Cognos.Insight | On both 2011 and 2021 |
| Information Builders | On both 2011 and 2021 |
| MicroStrategy Visual Insight | On both 2011 and 2021 |
| Qlik | On both 2011 and 2021 |
| SAP Visual Intelligence | On both 2011 and 2021 |
| SAS Visual Analytics | On both 2011 and 2021 |
| Tableau | On both 2011 and 2021 |
| TIBCO Spotfire | On both 2011 and 2021 |
| Oracle Endeca | On both 2011 and 2021 |
| Microsoft PowerPivot with PowerView | On both 2011 and 2021 |
| Amazon Web services | On 2021 List |
| Board | On 2021 List |
| Domo | On 2021 List |

* (2021). Positioning Technology Players Within A Specific Market. Research Methodologies Gartner Group.

| Name/Publisher of Software | From which list |
|---|---|
| Google Looker | On 2021 List |
| Info | On 2021 List |
| Pyramid Analytics | On 2021 List |
| Sisense | On 2021 List |
| ThoughtSpot | On 2021 List |
| Yellowfin | On 2021 List |
| Alibaba Cloud | On 2021 List |
| Actuate | No longer on list |
| Arcplan | No longer on list |
| Bitam | No longer on list |
| Board International | No longer on list |
| Corda Technologies | No longer on list |
| Jaspersoft | No longer on list |
| LogiXML | No longer on list |
| Panorama Software | No longer on list |
| Salient Management Company | No longer on list |
| Targit | No longer on list |

**Table 21. Magic Quadrant of Business Intelligence/Analytics software**

As you can see, only about half of the names on 2021 list were contenders in 2011, just ten years earlier. Half the list dropped off and more than half were new. So while you might want to recognize the names of software in the field of analytics, it is more important for you to recognize the category of software so that should you need business analytics, you have somewhere to start. Data Analytics software is sometimes called Business Intelligence or Competitive Analysis software. Regardless of what it is called, it makes sense to choose the software that is already associated with the transaction system where the bulk of the data is located. That's why you will see the top publishers of ERP, CRM, SCM and other OLTP systems in the list (and showing up more often than not in both Magic Quadrants). Since it really isn't the tool (despite what these software vendors might have you believe) but the talent of the analyst that can provide the business intelligence, which one you choose doesn't really matter. And they all basically do the same foundational tasks.

## How To Tell If Its Good Data

Data that comes from a variety of sources can be categorize on many different levels. These levels include:

- Data source reliability
- Data content accuracy
- Data accessibility
- Data security and data privacy
- Data richness
- Data consistency
- Data currency/data timeliness
- Data granularity
- Data validity and data relevancy

The reliability of the data source is important. An ERP system for example, would be considered more reliable source than social media. A source is reliable if the information can be tracked back to the original original source, and the person or persons responsible for that source are known.

Related to reliability is content accuracy. How sure are we that the data accurately reflects the underlying reality? Sales data is more accurate than order data because someone paid money for the item, and that won't generally happen unless it is a real sale. Of course, social media is not known for its accuracy, and neither is customer relationship management systems where people might provide fake information in an attempt to avoid being pressured into buying something they don't want.

**When is a Sale Not A Sale?**

Even sales data can be unreliable, however. I personally know more than one author who, instead of buying books directly from their publisher, went on to Amazon and purchased a lot of copies of their book in one day. This puts Amazon's heuristics into a tizzy and raises the level of the book to the best seller's list – at least for a short time. Once the expected recurrence of the sales doesn't happen, the book will slowly sink lower on the list until it is no longer a best seller, but by that time the author has already been noted as a best seller.

I discovered how Amazon's heuristics work when one of my books suddenly found it's way to the best seller list. (And no, I didn't purchase any copies of my own book.) But because it was a textbook, when a lot of students would purchase the book at the beginning of September, it bumped up the status to best seller because the heuristic isn't smart enough to figure out that it's a textbook. It figures it's going to see the same level of sales every week for the whole year and accounts for it accordingly. This is an example of when big data can give the exact wrong impression. By the end of the year the book has sunk back to its low status as a poor seller – until the next September when it shoots up to the top again for a few months.

Accessibility does not just mean that we have access to the data, but also whether the *format* that data is in is accessible. For example, a Word document or PDF with names and addresses is just about useless because the information is not provided in a format that is accessible to the computer. The information needs to be parsed into firstname, lastname, street, city, etc. A CSV (comma separated variable) file is a more accessible standard for transferring data from one system to another, but the data must be available in short discrete fields that don't contain control codes or special characters like quotes. If the data contains pictures, URLs or websites, emoticons, paragraphs of data or formatting codes such as line breaks, italics, boldface, etc – it is not necessarily going to be accessible to the system.

One of the bugaboos with data is keeping it safe and private. Information such as people's social security numbers, account numbers, credit card info, etc. are well known targets for data thieves. Lesser known, however, is that people's names, emails, phone numbers, birthdates, and relatives' names can also be very risky because data thieves can also use this information to steal someone's identity. People are constantly worried about keeping their data private, and yet also want to have the convenience of credit card information already stored and putting their birthday in social media so that others can wish them well. It's a tough balance.

Richness refers to how much detail is included in the data. Having someone's username and phone number is one thing, but to also have their name, address, what they purchased, how much they spent – well, that's a much richer data set. Getting pictures and locations is even richer. Social media information is generally much more "rich" than financial or sales data in ERP and SCM systems.

Data consistency is important and difficult to get. We previously described some examples regarding consistency of inventory data. Sometimes a company will hire a new person who enters the data in a completely different way than the previous person. Neither is right or wrong, but it would be inconsistent. Inconsistent data is dirty data, and often it must be manipulated to make it consistent. Going back to the candle example, an additional field called "color category" could be added so that magenta, maroon, and scarlet could be grouped under the color category of "red" so that "red" and "blue" can be compared.

Timeliness is another important aspect of data. Old data is often useless data. Imagine a company trying to sell a data set of people who looked up phone numbers in a phone book. Would this information be helpful or relevant today? Very few people use phone books anymore, so if the data is no longer current, it is no longer useful.

Granularity is one of the most difficult aspects of data. As described earlier, granularity is the level of detail which the data contains. More data with more fields and additional information is more granular than summary data with fewer fields of additional information. How granular the data should be is entirely dependent upon the questions that you want answered. Again – part of the art of analytics is knowing how much granularity you need for a given situation.

The final two are validity and relevancy. Different from accuracy, validity is whether or not the types of information are appropriate for the types of questions being asked. And relevancy is whether the data matters. For example, if we were trying to ascertain if people were willing to purchase a new kind of fitness watch, whether or not they have already purchased a fitness watch would be relevant. Whether they liked ice cream would probably not be relevant (though how much ice cream they eat might actually be relevant).

## 9-1. Interactive

1. **Class Discussion:** Imagine that you are the owner of a large business in an area where a hurricane is about it hit. What kinds of actions should you take? What are the potential problems? What are the potential solutions?
2. **Class Discussion**: What might be the problem with planning too much for the event of a disaster? What would the disadvantages of over-planning be? How can you know when you've crossed the line?
3. **Class Discussion**: Do you think outsourcing is a good idea or a bad idea for a business? If you were a business owner, would you consider it? Why or why not?
4. **Class Discussion**: How can you tell an authentic request for information or help from a scam or phishing expedition? How many fraudulent emails do you usually get a day? Have you ever responded to a fraudulent email? What happened? Have you ever been the victim of malware? What happened?
5. **Independent Activity**: What would you do in the event that you could no longer access any of your digital data? How much of your personal data is backed up? How recent is the backup? Is it a full, incremental, or differential backup? How much is backed up and in a secure location apart from the location of the original data?
6. **Independent Activity**: Make a list all the usernames and passwords you have. (Do NOT share the list with anyone – ever.) How strong are your passwords? How many do you have? How many of them are the same? How many systems or websites do you use that share the same username and password? What would you do if someone were able to get into all of your sites and systems and pretend to be you? Have you ever given someone else your username or password? Why?
7. **Independent Activity:** Take a look through a magazine, pick a product at random, and count the words that give you meaningful information and the words that are just marketing fluff. What's the Hypometer score? Try it again for an advertisement for an IT product or service you've just recently purchased.
8. **Independent Activity:** Review a few of the websites that purport to talk about business analysis and big data. How many of the systems discussed in this chapter are still around? What do they say they do?

## 9-1. Self Study Test Questions:

1. What is a Systems Analyst? What is a Systems Approach? What is Systems Analysis?
2. What other people might work with a systems analyst?
3. What are some other names for programmers?
4. What is Systems Development?
5. What two classifications are there for systems development methods? How do they differ?
6. What are at least four of the systems development methods? How do they differ?
7. Why do some companies do systems acquisition instead of systems development?
8. What are the phases of systems acquisition?
9. How can we avoid hype?
10. Why should we minimize customizations that require changes to source code?

11. What kinds of fees might be involved in purchasing complex enterprise software?
12. How much of a project budget should be spent in training?
13. What is meant by software support? To the vendor? To the business? To the user?
14. What are the advantages and disadvantages of outsourcing?
15. What is meant by cloud computing?
16. What is BCP? Why should a business have one?
17. What are the three types of backup? What is the difference between an image and a full backup?
18. Why is purchasing technology is fraught with difficulty?
19. How do we know that the field of information technology is immature.
20. When do you think the specialized terminology of IT will enter general domains of knowledge?
21. How would you characterize public infrastructure – still in infancy, evolving, or mature?
22. Are IT components integrated or must they be purchased separately and assembled by the buyer?
23. Why are the vendors of IT products and services are often subject to volatile finances?
24. How do you recognize when IT products and services are hype?
25. Why do IT products and services have amazing capabilities but strange limitations?
26. Can IT products make judgments or meaningfully relate two items? Why or why not?
27. Can IT products speak natural language? Can they follow rules of thumb?
28. Why can't IT products clean up dirty data or fix process problems?
29. Why do IT buyers often have unrealistic expectations about IT products?
30. What about IT concepts makes communication about products and services difficult?
31. What accounting rules and standards have not kept up with IT project realities?
32. How has the lifecycle of IT products changed in recent year?
33. Why has getting technical information about IT products and services become more and more difficult?
34. What steps can help you make good decisions about IT?
35. Why should you develop good relationships with IT vendors?
36. Why should you become technically literate before purchasing IT products and services?
37. Why should you define enterprise-wide standards for hardware, software, and processes?
38. How do you maximize periods of stability, and minimize periods of instability?
39. Which phase of the product life cycle should you buy hardware and software?
40. Why should you buy hardware and software that has been popular and stable?
41. Why should you buy in generations (all components manufactured or developed around the same time frame)?
42. How do you develop long-term partnerships with vendors of integrity?
43. Why should you ignoring hype and stay off the bleeding edge?

10

# Chapter 10: People of Organization

Overview: In this chapter, readers will learn the following topics:

- ⇨ Why Non-IT People find IT So Difficult
- ⇨ What We Need to Know About Training
- ⇨ Why IT People Are So Difficult
- ⇨ How to Manage IT People More Effectively

## Why Non-IT People Find IT So Difficult

A typical person, a non-IT person, often considers the IT people that they meet in companies or on technical support calls to be *difficult.* They actually have a point. Between the learning curve, the speed with which technology changes, and the immaturity of the field, non-IT people are at a disadvantage.

***Everyone Has a Learning Curve***

I know how difficult it is to understand information technology because I had that same difficulty myself—in the beginning. When I worked on my bachelor's degree to become a teacher, not one word about computers was mentioned. Nonetheless, in my very first job, in 1982, I was given a computer (an Apple IIe). Being the dutiful teacher I was, I took a class in how to use computers. But when I tried to apply the knowledge to the computer in my classroom, it didn't work. I took another class. I still couldn't use the computer. A third class—and I was still completely lost at sea when I tried to use the computer in my classroom. My father was a computer programmer, but the mainframes he worked on were completely different from the beige box on my desk. I was highly frustrated.

I was so dumb that when I borrowed PFS Filer (a database) from the technology resource room, I couldn't understand why I couldn't edit all the student names and addresses I had entered after I had returned the program. The concepts of application and data and storage were foreign to me. It wasn't until the fourth computer class that I figured out what I was missing. Finally I had met a teacher who taught me about word processing, spreadsheets, and databases, and understood how they worked instead of trying to teach me how to program using a computer language (which all of my previous computer courses had done). This experience, and my subsequent career in becoming an entrepreneur and starting a business to teach people how to use computers, made me think quite a bit about information technology and how we attain knowledge about them.

### *Assessing Technological Skill*

There are two main characteristics (shown in Figure 94, which is a copy of the same illustration found in Figure 3 in the first chapter) that identify any particular person's ability to work with computer systems: experience and ability.

I remind you of the question we posed in the earlier chapter: How long does it take a person to learn to play a musical instrument? A day? A week? A month? A year? More?

Chances are you said "more."

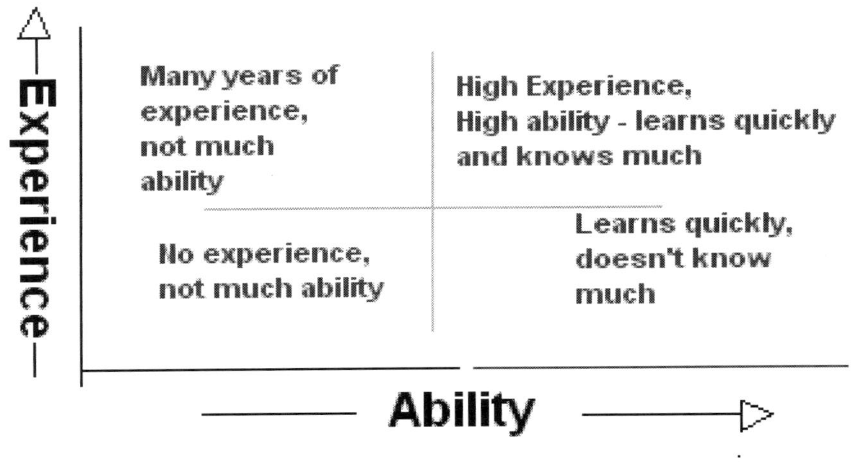


**Figure 94. Experience and Ability.**

***Experience and Ability Applied***

I took piano lessons for 10 years. I practiced enthusiastically every day. The result? Well, let's just say that even my friends grimace when they hear my plodding attempts to play a piano piece for the first time.

It doesn't matter how much I practice or how many years I study, it takes me about six months to learn to play a new piece so that it doesn't elicit the nails-on-chalkboard response from other people. And it better be an easy piece or I can't play it at all. I simply have no talent at playing the piano. If I work on the same piece for a year I might even sound good enough that the uninitiated would think that I was pretty good. But in truth, when it comes to musical instruments, I have high experience but low ability.

My former husband is just the opposite. He can pick up any musical instrument in the world—with no prior practice or instruction—and figure out how to play it very quickly. He's never had a piano lesson in his life, but unlike my pedantic pounding, his songs are beautiful music. His ability with music is extremely high.

While I started off slow, technology is a little different for me now. I can sit down to any computer application program, and within days or weeks be using it like I was the original developer. In the 1980s, when I taught individual CEOs how to use the computer programs on their computers, I often would walk in to the lesson having never seen the programs before. But I could always figure it out as we went along. By the end of the lesson, my CEO client would think I'd been using the program for years. I have high ability with technology.

But I recognize that others are not so lucky. I help many of my friends and family with technology, so I know that the norm is to have no idea of what's under the covers or how to make it work smoothly and efficiently.

For people who have low ability with technology, their experience on one system might make it appear that they are very good in technology, but their experience is not able to be generalized. I can play the piano, but I can't play a guitar or a flute. Someone like my former husband, with high ability, can play any instrument. I can use any technology program because of my ability, not just my experience.

If we don't have some way of assessing people's ability with technology, the hardware and software chosen will be more appropriate to the IT people who make the decision. Because IT people tend to be able to quickly and easily come up to speed on a new technology, they often make the assumption that most people are the same way. Nothing could be further from the truth. For 90 percent of the population, productivity on a new system occurs after years of using the system. Training can shorten this time period, but no matter what it takes most people time to become productive on a new system.

So here's the problem: What happens Why Nwhen the system changes every two years? For most people, they just start feeling comfortable on a system—just start to become productive—and the system changes and they are back to square one. The speed of change in technology is the biggest reason why people with low ability find IT so difficult.

Why does it take so long? Studies of knowledge have led many experts in the field to believe that the issue is one of indexing—making connections to other knowledge, even if this other knowledge seems unrelated to the current information being learned. People who pick up knowledge quickly either have already built indexes to hold that knowledge, or they know how to create their indexes in order to connect that knowledge. The daunting task of building the indexes is time consuming and more difficult the longer you have lived without them. The next generation of workers, of course, developed the indexes as children growing up. They are way ahead of today's generation of business leaders when it comes to technology.

Furthermore, using a computer is a combination of both intellectual and kinesthetic skills (both thinking and eye-hand coordination). Like learning to drive a car, most people learn to use a computer by using a computer. You can't learn it from a book. The more people use technology, the more quickly they pick up new technologies. The more similarities between new technologies and the technology they have already experienced, the more quickly the new methods will be learned.

People with high experience and high ability tend to gravitate toward highly technical jobs. We will probably find more of them among the information technology divisions of the company. They tend to be the most flexible regarding information technology. They will learn new technologies quickly, and they will be able to rely upon past experience to help them make the most of any information technology they use.

High-experience & high-ability people in technology, however, are not appropriate for all jobs. They tend to get bored easily, and do not do well in jobs that require attention to detail or unchanging tasks. They also don't tend to do as well managing people, or where a high degree of collaboration is necessary.

Specific jobs using a single system are sometimes better suited to people with lower ability. For example, an accounts payable position entails using the same system over and over again in the same way. Another example would be someone who enters shipments into a warehouse receipts program every day. In these cases high experience in the system used would be a bigger plus than general high ability. As a matter of fact, this combination can be spectacular. If a person with low-technology ability gets into a groove with a system, his or her personal productivity goes into a super-summit mode. It would be very hard for a competitor to match that level of productivity.

## Understanding the Nature of the Gap

To be personally more productive, we have to do more than learn just the basics. We need to learn all the little tricks and techniques for using a system more effectively. Furthermore, we cannot assume that our employees are taking it upon themselves to become more knowledgeable and experienced in any particular technology. When queried, people will usually say that they already know how to use a technology. Even if they realize they have much to learn, they find it difficult to admit when they don't know something, and often don't want to spend the time to learn more. This is human nature.

Let's consider an example. Imagine that we are running a large business in widget distribution. Now imagine that occasionally we hire someone who knows how to use the cell phones that we provide, but for the most part, the people we hire have never had to use a cell phone before. They pretend, of course. They pick up the phone, hold it to their ear and talk, imitating the people who actually know how to use the cell phone. But they don't actually know how to dial the phone and press the button to launch the call to talk with someone. Should we just ignore the fact that they don't know how to use the cell phone? Of course not. We would insist that they learn how to use the device properly so that they can do their jobs more effectively. We would bring in special trainers, if need be, because learning to use the cell phone is such a basic requirement of the job.

Of course, in today's environment almost everyone (except really old people) already knows how to use a cell phone. But what if they didn't? Imagine you are one of the people who doesn't know how to use a cell phone. You go into the office, and everyone else acts like using the cell phone is so easy a toddler could do it. They all seem to have no trouble using it. Would you go to any of those coworkers and say, "How do you make a call? How do you know what number to enter? What should you do when it rings?" Of course not. You might be embarrassed to admit that you don't know, and you won't want to face sneers or jeers from people who do know (although they are no better than you; they just had someone show them already).

But we can't allow the gap to continue to widen. As time goes on and the majority of people learn how to use this technology, the problem will get worse for the few who haven't yet learned it. The fear of appearing dumb prevents people from learning more about how to utilize hardware and software more effectively. Unfortunately, ignoring the problem tends to backfire. An obvious apocryphal example of the CEO who insists on the fancy laptop—but who has never opened it up. The most loved Dilbert cartoon is about a laptop that Dilbert gave his pointy-haired boss. Dilbert carefully explained that it had to be shaken upside-down in order to reboot. Amazingly enough, it looked just like an Etch A Sketch.

The less obvious but more common example is the person who enters the column of numbers into a spreadsheet, and then uses a calculator to add up the column of numbers and enters in the total. (You would be surprised by the number of people who do this!) Or the people who press the [Enter] key at the end of each line when they type into a word processor—especially when they want the second line indented (and they don't know how to use the indent function).

As entrepreneurs and technology leaders, it's important that we recognize the huge gap between those who have been using information technology regularly and those who are relatively new to the process or use it infrequently. Information technology is easy—when you already know how it works. Information technology can be devilishly difficult—when you don't already know how it works.

It often takes quite a bit of creative encouragement to get people to learn it. As noted, due to embarrassment or learned helplessness, few people will ask for help in learning to use a personal computer more effectively. An entrepreneur or business owner can cultivate a business culture that encourages learning in order to offset this tendency.

First, it is up to entrepreneurs and business owners to get over their own humps.

***The Executive Program of Computer Training***

I know how difficult it is for executives to get properly trained on technology. In the 1980s, a great deal of my computer training business was private, confidential, individualized instruction for the CEOs of many of the larger businesses in the area. These senior leaders could not ask their employees to teach them without losing status. They could not take a class (many of which are not paced properly for highly intelligent leaders who didn't have any context for information technology and therefore needed a completely different approach to learning it). What they desperately needed, and what I gave them back then, was a customized program of instruction, spread out over the course of six months to a year, in blocks of 90 minutes, two or three times a week. I called it "The Executive Program," and it was many times more expensive than any other class instruction. But the executives who went through it back in the 1980s are undoubtedly much more well-informed today about information technology topics because they got over the hump early. Those who didn't are now 40 years older, and still on the wrong side of the information technology hump. And now they face competition from youngsters who grew up with computers.

Once they've gotten over their own hump (and even if they haven't), it is up to the leader of an organization to establish a culture where people are not made to feel inferior because they don't yet know how to do some function on the computer, while at the same time demonstrating the importance of the knowledge. It is up to the leader of the organization to encourage people to ask questions and figure out a more effective and efficient way to do everything they currently do.

It makes the most sense for us, as business leaders, to take responsibility, find out where our employees are in their level of knowledge, and train them on the gap between what they know and what they don't know.

It is also important to understand the extensive time involved in the process. Information technology is not something that can be learned in a day-long workshop. Not even a week-long training class. Not even a month-long course. Learning to use information technology effectively, when starting from square one, takes years and years.

Most important, it is up to the leader of an organization to recognize that while people are learning new technology, they must be given leeway regarding their productivity. We can only expect drastically increased productivity once they have fully learned the new technology.

## Making the Assessment

I wish I could say that there is a specific test that we could give prospective employees to determine their technology ability and experience. The truth is that an assessment has to be based upon the particular needs of a particular job. I've found that, in most cases, an informal assessment by the supervisor works better than a formal paper-and-pencil assessment. Even low-ability employees are able to shop on-line and belong to on-line communities such as LinkedIn or YouTube or Facebook. But high-ability employees will be the ones with an integrated Google contacts which they can view on their phone or their computer. They are the ones with several email accounts, a mixture of POP3 and IMAP.

Chances are, you can tell high-ability employees by their lack of fear of learning new systems. You might want to ask employees if they would like the opportunity to learn a new software. Those who are willing to volunteer are more likely to be high ability.

Experience is a bit more difficult. Experience can be both wide and deep. If employees know how to use 10 different computer programs, their experience is wider than if they only knows two different programs. But if employees have used only one or two programs for the past 15 years, their experience is deep, not wide.

Level of experience can be estimated by simply observing the employee using the program. An employee with strong knowledge of a program will work very quickly and take many shortcuts; he or she will seem to fly through tasks. An employee with a low level of knowledge will use the menu (often the slowest method), read everything on the screen, and take much longer to accomplish the tasks. With high-experience people, an observer will have trouble keeping up with how they got from screen A to screen B. With low-experience people, an observer might feel slightly impatient because they seem to take forever to get anywhere.

One way to create an assessment for new people is to first work with someone who has high ability and high experience in one of the programs utilized by your business. Watch what they do and how long it takes them to do things. Then assess others' capabilities against those measurements.

Here's an exercise that would assess, for example, the technology skills for the word processor: Go into a document that someone within your company has sent you recently and change whatever option you must to reveal the invisible codes. (In Microsoft Word, it's usually a button with a ¶ on it.) Look at a paragraph that's been indented. If people have a ¶ at the end of each line and have manually lined up the indent with either spaces (.) or tabs (→) then they never learned to properly use the word processor, and they are wasting time each day lining things up manually. Another test: Did they use a template, or did they manually make all the changes needed to the normal template? (You can tell by going into the File—Properties choice and seeing if the template says "normal.dot" or something else.) Do they use Styles, or do they manually change the format for every word and paragraph? Watch your employees copy and paste. Do they painstakingly go back to the main menu for everything, or do they use the context mouse buttons or shortcut keys (Ctrl+C to copy and Ctrl+V to paste)?

You may think that these are small things, but all the hundreds of small things add up to significant difference in productivity over time—hundreds of hours a year. If your people are not as educated in using the software as they could be, they are wasting valuable time every day.

## Accepting the Learning Curve ROI

Many years ago, I worked with a client whose secretary, Rosemary, typed everything on a word processor, but she never saved anything—ever. She typed it, and then exited out of the word processing program because she didn't see the need to save. In other words, she was using the word processor as if it were a typewriter. For her it was quicker and easier to retype the document than to learn how to retrieve a saved document because Rosemary could type up a one-page letter in five minutes. It would probably take an hour for someone to show her how to save and name a document the very first time. This short-term hit to productivity was why the fairly intelligent manager, Pam, continued to allow her secretary to retype the documents instead of saving and retrieving them (not necessarily a bad thing). Pam wasn't willing to wait an hour for a document, and though she knew that Rosemary wasn't using the system properly, she wasn't willing to demonstrate the proper way, or encourage Rosemary to take a class so that she could learn the proper way. Pam did not even complain about the fact that it took much longer to get a change made to a document because Rosemary had to retype it from scratch each and every time. In Pam's mind, the difference was minimal and few enough documents needed to be changed.

The problem is that the company is losing out. By enabling Rosemary to use the word processor like a typewriter, the company is losing the electronic history of documents that would be helpful to the person who takes over for Rosemary after she retires. True, the ROI on a single letter is negative because Rosemary could type, print, and exit in less time than it would take to learn to save. However, the cost of copying, organizing, and storing copies of all those letters would be a drain on the entire company. Furthermore, the difficulty in making a single change may mean that the company was avoiding needed changes unnecessarily. When looking at the whole picture instead of just one document, the ROI on learning to save documents makes a lot of sense.

Of course, there are times when we have to drop our quest to do it a better way and simply get it done any way we know how. It is not easy to know where to draw the line. In a few cases, the front-end learning curve is not worth it for a single event. It makes little sense, for example, to learn how to mail merge (automatically create a letter for each name and address in a separate list) if we only send out one mail merge every five years to 20 people. In that case, we should simply save a template and enter the name and address manually. But most times, learning to mail merge is well worth the learning curve because once we've learned how to do something, we will find 10 others ways to utilize that knowledge to make things more efficient. It may be, for example, that we only send out one mailing every five years because it's so hard to do. Once we've mastered mail merge and it's a five-minute task instead of a 10-hour task, we may well use it several times throughout the year—increasing sales and profits and making 200 or 300 times return on our investment.

There are also times when we have to give up trying to change the people in our employ and just let them do things either manually or using whatever system they prefer. It is a smart strategy to wait until the extremely reliable and fast employee retires to try and implement a new system because chances are, we will completely lose the productivity of that employee if we switch systems and force him or her to use it.

Unfortunately, because it's difficult to quantify the return on the outset, often managers are unable to take that initial hit to productivity during the learning curve. In many cases, managers aren't really sure what's worth it and what's not. The difficulty is that the larger the gap between the employees who can easily use information technology and the employees who find it harder to use, the worse it gets. Not only will the people themselves be less productive than the technically knowledgeable ones, but there will also be more resistance to changes and new technologies.

## Overcoming Initial Resistance

There is always initial resistance to any new system. However, once the benefits of technology become obvious, no one ever goes back. When was the last time you saw an office filled with typewriters, index cards filing systems, telex systems, or rotary dialing telephones? Once people use word processors, contact management, e-mail, or spreadsheets, they rarely go back to doing it the old way. Once people discover how easy it is to find information on the World Wide Web, they never go back to wasting time trudging through library stacks. But first someone must show them how to use the word processor, contact manager, e-mail, spreadsheet, or Web browser. And someone must encourage them to use these tools appropriately.

Overcoming resistance to new technologies can sometimes require creativity and effort. Overcoming resistance is always possible when the benefits of the technology are real and experienced firsthand by the users. But the perception of benefit cannot be forced.

Jane and Leslie, the COO/CIO team for a network of hospitals in California, tried five different ways to implement a computerized physician order entry (CPOE) system. They tried four different vendors. They tried PCs, tablets, and PDAs. No matter what they tried, the physicians demanded that the system be removed.

Why? It took those physician five seconds to fill out a prescription on paper. The fastest physicians took two to three minutes for a computerized version and most took up to fifteen minutes. According to Jane, the doctors were frustrated, and felt that the system was stealing money out of their pockets because of the crippling impact on their productivity. Even when the system was finally implemented (due to pressure from the CMS which mandated that health systems use computerized prescription systems), it took about a year for physicians to recover to 90 percent of their original productivity.

One of the most difficult areas to deal with regarding resistance is sales-support software. Sales administrators would love for their salespeople to utilize information technology to the max because it would give them more control over the process (they believe) and enable salespeople to make more sales in less time. Certainly, there are some information technologies that are very helpful to salespeople. Wireless communication, cell phones, dynamic inventory information, and remote sales order capability can increase sales. But customer relationship management (CRM) programs often meet with strong resistance from salespeople. Like the physician who loses time on every prescription, CRM programs often interfere with the sales process and steal important relationship-building time away from salespeople. I've seen cases where the best salespeople in a company refused to use administratively mandated programs, while the worst salespeople in a company used them eagerly. If the benefits were real, the opposite would be true.

The best way to overcome resistance is to have a large number of users, not just a token few, involved in the decision making regarding any new system. George, CEO of a large regional teaching hospital, learned this lesson the hard way.

Ten years ago the administration tried to make the emergency room (ER) entirely paper free. The project planning lasted about six months and was done mostly by administrators, and the system was installed within another six months. George mandated its use by all doctors and nurses in the ER, but most of the doctors and nurses refused to use it. The hospital eventually scrapped the project.

For a more recent medical administration automation (MAA) project, George used a different tactic. There was still a high level of resistance initially. But this time the administrators involved many doctors and nurses in the project planning. They then spent two years in planning and then another year in piloting the system on various floors. Unlike the paperless emergency room system, this time they also included a great deal of ongoing training and support in the plan.

The results of the pilot clearly told the tale. Staff who had worked on the floors *with* the new technology were refusing to go back to work on the floors *without* the new technology. Once the doctors and nurses experienced the benefits, they became fans of the new system. Word spread to others, and for the most part resistance disappeared.

In summary, there are four keys to overcoming resistance to new technology:

- Make sure the users of the system are the ones who get to make the plan and choose the system (not administrators or information technology folk).
- Spend more time planning than implementing.
- Allow for a generous start-up time while the new technology is being learned without the pressure of quotas or other measurements based upon pre-technology processes.
- Make sure that there is ongoing support and training—not just at the beginning of the implementation, but for years afterwards, too.

## What We Need To Know About Training

Research is pretty clear that trying to save money by limiting training is not very smart. Organizations that have invested in new financial systems, but did not invest in training, did not do as well as organizations that invested in financial systems and the training for them.[*] The key was the percentage of the staff involved in the training. In one study, companies that involved higher percentages of their staff in training had higher returns from their information technology investment. Another study found that organizations that spend roughly 20 percent of project funds on training succeeded, while organizations that allocate a mere 4 percent to 6 percent to training (a much more common amount) tended to fail more often.[†]

You may think that this is common sense—I certainly do—but I've been surprised, again and again, by decisions of senior leaders to cut the training budget, or to plan only for initial and not ongoing training, or choose to train only a few people.

Why do entrepreneurs make the decision to cut the training budget so often? Why is the training budget one of the first things to go when money gets tight?

For one thing, training quality can vary widely. If one company purchases a training tutorial for 5,000 employees at a cost of $100,000, and another company custom-designs and implements live training for five different levels of employees with an experienced trainer at a cost of $100,000, are they both equivalently trained? My experience says "no," but there are lots of vendors of on-line tutorials who will disagree with me.

For another thing, it is rather hard to trace specific financial returns to training. The cost of training can get very high. As noted earlier, 20 percent is not too much to spend, yet very few companies invest that much.

Imagine that you are a decision maker about to commit $5 million to implement a new system. The project manager is about to launch the project including a 20 percent budget for training: $1 million. Imagine, when being presented to your boss, you get the following reaction: "What do you mean we need to budget $1 million for training—are you crazy?" The answer is to remember the value of a failed project. Would you rather invest $5 million in a successful project or $4 million in an unsuccessful project? **Implementing a new system without paying** a significant amount **for training is like buying a car and then refusing to purchase any gasoline for it**.

The third reason is how training appears in an ROI calculation. According to generally accepted accounting principles (GAAP), training should be listed as a one-time expense. This might put the ROI on the negative side, and doesn't accurately reflect the valuable ongoing benefits of training. One way to adjust for this problem is to include the training dollars in maintenance kits, documentation costs, and upgrade costs, all of which are capitalized and depreciated over time. The ability to capitalize the training costs is why so many vendors are willing and able to bundle

---

[*] Erik Brynjolfsson, now a doctoral graduate from MIT Sloan School of Management, did a study on "Intangible Assets: How the Interaction of Computers and Organizational Structure Affects Stock Market Valuations." A paper based on his dissertation was published in *Brookings Papers on Economic Activity,* compiled by Laura Hitt Brynjolfsson from the Wharton School at the University of Pennsylvania, and Shinkyu Yang from Stern School at New York University.

[†] Nay, Douglas L..2003. An Empirical Study Of The Effects Of Enterprise Solutions Of Profitability And Internationalization. PhD diss., Lubin School of Business.

training dollars with the software and hardware—they know that they wouldn't get those training dollars any other way. While not illegal to bundle training costs, it is better, in my opinion, to have the training dollars spelled out. But if the financial arm of a company places limitations on the amount of training because of its impact to the return on investment calculation, then I see nothing wrong with encouraging a vendor to bundle the costs. Good training is, after all, one of the most valuable components in the purchase of a new system.

Of course, getting training from the vendor of the hardware and software may work fine—but it also may have problems. For one thing, because of the difficulty of getting recognition of the importance of training, hardware and software vendors know that they have a captured training market. That means the training from the vendor of the hardware or software often simply isn't very good. Vendors don't have to pay top dollar for really great trainers or put as much effort into the quality of their training. They don't have to try so hard because they can fill up their classes with trainees who are getting the training, in their minds, "for free." Unfortunately, when the quality of the training goes down, the value of the training goes down as well. It's like trying to run a car on a mixture of gas and water. The magical effect that training has on making systems pay off disappears when the training is not of a high enough quality.

I'm not saying all vendors have poor training. Some do a very high-quality job. But you can't always count on that—so it makes sense to check it out before you accept the training as part of the package.

## *Just-In-Time Training*

Another key to employee productivity is just-in-time training. Rather than sending people off to day-long or week-long workshops, it makes much more sense to ensure that anytime employees needs a particular set of training, they can get it. Need to use the invoicing module for the first time next Friday? Schedule a few hours of training on Wednesday or Thursday. Need to send out 150 letters to people using a mail merge next Tuesday? Schedule a few hours of training on Monday.

People tend to learn more (1) when the training is immediately relevant and (2) when the training is hands-on. Just-in-time training can ensure both of these things in a cost-effective manner. And with new technologies such as Zoom for web conferencing that allows people to share their screens or trainers to quickly record little training videos for individual needs, high quality training is more cost effective than ever before.

## *How to Tell the Quality of the Training*

So how can we tell if the training we are considering is high quality? First, it relies upon live trainers—and trainers who do both training and consulting for a living. Trainers who do nothing but training rarely have the real-life experience that is the essential secret ingredient to high-quality live training.

Second, the training is flexible enough to accommodate the needs of the people taking the training. When a training group is relatively large, it's impossible to meet all the needs of everyone. But sometimes, even if the training group is small, the curriculum requirements restrict the ability of the trainer to answer questions and get into the kind of details that will really help the trainees. So courses with a too-strict curriculum are actually detrimental to high-quality training.

You might think that I would add high-quality materials to this list, but in practice I've found that although materials can add to the quality of training, they are not essential for high-quality training. Furthermore, the high cost of material development often outstrips its benefit, especially in the constantly changing world of IT. The experience and talents of the trainer are really the deciding factor between an okay training and great training. A great trainer can add great value just by sharing experiences. A poor trainer can have the best materials in the world and it won't help the quality.

Additionally, in my experience, there is no substitute for live training with an experienced and talented teacher. I have found many tutorials to be a complete waste of money. Five hours of a lock-step tutorial aren't nearly as effective as 10 minutes with a high-quality trainer.

### *Matching Hardware and Software to Ability and Experience Levels*

Experience and Ability characteristics should also figure prominently when we make a decision on hardware and software. To mitigate the effect of the difficulty in learning new technologies, we can assess people's skill in technology and match the hardware and software types to their ability levels.

When dealing with employees with high ability, we can choose software with more power. Software has more power when it is possible to program within the program with templates, macros, ad-hoc reporting, and so on. Word, Excel, and Access are all much more powerful programs, for example, than PowerPoint, Works, or Encarta. When dealing with employees with low ability, ease of use should be the guiding factor when choosing a program. Easy-to-use programs are sometimes not as powerful, but they are much more effective among people who have less need of power. Most people will never use the advanced features of software anyway, so why confuse them with the option? Often, a good pairing comes from someone with high ability customizing the programs for people with lower ability.

Problems can occur if there is a mismatch. If we give simple technology with few options to people with a great deal of ability and experience, we limit their productivity. They will be frustrated at what they know they could do if they were only given the proper tools. At the same time, we can't give sophisticated and complex systems to people with low ability or experience, or the capabilities will be wasted and the people will still not be able to get the information they require. In the end, balance is the key.

## Why IT People Are So Difficult

Let us now turn our attention to that 10 percent of the people who actually develop, install, and maintain all this information technology. If it hasn't become apparent to you yet, let me point out that technology people are not the same as non-technology people. The stereotype of the geek, pockets filled with gadgets, hunched head-down at the computer, effortlessly gliding around the virtual world (while a face-to-face meeting would tie the tongue and put knots in the stomach)—is not always far from the truth. The previous section was about general staff—administrative, sales, marketing, and operations employees. This chapter focuses specifically on managing information technology staff, as well as IT vendor personnel. And while we know that there are always exceptions to the rule, and that generalizations are never 100% accurate, for the purposes of understanding how to manage IT people, we are going make some generalizations about them in this section.

First, IT people seem to have completely different motivations from most businesspeople. Second, they often don't have the same people skills as non-technology people. Third, they sometimes pursue technical excellence and logic to the exclusion of common sense. Fourth, they may have developed a sense of entitlement, especially if they have achieved guru status in the IT world. In order to use their skills effectively, we will need to adjust for these differences.

### *Why Business Leaders Need To Know Their IT Staff*

Some small business owners can avoid the pitfalls of managing IT people directly, because their businesses are small enough that they don't actually have any hard-core IT people on staff. They either rely upon vendors, or struggle along without any real IT guidance. However, IT vendors are filled with IT people, so it would behoove all entrepreneurs and business owners to read and understand this section.

***Replacing the IT Gurus – or not.***

In my own background, two of the start-up companies in which I was involved were based upon the technical expertise of gurus (which is not me). One company was based upon the brilliance of enterprise network management engineers, and the other was based upon the superstar capabilities of Web site developers. In both cases, the success of the business was directly related to how well the leaders of the organization managed the gurus of the organization.

Once I was hired by the CIO and CEO of a midsized manufacturer because their director of IT was difficult. They wanted to fire him. They were unsure what would happen to the network, however, since neither the CEO nor the CIO knew anything about how the network was set up. All they knew was that this director of IT was spending too much money and giving them a hard time. He was disrespectful to them. He was ignoring their directives and going off and doing things on his own. So I came into the organization as a consultant ostensibly to write up a report on network security, but really I was there to learn as much as I could about the network so I could quickly secure the network when they fired the director of IT, and to help them find a replacement.

What I found was surprising. In every nook and cranny I checked, this director of IT had made the right decision. As I looked at the history of the network, I could see how he had taken a very backward system and, bit by bit, and for very low cost, transformed it into a high-performing, secure tool for the administrators of the corporation. He had also taken a look at the old-legacy manufacturing system and started down the road on an excellent replacement plan.

To top it off, he was not paid very well for his efforts. As I started the search for his replacement, I could see that it would cost three times his salary to find someone with the breadth and depth of his domain knowledge.

I delved a bit more deeply into the conflict he was having with the business leaders. Apparently, they had given him some directions he felt would compromise the network. They told him to ignore security protocols and switch over to an obsolete operating system. Initially he argued with them. He was not very tactful in his responses to their "boneheaded" commands. Finally he simply ignored them and didn't return their calls. But he was still working 60 hours a week ensuring that they had the best infrastructure they could possibly have at the lowest possible cost. I was able to convince the CIO and CEO to keep him, and he is still working there as I write this.

Many of today's entrepreneurs will find this section extremely useful because they may be trying to build a business around the skills and talents of a particular IT guru or team. Ethernet, Linux, enterprise network management, the public Internet—all of these innovations and hundreds more came out of virtual teams of technologists in a relatively short amount of time. They were built within a special IT culture of knowledge. But it was rarely the technology gurus themselves who did the business building. For most, there needed to be a partnership between IT people and non-IT people.

This story of the manufacturer senior leaders trying to replace an irreplaceable and effective IT guru demonstrates the pervasiveness of the difficulty when non-technologists try to lead technologists using standard hierarchical "I know better than you" command structures. It also demonstrates that good technologists often have poor people skills.

There are many experienced technologists in the field, but few who can both do the work and work with people. Often, executives have to make a choice. Is it more important to have superior technology at a low price? Or is it more important to have people who get along together and have substandard technology?

## Characteristics Of IT Folks And Gurus

As noted, many entrepreneurial businesses are dependent upon the skills of an IT guru. Unless they are managed right, gurus themselves can be the obstacle to their own success. They have a tendency toward arrogance, and often get involved in petty technological religious wars (like

AWS versus Azure, or Microsoft versus open source) rather than focusing on accomplishing the task at hand. Even if the business leader is not counting upon the specific talents of a guru for the growth of the business, all businesses need brilliant technological people to quickly fix issues.

Despite all the drawbacks brilliant IT people are often essential for success. Technological people of lesser skill may end up being nicer, but they also tend to spin the wheels of never-ending problems and may do more harm than good when it comes to fixing issues that impact productivity of the whole company. Gurus keep the computers running smoothly and solve every problem that comes up. We can't live profitably without them.

If your company has been lucky enough to find brilliant technology people, then you know what a big difference it makes in how productive your company can be when using technology. But if these same people have become a thorn in your side, don't despair. Don't make the mistake of letting them go. Smoothly running computer systems are often worth the risk of fracturing a few feathers. We just must learn how to manage them.

### *Techno-Jargon*

Very few people understand the language of information technology. Techno-jargon is the terminology that IT folks use to converse. Believe it or not, techno-jargon is not purposefully used to exclude non-IT people. Techno-jargon is used because IT folks must discuss issues at a level of detail that is beyond what most people know about IT, and the jargon provides the shades of meaning necessary to have accurate and meaningful conversations. A doctor can't just say the leg is broken; there are several bones in the leg that might be affected. A doctor, when talking to another doctor, would need to specify an intertrochanteric fracture of the femur bone.

Furthermore, the jargon gap is due, in part, to a time lag before the concepts permeate the culture. The vocabulary is not well known among the general business population yet. Businesspeople, after all, have just recently even begun to think of technology as part of the operation, so there is a gap between their knowledge and the vocabulary. The gap will eventually disappear and what was once considered lingo will be part of the general knowledge.

Don't believe it?

Well, imagine that you have just been magically transported back 95 years. Walk around a typical small town and talk to the people who live and work there. Ask someone to tell you what a parking brake is. Try to explain to someone how a carburetor works. Look around for a gas station or a car mechanic. When you don't find them, ask a few people where they are, and note the look of puzzlement on their faces. How many people would you meet who would know that you must change the oil in your car every few thousand miles? How many would know that it makes no sense to put your foot on the gas and the brake at the same time?

Back in the early 1910s, information about automobiles was not generally known. The terminology was specialized, and only known to the few who had built their own cars. Those who actually owned these cars were different from everyone else—so they likely hung out together and developed their own specialized culture. When problems arose they would check with their buddies instead of a commercial establishment. Garage mechanic was not yet a job classification.

Think about it. Today, most people know what a car is, what it looks like, how to drive it, perhaps even how it works. They know that a car has a body, an engine, and that it runs on gasoline. They know how to turn a steering wheel, they know that cars can't be driven under water, and they know that a Porsche is faster than a Hyundai. They know the oil has to get changed regularly, and they know that if they don't fill up the tank before they run out of gas they will get stuck somewhere. Think about the amount of general knowledge there is about automobiles among everyone—from children to senior citizens. Words that were techno-jargon in the 1920s are now commonly understood terms.

Now try to imagine what it would be like to purchase a car without understanding any of those terms. Can you imagine looking at a brochure and not understanding what is meant by a V-6 engine or power steering or anti-lock brakes?

IT folks understand buzzwords and jargon, and have a hard time explaining them to people who don't understand them. IT folks see many business leaders making decisions about technology products and services without understanding what they are buying, what components are included, or how it all works.

How many business leaders have you encountered who would know that trying to put Windows 10 on an AS400 machine is akin to putting the engine of a Volkswagen Bug inside the body of an Oldsmobile Regency? How many people know that occasionally you have to optimize the hard drive space, and re-index the databases? Or that you can't run an enterprise network management software on the same machine as a relational database? Any IT person would recognize all of these terms and would know, without having to think about it, all of these general IT facts. Most businesspeople would not, but they make business decisions that are influenced by all these facts.

That isn't to say we haven't made progress—we have come a long way. Today, terms such as *Web site, Internet, e-mail, personal computer, printer,* and *flash drive* are pretty commonly understood. Most people can name Microsoft, Hewlett-Packard, and Google as large technology companies. More than 20 years ago that might not have been true.

But we still have far to go. How many CFOs can explain the difference between an operating system and an application? How many CEOs can differentiate between a BPM system, a CRM system, or an ERP system? How many know that virtual private networks are neither virtual, nor private, nor networks? How many know that Quality of Service has nothing to do with quality or service?

Suffice it to say that IT folks sometimes feel like they are speaking English while everyone else is limited to Latin.

> Just a reminder of the acronyms:
>
> An ***operating system*** is the low-level instructions that interface between the programs and the hardware. The ***applications*** are the programs like Office, SAP, Oracle, Outlook, etc.
>
> ***BPM*** is business process management software. ***CRM*** is customer relationship management software. ***ERP*** is enterprise resource planning software.
>
> A ***VPN*** is called a virtual private network, but it is really just a way of encrypting information for privacy because the Internet has no built-in security. Using a VPN is akin to putting all of your messages in a locked Kevlar bank envelope before sending them through the U.S. postal mail.
>
> ***Quality of Service*** (QoS) refers to being able to distinguish between different types of messages on a private Internet. QoS does not exist on the public Internet. On an Internet, if a videoconference system and an e-mail system share the same network, installing QoS on the routers will allow the routers to be programmed to hold the e-mail message traffic while letting the videoconferencing traffic go through uninterrupted.

## Logic and People Skills

If you have ever tried to manage someone who has superb technology skills and knows it, you may recognize that technological brilliance often precludes the ability to work well with others in a team. They get testy when forced to deal with real problems that interfere with the elegance of their systems. According to them, if everyone would just conform to the logical world of computers, everything would be peachy keen—no one would be frustrated. They often feel that they are being forced to work in a world filled with bumbling dunderheads who don't know a kernel command from a universal resource locator call. They are frustrated at dealing with people who can't memorize command-line syntax or even troubleshoot a corrupted driver. Most people use Windows and mice to click on menus, but real IT folks like to enter the commands directly because if they know the syntax they can accomplish many things more quickly and more powerfully than if they are limited to a menu.

> In Unix computer systems, the central core is called a ***kernel***. A command directed to the hardware directly through the core programming language is called a ***kernel command***.
>
> A ***universal resource locator*** (URL) is the address of a Web page. To make a URL call is to write a command to read the contents of a Web page, as in http://google.com, which is the URL call to the home page of the Google Web site.
>
> ***Command line syntax*** is following the rules for writing commands. For example, *dir/w* is a command to provide a short wide list of all the files in a folder. *Dir\w*, for example, would be the wrong syntax.
>
> A ***driver*** is a file with instructions for using a peripheral like a printer or a flash card. Because a driver is written by a manufacturer to work with the operating system and hardware of other manufacturers, they often don't work correctly or get corrupted.

You will recognize the true gurus because they will want to automate everything. They are exasperated by a world that doesn't see the value of technology the way they do. They will spend six hours writing a program to automate a two-minute manual task done four times a year.*

People who get along superbly well with computers often do not get along as well with other people. The rules of the computer world are complex, but stable, logical, and predictable—easily understood once you've cracked the code. The real world of relationships, on the other hand, is fraught with illogical and unpredictable behavior. In a computer system, variables are easily controlled. People are not so easily controlled. Face-to-face encounters with people demand flexibility. Rejection is a potential risk to one's self-esteem. Some technology people take a preemptive strike against rejection by acting rude and insolent from the get-go. They seem to have no respect for authority or hierarchy.

To deal with the lack of people skills, the most important prerequisite is to be understanding of the behavior. Think about it this way: Non-IT people think IT is complex and difficult, and people are easy to get along with if you are friendly and helpful to them. IT people think that IT is easy to get along with if you just understand the logic, and people are complex and difficult to deal with. Just as non-IT people would appreciate a little more patience when they have trouble with IT, IT people would appreciate a little more patience and understanding when they have trouble with people. Although rude behavior should never be tolerated, a certain amount of gruffness should be expected.

Sometimes it helps to minimize contact between IT people and non-IT people. The less time the IT person needs to spend dealing with non-IT people, the fewer problems the business will have.

## Non-Business Motivations

Typical business leaders, with their core talents at managing people and making strategic business decisions that produce profits, are often totally flummoxed by the dedicated IT people who aren't governed by the same internal motivations as most people. Just what makes them tick?

---

* You may think I'm exaggerating, but I've seen it myself. Actually even though I'm not really an IT guru, in my early programming days I was guilty of doing the same thing!

The best technologists are typically highly intelligent independent thinkers, dedicated to doing a good job. As noted, they also tend to be poor communicators who don't play well with others. But there is an exception: IT people congregate with other IT people, who then provide their true motivation for work—recognition. The best IT people wish to impress their virtual peers.

The rules governing the dynamics of technology teams are simple—the most elegant algorithm or the most in-depth insight wins. Hierarchy is established immediately after public demonstrations of domain knowledge, which determines which role will be played by which guru (as tacitly agreed upon by all members of the team). Only performance matters. No one cares about looks or congeniality. Each takes turns showing off to each other.

The motivations of suits (a derogatory name IT people often use to connote businesspeople) are often unappealing to IT people. One time I carefully set up policies to govern how my IT staff would earn raises, or corner offices, or offices with windows. I shouldn't have wasted my time. It turned out my IT staff couldn't care less about money. Or a corner office. Or any of the normal trappings of success. What they care about is knowledge and getting information technology to do things. All 10 IT folks chose to stay in their windowless cubicles in a shared space rather than be separated off in their own windowed offices! (They did take the raises, of course, but personally I don't think they were very motivated by them.)

Obviously the answer to this issue is to recognize that the trappings of success that motivate businesspeople may not work for IT people. We may need to spend more time and effort discerning their true motivations.

### Aversion to Multidisciplinary Teams

Between the lack of interpersonal skills and a different base of motivation, we can see why teams that mix businesspeople and technology people often don't work. For one thing, businesspeople can't immediately ascertain who has the most talent, so they have a hard time understanding the IT hierarchy. A brilliant algorithm won't necessarily increase sales or lower costs, so it is not important to businesspeople. As a result, they often don't pay the proper respect to the achievements of the IT folk. A non-IT person might give more respect to the nicer IT person (who doesn't know as much) than to the truly brilliant IT person who has the best chance of solving their IT problems. That's akin to treating the janitor like royalty and ignoring or insulting the VP in the office.

We often must find alternatives for IT folk that don't require them to interact and play nice with others on a non-IT team.

One common strategy that doesn't work well is the matrixing solution. A *matrix organization* is one in which project teams band and disband, and in which a traditional hierarchy is hard to discern. In a matrix organization, IT personnel report to several different managers, and are responsible for prioritizing among the requests of them all. My experience is that matrixed IT people end up pleasing no one. They are put in a position where they can surreptitiously spend all their time working on esoteric technical issues without any real-life application, telling each person demanding their time that they are working on a problem for someone else. Instead of solving problems, they create new ones.

One strategy I've found that does works well is to appoint a liaison to work directly with IT people so they don't have to attend team meetings. Also pairing a non-IT person with an IT guru tends to work well. The non-IT person must have enough IT knowledge to gain the respect of the IT person, but they don't have to be super technical (just tolerant). The non-IT person can do the documentation and training for the systems while the IT person can develop, configure, and install the systems.

### The Hero and the Challenge

I don't know many IT people who have *not* experienced the thrill of being the hero. Because there aren't that many people who understand IT, often IT people can quickly and easily fix problems that completely mystifies everyone else. I can tell you, there is no better feeling in the world than to be the knight in shining armor to an office full of non-IT people.

However, I've noticed a very bad trend. IT people, in seeking opportunities to repeat this feeling, may subconsciously manufacture crises that they are then called in to fix. Even if not precisely manufactured, IT people may not do all they can to ensure smooth-running systems. They scorn documentation and often will not share their secrets of configuration or programming. They do not have any motivation to ensure that systems can run perfectly without them. Just as businesspeople can become addicted to the urgent rather than the important, IT people often become addicted to the crisis solution rather than the stable system.

Face it; the stable smooth-running system is no fun. Once a system is working perfectly, most IT people will feel compelled to tinker with it, ostensibly to incrementally improve it, but resulting in constant changes that require further tinkering to fix. Vendors encourage this behavior by continually providing updates and upgrades they must install.

The truth is, a smooth-running system needs no maintenance, no upgrades, no work. With the exception of bona fide security patches, there should be no need to install or update anything. However, there is no glory in keeping a smoothly running system going. IT people can't be the hero that rushes in to save the day if everything runs smoothly. They might be ignored. Or worse—the leaders might get the mistaken impression that IT folk are not needed and fire them! I've seen this happen. If systems run too smoothly, you might forget that the IT folks who got the systems running smoothly are the *only* people who can keep them running smoothly. IT is still too complex for a second IT person to come in and take over. In other words, the configuration of a smoothly running system is unique, and would take months (if not years) to fully understand. Organizations cannot afford to eliminate their IT people (or their IT vendors).

But how do you get IT people currently addicted to being the hero to change their focus and instead ensure smoothly running systems? The answer is to build smoothly running systems into both their formal performance review and an informal pat-on-the-back habit. Keep measurements of how long the system has been running without a problem, how many backups have been done without incident. Don't count help desk calls (because that will discourage people from calling) but conduct surveys and count satisfied users.

To keep IT people busy while the systems are running smoothly, direct them toward the future of the organization. Provide them with a budget and a test lab. Have them keep on top of the standards bodies and install and test each new product or service that comes out in the domain of the Hedgehog Concept of the company. Feature their results in a quarterly or annual presentation to the senior leaders. Get them involved in planning the next generation (5 to 10 years out) of IT systems. Most of all, treat them with a lot of respect for their knowledge and skills.

### *Not-Invented-Here Syndrome*

The not-invented-here (NIH) syndrome is not unique to information technology, but it is much more of a force, I believe, in causing troubles for good relationships with IT folks.

People with the NIH syndrome believe that they are the *only ones* can develop the solution to the problem or figure out the best way to handle a process. Gurus are especially prone to this disease—often developing entire libraries of code rather than reusing code that another developer publishes.

Doesn't that mean the company owns more of the technology then? One would think that if a really talented technologist works for a business and develops a library of procedures and code while in the employ of that business, the business would be able to continue to utilize the code after the technologist leaves. Certainly they have the legal right to do so. But procedures and codes are often completely unintelligible to anyone except the original developer who wrote them. They are rarely transferable to another system. Furthermore, system configurations include thousands of variables that are only known to those who set them up. Losing a top technologist is often a major blow for a small entrepreneurial company.

The NIH impulse also leads organizations to hire their own programmers or develop their own customized software rather than take advantage of third-party or commercially developed packages. Too often, in the buy or build question, the IT people will always recommend the build road despite the fact that the buy road is more cost effective and leads to smoother computing.

### Protector of the Gap

Like most people, IT folks like to be needed. But the language and general knowledge gap between IT and most people enables them to take advantage of the situation to their own benefit; sometimes subconsciously, sometimes not.

***Don't Tell Them How Easy It Is***

Early on in my career, I was talking with a programmer who was charging the owner of a corporation quite a bit of money to code a program to calculate a sales index which needed to be generated every day. The formula was very complex, and took the owner about 15 minutes to calculate every day. Because the owner put a high value on his time, the ROI on the programming project was positive even though coding the program cost thousands of dollars. In my initial meeting with the owner, he beamed with pride at his use of information technology to save time, and he boasted about getting it so cheaply since the programmer was charging less than the going rate for programming.

When the programmer showed me what he was doing, I realized that the entire program was unnecessary. I asked him why he didn't simply teach the owner how to enter the formula into a spreadsheet. It would take only five minutes to enter the formula, and the answer would appear instantaneously each morning when the owner entered the daily numbers since the formula would automatically recalculate. The programmer's initial reaction was knee-jerk. "You can't tell them how easy it is to use a spreadsheet! Then they wouldn't need to pay me to program for them!" Since the owner was my client, I went ahead and explained how to use Lotus 123 (the most common spreadsheet at the time) to do the calculation. The programmer was not happy with me that he lost the contract, but once he thought about it, realized it was the right thing to do. (Because I praised his skills in front of the owner, the programmer was hired for the next—legitimate—programming job, so in the end he didn't lose anything.)

We can't expect people to work against their own best interests. I don't blame this programmer for his reaction; his livelihood was at stake. It is this same impulse, that made many a programmer in the 80's deny that microcomputers and spreadsheets would ever amount to anything. It's the same impulse today that causes Web developers to emphasize the complexity in developing a Web page. I'm not surprised by how difficult it is to get IT people to explain technology to a businessperson; their salary depends upon the businessperson not understanding IT.

### Mutual Mystification

The gap of knowledge makes "mutual mystification" occur often. Mutual mystification is when a businessperson and an IT person are talking to each other, but are completely unaware they are talking at cross-purposes or talking about two completely different things. In the last chapter, I shared this exchange between a Vice President at a bank and the lead Sales Engineer for an enterprise-network-management software product.

VP: **Can this software check the entry of the clerk so that the transfer always goes through?**
Sales Engineer: **Absolutely—yes. Every time a packet gets sent, an acknowledgement is returned so that the sending computer knows if it was successful. The transfer always goes through**.

The VP ended up thinking the sales engineer answered the question in the affirmative. In reality, nothing could have been further from the truth. After they spoke in that manner for about 30 more minutes, I finally jumped in to clarify and translate for the Sales Engineer what the VP meant. Once the sales engineer understood what the VP was asking, he admitted that the program he was selling was not designed to do that. But I can imagine this type of conversation happens all the time—without someone there to translate.

Imagine asking a road builder working on a new road, "When the road is finished, can the cars on it be painted blue? Will they ride smoothly?" The road builder would answer, "Yes, of course," even though the road builder has no idea what color the cars will be or if they will ride smoothly. The road builder knows that the work being done will enable a smooth ride for cars (with good shock absorbers, of course) of any color. Just because the road builder said that blue cars would ride smoothly did not mean that the road builder would provide the cars, nor that they would all be blue.

Such cross-communication is why entrepreneurs and business owners, upon finding out that the software they just implemented won't do a required task, will often say, "But I asked my IT people if this software did this task and they said it did!" The businessperson and the IT person are mutually mystified when they are disappointed in the results.

### *Technology Wars*

The topics constantly change, but there is always a technology war going on between vendors or between standards or between ways of doing things. I refer to technology wars as religious wars because of the passion and fervor of the technologists who wage them. IT people are almost always on one side of the war or another, and it impacts their ability to do their jobs.

In the past, technology people have lined up on various sides regarding a variety of "technology wars", including…

- mainframe versus microcomputer (microcomputers proliferated unchecked);
- IBM DOS versus Windows 3.1 (Windows 3.1 clearly won);
- Betamax versus VHS (VHS ruled until DVDs and streaming took over);
- Ethernet versus Token Ring (Ethernet won);
- NetWare versus NT (NT was the clear winner);
- Sybase versus Oracle (Oracle came out on top);
- NetBIOS versus TCP/IP (TCP/IP won);
- Yahoo versus Google (Google won);
- Netscape versus Explorer (Explorer won);
- Windows Server versus Open Source servers (Linux, the open source alternative, won);
- Windows IIS web server versus Open Source servers (Apache and NGINX won);
- Smartcards vs Digital Wallets (Digital Wallets won);
- Cryptocurrency versus traditional banking (traditional banking won);
- AWS versus Azure versus GWS (the war is still going on).
- Google Drive, Apple's iCloud, Dropbox, or Microsoft OneDrive (also still going on)

The winner takes the market share lead, and the loser resolves to a niche market or disappears altogether. Choosing the wrong technology in the midst of the war when the outcome is unknown is fraught with problems. For example, back in the early 1990s the IT director of a small manufacturer had heard about Microsoft NT (which was new then), so he hired an NT guru into what *had been* a NetWare-only shop. The anguished CEO called me a few months later when problems arose—all the NetWare staff had quit, and the new guy didn't know Novell. The senior managers had not even been aware that bringing in an NT expert would signal the end of Novell to the existing staff. The senior leaders had naively assumed that the IT staff would welcome a different operating system and be happy to learn it from the new guru. (Hint: IT people, having invested untold years in the systems they have already learned, are often loathe to learn new ones at someone else's request.)

The biggest current battle is among clouds. The current battles in play are Google Drive, Apple's iCloud, Dropbox, and Microsoft OneDrive. No one know which cloud service will be victorious in the end because the battle just recently heated up and skirmishes are raging. In three or four years, one will rise and everyone will eventually migrate to that one*.

Mark, a client and friend, is the CIO of a large nonprofit organization. Because Microsoft gives huge discounts to nonprofit organizations, Mark has a Microsoft shop with Exchange, Outlook, Office, SharePoint, and so on. But his technical employees are all staunch open source advocates. Instead of spending their time figuring out how to use all the Microsoft products, they spend their time developing alternatives to those products using open source tools on Linux systems. As a result, there are constant problems on the network and nothing seems to work smoothly. I suggested to Mark that he should choose one or the other (Microsoft or open source), but he does not see the current battle as the cause of his problems.

Though we don't want our businesses to get involved in the wars, it is important to know what battles are being waged and where your IT people stand in it. Otherwise, the skirmishes can hurt business.

## Managing IT People Effectively

Remember that it doesn't do any good to rail against the perceived faults of IT people. Those faults may be the reason IT people are so good at what they do. People who are personable, friendly, profit-oriented, and good team members rarely develop the logical, dedicated focus necessary for high levels of IT knowledge.

Once we recognize the characteristics, we can manage technology people effectively. To summarize from this section, follow these guidelines:

- Give the IT people recognition for their role.
- Focus them on what they do best, and supplement them on what they don't do well.
- Give IT people leading-edge technology in isolated labs.
- Provide IT people with liaisons to teams instead of forcing them to be part of the team.
- Prevent crises by rewarding a smoothly operating system.
- Understand the technology wars in which gurus are involved, and be willing to live with the consequences of taking sides.

As noted earlier, IT people thrive on recognition. They usually don't care much about money or status—they care about recognition among their peers for their knowledge. They want to give their opinions, and they want to be taken seriously. We don't necessarily have to follow every suggestion they make, but they will fume if they are not asked. So smart business leaders discuss all potential strategies with their IT people before decisions are made.

Second, IT people will work hard, harder than most other people, but they will not do what you want them to do (like plan and organize their work, write up documentation, or communicate with others). Instead, they work harder to further their own knowledge and push the limit of the technological capabilities of the systems. Don't expect them to understand the business needs or appreciate profitability as a reward. Establish guidelines so that they focus on what is helpful to the business and minimize administrative functions for them as much as possible.

Third, gurus will not be happy spending day after day on maintaining older technologies. The importance of maintaining stable, smoothly running systems that require no active work is at conflict with finding exceptional IT people who hate monotony. We can attract and keep gurus while preventing them from using the operational network as their personal lab by providing them with a separate budget and letting them spend a portion of their time investigating various options for the future. There is nothing a guru loves more than playing with the latest and greatest technology. If you can get the guru to focus on your Hedgehog Concept and think about

* During technology wars, it is best not to get invested in any one technology but keep your options open. For example, I use all three of the non-Apple clouds (and I would use that one as well if I were still using my iPad regularly, but my Android phablet is more convenient).

innovation, you may receive a return that is many times the value of the time and money they spend on it. This will also help with the hero-seeking issue. An IT person may become a hero by finding that perfect information technology that aligns with the Hedgehog Concept of the business. That works much better for a business leader than having systems break all the time as IT people tinker with them.

Fourth, don't expect good team member skills. Find other people to manage projects and write up documentation. IT people, gurus especially, will be happier if they can focus their intellect on furthering the technological limits of your company in the core business areas and keeping the systems running smoothly. If you aim them right, their happiness means more profits for you.

Fifth, build in measures that reward no-problem systems. Do not overly reward the knight in shining armor who comes into work in the middle of the night to solve a computer problem. Show more appreciation to the technologists who prevent such crises. Otherwise, quest for recognition will lead to constant crises that IT people are only too happy to solve.

Sixth, understand the current technological wars going on and understand which side of the battle your IT people are on. It would be best to find IT people who don't strongly take sides in the current technology war, but that's rare. The next best thing is to know the technical direction of your business and choose IT people whose opinions are compatible with it. If they are on the wrong side from the technology direction of the company, find others. Don't expect IT people on different sides of a war to get along.

Talented IT staff can easily fulfill a very valuable role that brings significant strategic value to your business—but only if you manage them well. By following these few simple guidelines, you can maximize the skills and talents of your IT employees. They will be happier and do more for your business in return.

## 10-1. Interactive

1. **Class Discussion:** Imagine that you are the new owner of a large business where the previous owner had hired people on both sides of a current technology war. How would you handle it?
2. **Class Discussion**: If you are more of a business person, consider the last 3 or 4 technology people with whom you've interacted. How would you assess their technology ability? How would you assess their people ability? If you are more of an IT person, consider the same two questions for the last 3 or 4 business people with whom you've interacted.
3. **Class Discussion**: What are your leanings in the current technology battle over the cloud space? What do you think the cloud landscape will look like 5 years from now?
4. **Individual Activity**: Would you consider your information technology system stable and smooth running, or fraught with problems and issues? Why? What can you do about it?
5. **Individual Activity**: Do a test on your own word processing documents. Reveal the invisible codes and assess yourself. Do you use styles and templates? Are your paragraphs properly formatted with indents and paragraph markers only at the end? Do you press the [Enter] key and insert a paragraph marker instead of using the spacing between paragraph capability? Do you use automatically generated tables of contents? If the answer for any of these is **no**, then do that activities from Chapter 7 or get someone to show you how to do these things. Styles and templates, for example, are much more efficient than manual formatting. But their use demonstrates an advanced knowledge or ability because they are known and used by very few people.

## 10-1. Self Study Test Questions:

1. How can you assessing people's ability and experience in technology?
2. What ways can you encourage people to become more productive in information technology?
3. How do you narrow the gap between those who know, and those who don't?
4. What is the ROI learning curve?
5. How can you overcome initial resistance to a new technology?
6. What defines good training? How do you increase the quality of training?
7. How do information technology professionals differ from typical business professionals?
8. How do IT people obtain recognition?
9. Why should we "coddle" IT people by supplementing them on their weak people skills?
10. How can we reduce or prevent crises in our information technology?
11. What do we mean by a technology war? How should we handle them?
12. What is the purpose of knowing what technology wars are currently raging?
13. Why should we give IT people their own budget to purchase leading edge technologies to test in isolated labs?

11

# Chapter 11: People - Foundations

Overview: In this chapter, readers will learn the following topics:

- History of Information Systems
- Internet Growth
- Technology Revolutions
  - Breaking the 640k Barrier
  - Standardizing the Internet and World Wide Web
  - Dot-Com Fever & Fiber Glut
  - Switch to Digital
  - First (Last) Mile Crisis.
  - Going Mobile
  - Videoconferencing
- Technology Guideline Laws
  - Loewy's Law
  - Moore's Law
  - Metcalfe's Law
  - Gilder's Law
  - Jakob's Law
  - Wirth's Law
  - McGuire's Law

## History of Information Technology

It is important to understand the technological events that formed the foundation for the IT world of today because they explain some seemingly strange situations. For example, have you ever wondered why the keys are placed where they are on the QWERTY keyboard (called that because the first five letters in the left corner are Q W E R T and Y)? In the early days of typewriters, when they were mechanical and manual (as seen in Figure 95), the typists could type faster than the mechanical arms could strike the paper, which caused the machine to jam. To avoid this, the keyboard was laid out so that the most commonly used letters were placed far away from each other and in locations more difficult for the typist to reach. Less commonly used letters were placed in the most easily reachable position. That way, the typists were forced to slow down. Although today's computers could easily keep up with the most nimble of typists, tradition is a very hard thing to break. Since most typists already know how to type using this inefficient keyboard, it would be difficult to change to a more efficient one*.

**Figure 95. An early mechanical typewriter.**

It would be helpful to have a basic timeline of major foundations for today's information technology. One interesting perspective on the foundation of information technology was

* Not that people haven't tried. Among others, the Dvorak keyboard was much more efficient, but it is very difficult to get people to change once they've learned something, and typing is a skill that takes years to master.

discussed in the PBS special *The Triumph of the Nerds,* which was shown on television in 1996. The timeline in this section is based loosely on the companion PBS website (modified for this textbook and used with permission).

## IT Foundations: Prehistory

There are many people in history with inventions that form the foundation for what is now the PC revolution. This section discusses a list of some of the most influential roughly in the order in which they appeared in the timeline. Some were only ideas for which we don't know exactly who made the invention. We've separated out the electronics (which started about 1941) into the next section.

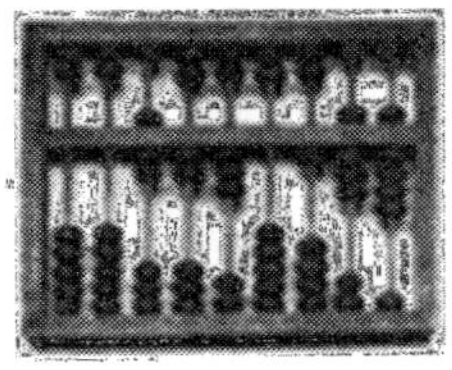

The **abacus**, a simple counting aid, may have been invented in Babylonia (now Iraq) in the fourth century B.C.E. (Figure 96).

**Figure 96. Abacus - c. 3000 B.C.**

The **Antikythera mechanism**, used for registering and predicting the motion of the stars and planets, is dated to the first century B.C. It was discovered on the Antikythera Islands off the coast of Greece in 1901. It may have been built by **Hipparchus** or his student **Posidonius,** but no one is sure.

**Arabic numerals** invented in the 6th century were introduced by **Muḥammad ibn Mūsā al-Khwārizmī** to Europe in the eighth and ninth centuries A.D. Roman numerals remained in use in some parts of Europe until the seventeenth century. The Arabic system introduced the concepts of the zero and fixed places for tens, hundreds, thousands, etc. It greatly simplified mathematical calculations.

**John Napier**, Baron of Merchiston, Scotland, invented **logs** in 1614. Logs allow multiplication and division to be reduced to addition and subtraction and were the basis for the slide rule, a hand-held device used by engineers in the '40s, '50s and '60s to do complex mathematical formulas.

**Wilhelm Schickard**, a professor at the University of Tübingen, Germany, built the **first mechanical calculator** in 1623. It calculated up to six digits, and carried digits across columns. It worked, but never made it beyond the prototype stage.

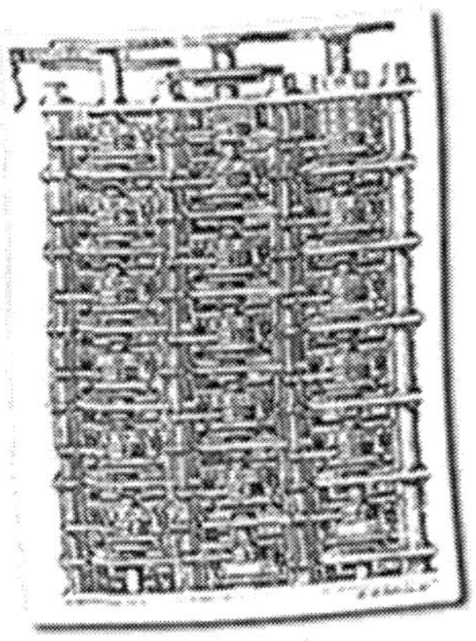

**Blaise Pascal** built a **mechanical calculator** in 1642. It has the capacity for eight digits, but has trouble carrying and its gears tend to jam.

**Joseph-Marie Jacquard** invented an automatic loom known as **Jacquard's Loom** (Figure 97) controlled by punch cards, which is often thought of as the first example of computerized instruction.

**Charles Babbage** conceived of a **Difference Engine** (also known as the Differential Engine) in 1820 or 1821, a massive steam-powered mechanical calculator designed to print astronomical tables. He attempted to build it over the course of the next 20 years, only to have the project cancelled by the British government in 1842. Babbage's next idea was the **Analytical Engine** - a mechanical computer that could solve any mathematical problem. It used punch cards similar to those used by the Jacquard's loom and could perform simple conditional operations.

**Figure 97. Jacquard's loom: The first example of software.**

While the Analytical Engine was important, it might have been lost to history if not for Babbage's protégé **Lady Augusta Ada Byron**, the Countess of Lovelace. She met Babbage in 1833 when she was just 17, but because of her unusual talent in math and science, she was asked to translate Babbage's articles. She describes the Analytical Engine as weaving "algebraic patterns just as the Jacquard loom weaves flowers and leaves." Her published analysis of the Analytical Engine is our best record of its programming potential*. In it she outlines the fundamentals of computer programming, including data analysis (such as an **if** statement ), **looping** (such as a *Go To* statement) and **memory addressing** (storing a variable to be used later).

**George Boole** was a 19th century mathematician who described all the possible ways that two logical tests could be evaluated utilizing the terms AND, OR, as well as NOT, called **Boolean Logic**. He also invented an entirely new way of thinking about algebra called Invariant Theory which was a foundation for Einstein's Theory of Relativity.

## IT Foundations: Electronics in 40's and 50's

It is interesting to note that when first computers were built, they were simply called electronic machines or calculators. The term *computer* referred to the **people** who add, subtract, multiply, and divide for businesses and governments, not to machines.

**Konrad Zuse**, a German engineer, completed the first general purpose programmable calculator in 1941. He pioneered the use of **binary math** and Boolean logic in electronic calculation.

**Tommy Flowers** invented **Colossus**, a British computer used for code-breaking, which became operational by December 1943.

**John Mauchly** and **J. Presper Eckert** of the University of Pennsylvania invented **ENIAC**, or Electronic Numerical Integrator and Computer, was developed by the Ballistics Research Laboratory in Maryland to assist in the preparation of firing tables for artillery. It was built at the University of Pennsylvania's Moore School of Electrical Engineering and completed in November 1945. They later went on to develop **UNIVAC**, the UNIVersal Automatic Computer shown in Figure 99, in 1951. It could store 12,000 digits in random access mercury-delay lines. The following year they developed **EDVAC**, for Electronic Discrete Variable Computer, which was completed under contract for the Ordnance Department in 1952.

**John Bardeen, William Shockley** and **Walter Brattain** of Bell Telephone Laboratories developed the **transistor** in 1947. Some transistors (and a dime to show scale) are displayed in Figure 98.

**Figure 98. Transistors.**

**Figure 99. UNIVAC (1951).**

German physicist and engineer **Werner Jacobi** developed and patented the first known **integrated transistor amplifier** in 1949.

* Lady Lovelace's work itself might have been forgotten had not B.V. Bowden, who republished them in *Faster Than Thought: A Symposium on Digital Computing Machines* in 1953. Found in Ada Lovelace: Biography published in April 2, 2014, updated in July 29, 2020. https://www.biography.com/scholar/ada-lovelace Last accessed October 26, 2020.

In 1952, **G.W. Dummer**, a radar expert from the British Royal Radar Establishment, proposed that electronic equipment be manufactured as a solid block with no connecting wires. The prototype he built didn't work and he received little support for his research. Nevertheless, **solid state electronics** would eventually see wide use.

In 1953 **Harwick Johnson** filed a patent for a prototype **Integrated Circuit**.

**Jack St. Clair Kilby** of Texas Instruments and **Robert Noyce** of Fairchild Semiconductor both announced the integrated circuit in 1959. The actions of both companies started a patent war which wasn't resolved until 1966.

The **IBM 360** was introduced in April of 1964 and quickly became the standard institutional computer. Developing it nearly destroyed IBM, however, because warring factions within the company could not get along to finish the product. **T. Vincent Learson** is credited with finding a way to get the competitive groups to work together instead of against each other. By the mid-'80s the 360 and its descendants will have generated more than $100 billion in revenue for IBM.

## *IT Foundations: Minicomputers in 60's*

In the 60's, a typical minicomputer cost about $20,000 and had no memory. Data was stored on reel-to-reel magnetic tapes that were manually loaded (mounted) on the drives. Programs were laboriously entered and run by using computer punch cards with square holes indicating binary code (harkening back to the cards used by Jacquard's loom).

The first inkling that computers could do more than math was when **Ivan Sutherland** demonstrated a program called **Sketchpad** on a TX-2 computer at MIT's Lincoln Labs in 1962. It allowed him to make engineering drawings with a light pen.

**Gordon Moore**, predicted that the number of components in an integrated circuit will double every year (**Moore's Law**)*. He noted that an integrated circuit that cost $1000 in 1959 went for less than $10 in 1965. Gordon Moore and Robert Noyce founded Intel in 1968 which went on to become the largest manufacturer of integrated circuits by the year 2000.

**Doug Engelbart** demonstrated in 1968 a **word processor**, an early **hypertext** system and a **collaborative application** (three now-common computer applications).

Xerox's chief scientist **Jacob E. "Jack" Goldman** hired **George Pake**, a physicist specializing in nuclear magnetic resonance, and created its Palo Alto Research Center - Xerox PARC - in 1969. Its mission: to explore the "architecture of information."

## *IT Foundation: Innovation and Personal Computers of the 70's and 80's*

The seventies and eighties were when microcomputers (i.e. personal computers) began to proliferate. Some large businesses and banks began to install computers and start electronic data processing departments. By the late eighties, businesses began to consider personal computers a valuable addition to their office equipment, beyond the more common larger computers.

**Bill Gates** and **Paul Allen** formed **Traf-O-Data** in 1971 to sell their computer traffic-analysis systems. Four years later they started **Microsoft** to develop BASIC for the Altair 8800, and six years after that they were approached by IBM to develop the BASIC programming language for its personal computer project. Microsoft Windows 1.0 shipped in November 1985. Microsoft's sales reached $1 billion by 1989.

**Figure 100. Apple II (1977).**

**Steve Jobs** and **Steve Wozniak** started building and selling "blue boxes" in southern California in 1971. A blue box allowed people

* Moore's Law is discussed in more detail in the next section, but was the first recognition that computer prices would drop regularly instead of going up regularly which was the more typical product trajectory.

to make long distance phone calls without paying for them. By 1977 Apple sold its Apple II (Figure 100) for $1,195, including 16K of RAM but no monitor. It used 5¼ inch floppy disks, but had no hard drive. By 1980, Apple had captured 50% of the personal computer market.

In 1972 **Gary Kildall** wrote **PL/M**, the first high-level programming language for the Intel microprocessor.

**Jonathan A. Titus** designed the **Mark-8**, "Your Personal Minicomputer," according to the July, 1974 cover of *Radio-Electronics. Popular Electronics* featured the MITS Altair 8800 on its cover, January 1975 and hailed it as the first "personal" computer. Thousands of orders for the 8800 rescued MITS from bankruptcy and spawned the start of the Homebrew Computer Club (Figure 101) which Steve Jobs and Steve Wozniak joined, creating and selling their first Apple computer.

**Figure 101. Homebrew Computer Club**

**Dan Bricklin** and **Bob Frankston** of Software Arts developed the first spreadsheet program for the Apple II, **VisiCalc**, by the spring of 1979. It was released in October and became an immediate success. Copies shipped per month rose from 500 to 12,000 between 1979 and 1981.

**Figure 103. IBM PC (1981).**

Not wanting to be left out of the new market, the **IBM PC** was released in August 1981, designed by a team lead by **Don Estridge***. The IBM PC would have an outsized influence on the field because of the deep customer base for IBM computers.

Apple was not to be outdone, however. In 1984 it released the Apple Macintosh which boasted a new kind of interface first seen at Xerox, a GUI (graphical user interface). It featured the new graphical interface and required a mouse. It used the 8-MHz, 32-bit Motorola 68000 CPU (instead of Intel), and had a built-in 9-inch B/W screen. It cost $2,495 and had no internal hard drive, only a single floppy drive which required the floppy disk to have a bootable system and the application on one disk. It boasted the advanced feature, however, of being able to eject the system disk so that data could be stored on a separate floppy disk – something a single-drive IBM PC could not do.

**Figure 102 Apple Macintosh (1984)**

## History of Storage Media

Storage media have changed drastically over the past twenty years. Early on, computers did not have hard drives at all, and only came with floppy drives.

Floppy drives were called that because the disk (made out of a cardboard square and a round internal disc of magnetic media) was physically floppy (and not very robust – you could ruin the entire thing by bending it, touching the open hole over the magnetic media, or getting it near a

* As an aside, my uncle, H. Terry Liffick, worked for IBM in Boca Raton, FL and was on the team that designed the IBM PC. He purported to have decided on the much-denigrated "chicklet" keyboard that the IBM PC Jr used before being replaced with a full standard keyboard. He said that it was thought that the IBM PC Jr would be used in homes by non-typist, so they didn't think it needed a full keyboard. They were wrong.

magnet)*. The first floppy disk was designed by **Alan Shugart** of IBM and was 8 inches and stored a whopping† 80 kilobytes. The next generation was a double density 5¼ inch floppy disk which could hold 360 kilobits of data on each side (you could take them out, turn them over, and play the other side for 720k of data total). During the early 90s the 5¼ inch floppies gave way to 3½ inch floppy disks which had the added of advantage of a metal cover over the magnetic media and could hold 1.44 megabytes of data. As files got larger because of colors, fonts, and graphics, various proprietary disks such as the **Iomega zip drive** (developed by **David Bailey** and **David Norton)** that could hold 100 megabytes of data came on the market. **Compact disc** drives, invented by **James T. Russell**, also became prevalent as CDs could hold 750 megabytes of data.

All of these data storage media fell by the wayside around 2005 because (finally) most computers supported a more flexible method of transferring data; the flash drive. The flash drive was introduced in the year 2000, and within five years all desktop computers could use flash drives without the installation of special drivers. The first flashdrive was first marketed under the name **Thumbdrive**, invented by **Henn Tan** and used the new standard interface designed by **Ajay Bhatt,** the **USB** (universal serial bus). They have also been called keydrives, memory sticks, USB drives or USB sticks, and a myriad of other names. You can buy them in a variety of sizes from 128 megabytes to 512 gigabytes. Because of their low cost, reliability, and flexibility, they are the most common way of transferring data from computer to computer when the computers are not connected through a network (in which case, most times it is easier to use the network to transfer the data).

## *IT Foundations: Networking*

Timesharing, the concept of linking a large numbers of users to a single computer via remote terminals, is, was developed by **Fernando Corbato** at MIT in the late '50s and early '60s. It was just a precursor for **Paul Baran** of RAND (originally stood for Research And Development, but is no longer an acronym) who developed the idea of distributed, **packet-switching networks**.

ARPANET (Advanced Research Projects Agency Network), directed by **Joseph Carl Robnett Licklider**, went online in 1969. **Bob Kahn** and **Vint Cerf** developed the basic ideas of the **Internet** in 1973. The TCP/IP (Transmission Control Protocol and Internet Protocol) specifications were released in 1974. TCP/IP was established as the standard for ARPANET in 1982.

In 1974 BBN Technologies (the Bolt, Beranek and Newman consulting firm of **Leo Beranek, Richard Bolt,** and **Robert Newman)** opened the first public packet-switched network - **Telenet**. A UUCP (Unix-to-Unix Copy) link between the University of North Carolina at Chapel Hill and Duke University set up by **Tom Truscott** and **Jim Ellis** established **USENET** (User's Network) in 1979. The first **MUD** (multi-user dungeon, an online gaming environment) was also developed in 1979, at the University of Essex by **Roy Trubshaw** and **Richard Bartle**.

In 1987 the number of network hosts in ARPANET broke 10,000. In 1989 that number topped 100,000. Congress, influenced by Senator **Al Gore**, moved to **privatize the Internet** so that it could be used for commercial purposes. Until 1995, only researchers, educators, and the government could utilize the Internet.

---

* While typically an apocryphal story, for me it was first-hand. In graduate school I had a friend in my computer applications class who, for three weeks in a row, came to class with her homework disk only to find that it was blank and all her homework was gone. It took the instructor a while to figure it out, but finally he asked her where she stored the disk, and she answered with all seriousness that she did what she always did with her homework papers for her other classes; put it on her refrigerator with a magnet so she wouldn't forget it.

† I know I shouldn't joke in a textbook, but the *whopping* is facetious. Although at the time, 80 kilobytes was a lot, now it is very small, hardly enough to store a single page letter with formatting.

**Timothy Berners-Lee** (Figure 104) and **Robert Cailliau** developed the **World Wide Web**. CERN (European Council for Nuclear Research) released the first Web server in 1991. Within a single year, the number of hosts on the newly commercialized Internet using the World Wide Web exceeded 1,000,000. The World Wide Web sported a growth rate of 341,634%* in service traffic in its third year, 1993.

**Figure 104. Timothy Berners-Lee**

## Internet Growth

Over the next twenty five years, the number of host computers on the Internet has grown exponentially, and has now surpassed a billion hosts, as can be seen in Figure 105.

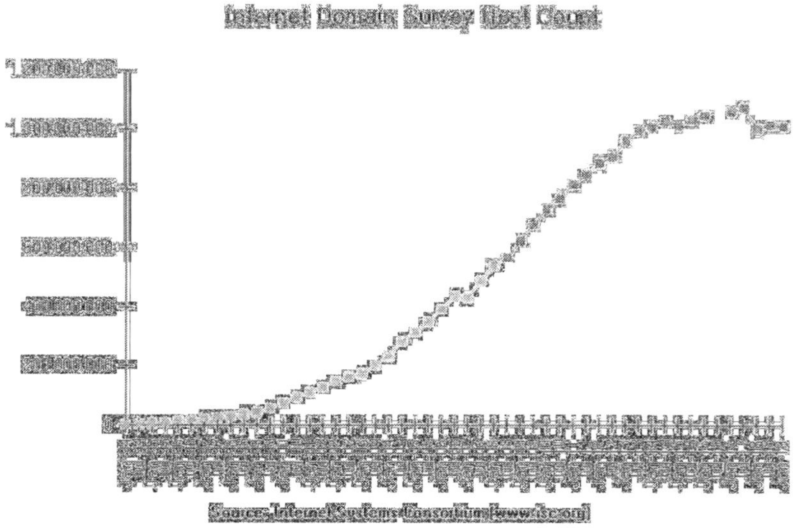


**Figure 105. Internet domain host count†.**

* No, that isn't a typo. That really is a growth rate of three hundred forty-one thousand, six hundred and thirty-four percent within 3 years. The World Wide Web had an astronomical growth rate, which caused the dot com boom and subsequent bust.

† Found on https://ftp.isc.org/www/survey/reports/current/hosts.png, posted January, 2020 by the Internet Systems Consortium. Last accessed October 27, 2020.

The number of users has grown even more. In the year 2000, less than 5% of the population had ever used the Internet to communicate. By 2010, almost 30% of the world's population uses the Internet, and in North America, the penetration is over 75% of the people, as can be seen in Figure 106.

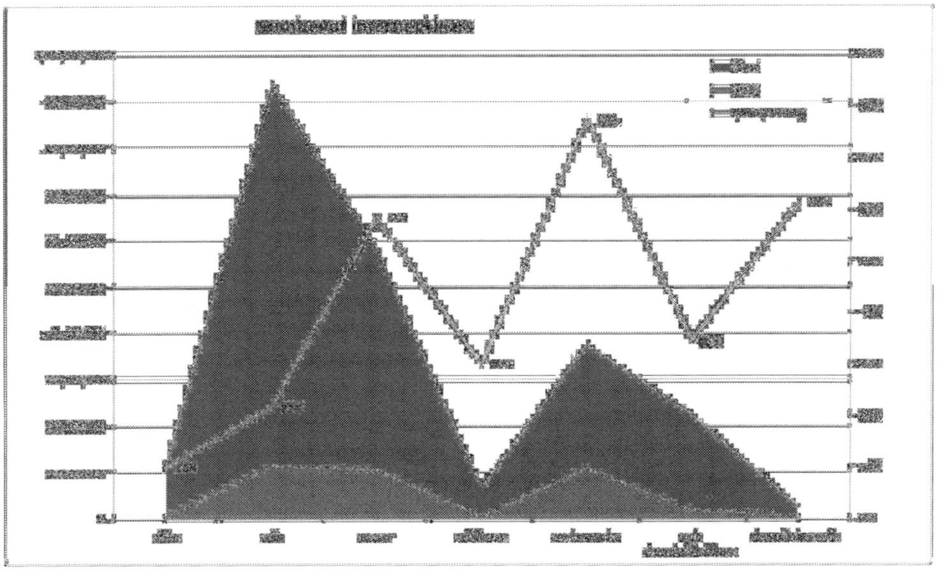

**Figure 106. Growth of Internet users, by Country.**

The current number of Internet users can be found in Figure 107 along with the percentage of users within each region. In North America, for example, 90% of the population uses the World Wide Web on top of the Internet.

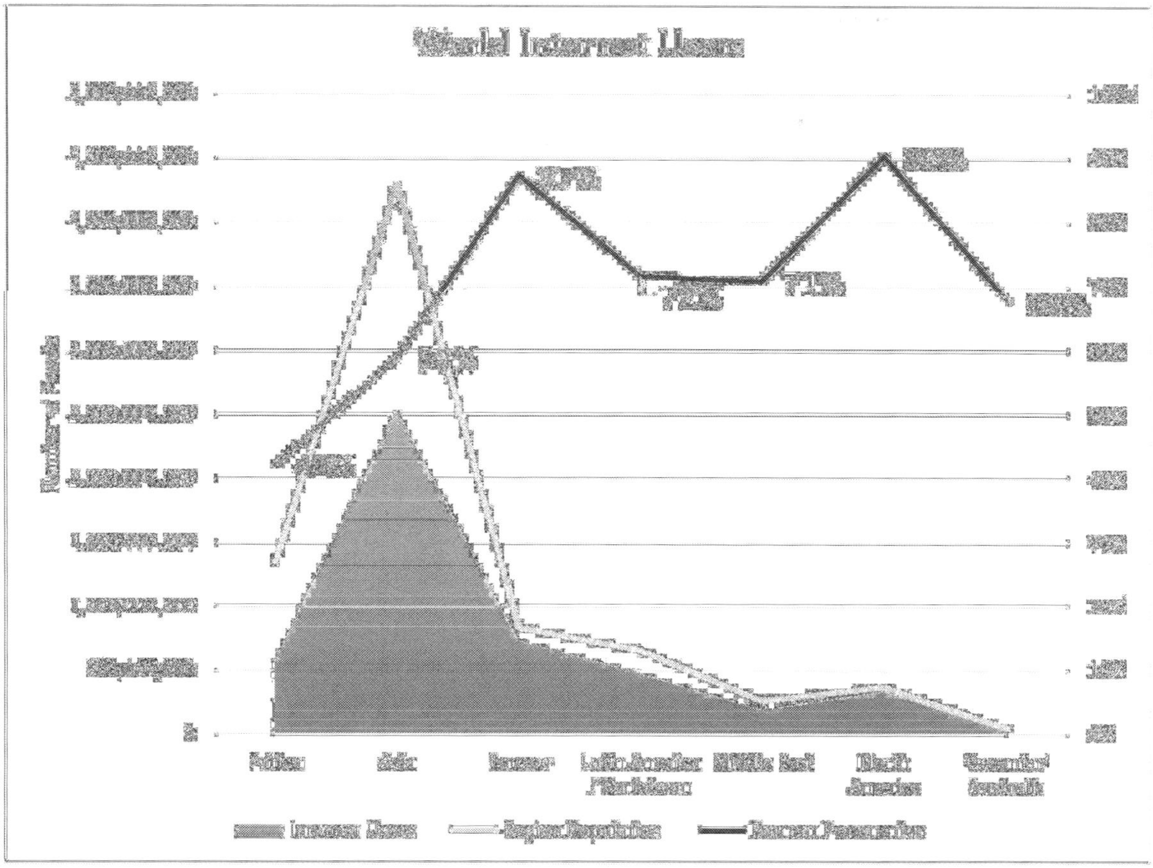

**Figure 107. World Internet Users 2020**

## Technology Revolutions Leading to Today

Many people believe that the introduction of the personal computer in the '80s "changed everything," much like the introduction of the telephone "changed everything" and the introduction of the combustion engine "changed everything."

However, in 1992, those with computers were still in the minority even though personal computers had been around for more than ten years. Computers were still mostly at work (not in the home). Computers at that time did very specific tasks and ran very specific programs. Except for a few hobbyists who played games and enjoyed learning to program, a computer was not a commonly seen item. In 1992, the first really stable computer hardware/software that could be used by everyday people (instead of only the technically adept) had just come out.*

Fast forward another twenty years, however, and it is a different world. More homes in the United States have computers than don't have computers. Email and websites are as essential to business as a telephone. Many businesses are virtual and don't even have a physical office building. All business is transacted online.

It is important to keep in mind that the availability of technology itself does not make the revolution; acceptance and ubiquity of the technology makes the revolution. And acceptance is up to people, not machines.

* Some consider the Intel x486 chip with Microsoft Windows 3.1 operating system the "first stable useful personal computer". Others may argue that the Macintosh was better, but the proprietary nature of Apple technology, (as compared to the "open hardware standards" architecture of the personal computers with the x486 chip), meant that the Macintosh would be considered a "niche" machine for education and art/publishing, whereas the PC was a general purpose business machine that could be used in thousands of situations. That did not change until the iPhone and the iPad, which made Apple explode out of its niche product status.

## Why You Need to Understand Revolutions

It would be a mistake (a common mistake, but a mistake just the same) to think that the sudden growth of personal computers came about solely because of the Internet. To better understand today's technology, it makes sense to look at the "sub-revolutions" that impacted the current state of information technology.

## Seven Sub-Revolutions

Basically, there were seven sub-revolutions after the introduction of the personal computer in the '80s.

- Breaking the 640k Barrier
- Standardizing the Internet and World Wide Web
- Dot-Com Fever & Fiber Glut
- First Mile Crisis (or Last Mile Crisis)
- Switch to Digital
- Going Mobile
- Videoconferencing

### *Breaking the 640k Barrier*

One of those little details that few outside the technical worlds of hardware and software understand is the impact, in the early days of computers, of the huge expense of memory and hard drive space. Programmers would specialize in getting programs to run in the typical 2 k bytes of memory available in the seventies and early eighties on the computers of the day.[*] Whole industries were developed out of the need to fit more programming in less and less space. Every byte saved made the computers cost less.

The focus on saving memory and storage set the stage for a technical limitation called *the 640k Barrier*. Early PCs had a built-in limitation of 640 kilobytes of usable active memory. When this limitation was established, a top-of-the-line, million-dollar computer had 64 k bytes of memory. It was reasonable to make the technical limit 10 times that amount for a personal computer . It was not thought that anyone would need that much memory[†]. Similarly, the typical hard drive was 10 megabytes, so the limit of hard drive space that could be addressed was established as 2,000,000 megabytes (or 2 gigabytes), again, much more than anyone thought would be needed by a personal computer. At the time, filling up 20 megabytes of hard drive space was unthinkable. A computer with 1 gigabyte of hard drive space cost over $5000. Most computers still stored data on large reels of magnetic tapes because hard drive space was simply too expensive. Personal computers used 5¼-inch floppy disks that stored less than 1 meg.

The problem was that, while 640k memory was more than enough in a text-based computer world, it wasn't even a blip in the memory needs of a graphic world. When operating systems like the Macintosh and Windows started adding little pictures (icons) and menus that could be clicked on with a mouse for the GUI, the existing memory and hard drive space requirements were simply not enough. The difference between the memory/storage needs of **text** and the memory/storage needs of **graphics** is huge. (There's another huge leap from graphics to video as well.)

For years the personal computer industry hobbled along, and people who wanted to use Windows or any other graphical user interface had to purchase memory managers and learn the difference

[*] A typical personal computer these days has 8,000,000 k bytes of memory (8 gigabytes). More details on memory and hard drives can be found an earlier chapter.

[†] You may hear reference to a quote by Bill Gates stating "Who in their right mind would ever need more than 640k of RAM!?". Bill Gates, however, denies that he ever said that.

between "real" memory and "virtual" memory. Since every peripheral used up memory, there was a severe limitation on how many devices could be attached to computers, and attaching a new device usually meant days of trial and error in changing "interrupts" and remapping the existing memory to accommodate the new device. Owning a computer required a great deal of knowledge and training.

Then came Windows 95*. Windows 95 introduced two features, *Virtual Memory Management* and *Plug and Play*, that made it more user friendly than any other existing operating system (with the exception of the Macintosh which was always friendly, but was proprietary†, too expensive and not considered a "business" machine.)

Virtual Memory Management was a feature whereby the operating system itself managed the memory so that the 640k Barrier, in the eyes of the end user, disappeared. (It was actually still there, but because part of the hard drive could be used as if it were memory, the technical barrier disappeared from the point of view of the software, so many more programs could be run at the same time.)

Plug and Play device installation meant that when a new device was attached to a computer, Windows 95 would try to identify it and install the proper drivers so that the new device would work.‡ Instead of days of frustration and hours on the phone with technical support, installing a new device became a few minutes of time.

Why would these two seemingly tiny technical details make such a difference? Because it made it possible for someone without any technical training at all to buy, install, and use a personal computer. Huge numbers of people went ahead and did just that. This PC proliferation established a base of people who were in position to move to the next level: the Internet and the World Wide Web.

### Standardizing the Internet and World Wide Web

With a large installed base of personal computers, the stage was set for people to be able to communicate with each other using these computers. Although small private networks had been around for years, in 1992 connecting with another computer was still, by and large, something that only technical people did. Computers had to be connected directly to each other, through wires or leased lines, point to point. There were hundreds of different networking protocols§. Each protocol had its own way of doing things, its own rules for communication between computers. Each protocol needed its own type of device and connection, and few of them interconnected with each other. There were special devices (called bridges and switches) to convert from one type of network into another type of network. But in 1992 the network of one company almost never communicated with the network of another company.

Of course, in 1992 the Internet existed, and it was interconnected. But only researchers, educators, and government employees were allowed to use the Internet. Then Al Gore convinced Congress in the early '90s that the US government should hand over management of the Internet to private

* Windows 95 was an *operating system* for a personal computers. An operating system provides the instructions for the system; how the keyboard works, how the hard drive works, how the monitor works, etc.

† A *proprietary* system was completely owned and controlled by one company, enabling that company to set higher prices because third parties were excluded from competing with them.

‡ A *driver* is a tiny file that contains the instructions for the device so that the operating system knows how to access that particular device like printers and cameras and monitors, etc. Many drivers come with the operating system, but some need to be provided by the device manufacturer.

§ A *protocol* is an agreed upon way of doing things. In networking, the protocol defined how the computers connected, how they sent messages back and forth, and how they disconnected. At the time, protocols were *circuit based*, which meant that one computer connected to another, and they sent messages back and forth, and didn't break the circuit until all the communication was completed. Today's networks are *packet based*. This is discussed in detail in *Chapter 2: Networking Hardware* commercial ventures and allowed commercial businesses to use it. At the same time, Timothy Berners-Lee and Robert Cailliau the system of hyperlinking (the World Wide Web) that made using the Internet easy. Because of the Internet and the World Wide Web, suddenly someone with no technical training at all could connect to any other computer on the Internet network, and could retrieve information from any of those computers.* All they had to know how to do was click a mouse.

### Dot-Com Fever and Fiber Glut

In the latter part of the '90s, the business world took notice of new technologies popping up. One of the first companies to take advantage of the personal computers on people's desks and the World Wide Web was eBay, which incorporated in 1995. eBay put buyers in touch with sellers (but did not sell or buy anything itself). This marked the launch of a period that many refer to as the Internet Boom, Technology Bubble or the Dot Com Fever. It lasted until February 2001. During this time, many people saw the power of the Internet and information technology in a new light.

At the time, the media announced (with a straight face) the "new economy," where sizzle was more important than steak. Many invested heavily in any technology company that sounded good. After all - the business they invested in might be the next Microsoft! Venture capital was easy to get, and did not require years of profits as had been required in the past. Thousands of companies, financed with venture capital, started on the road to make millions in the revolutionary new world where every home had a computer and was connected to the Internet, and every business network could easily connect to every other business network.

While the vision was attractive, the reality was that it would take years for the new technology to stabilize enough for business ventures with actual profits to appear. Some people did urge caution. In a now-famous 1998 speech, Federal Reserve Chairman Alan Greenspan warned the financial world of impending collapse by calling the over-investing in information technology "irrational exuberance."

As might have been expected, the boom did not last forever. Indeed, the dot-com frenzy was followed by a deep crash. The technology bubble had burst. Instead of wild speculation and easy-to-find investors, people were losing money in the millions and venture capital had dried up completely. Companies went out of business by the thousands as the "new economy" proved to be more shadow than substance.

The final nail in the coffin of the boom times was the terrorist destruction of the World Trade Center on September 11, 2001. The entire business world, especially information technology industries, went into a recession that lasted years. (Of course, the bursting of the technology bubble was just a blip on the radar compared to the financial crisis that started in October 2008 that started with Black Week†, the start of an eight-year-long world-wide depression. And the recent 2020 economic issue due to the Corona-19 Virus Pandemic may end up being even worse that that one!)

*Well, it felt like suddenly, but actually it took many years. The Internet started functioning in 1982, and the World Wide Web was invented in 1990. More details on both is in *Chapter 4 and 5.*

† Popular press has dubbed the week of October 6, 2008 as Black Week, in reference to Black Tuesday (October 29, 1929) which harkened the coming depression and Black Monday (October 19, 1987) when stock market declined by 25% in one day. The 2008 financial crisis has been dubbed The Great Recession (Kwoh, Leslie. 2010. "Economic slump called 'Great Recession' by Associated Press manual." New Jersey). According to noted economist Brian Beaulieu, "There is no great rebound on the other side of this downturn, only a long, slow climb out." (Diamond, Paul. 2009. Economic Forecast: Time to Hold 'Em, Not Fold 'Em. In The CEO's Guide to Doing Business In a Recession: Vistage International. pg 2).

This investment led to the next phase, the Fiber Glut. Before the Internet boom ended, one of the industries that received the most funding was *fiber optics.*[*] Newly formed fiber optics companies bought up rights-of-way and laid fiber optic cable in preparation for the coming need for connections between cities. All those newly formed companies had hoped to take advantage of the coming need for high-quality connectivity, but were instead sold for pennies on the dollar, lowering the cost for connecting to the Internet. This provided real competition to the un-competitive leased data line business that previously only the local phone company could provide.[†] In response, the phone companies also increased their fiber capacity to prevent being left behind. The quality and capacity of the highways between cities improved drastically during the years of the technology bubble.

This fiber glut also contributed to the increasing globalization of services. Because of cheap fiber crossing the oceans, it became feasible to outsource tasks and functions like programming, transcriptions, and customer service to India, China, Russia and dozens of other countries.[‡]

The problem was that though connectivity between cities was cheaper, getting from the fiber highways to the individual homes and businesses where people could use the broadband capacity was still expensive. We can think of it as building superhighways from country to country and city to city, but without any entry and exit ramps. The fiber glut, therefore, was the foundation for the next sub-revolution, known in technology circles as the *Last Mile Crisis*.

### Last Mile Crisis (or First Mile Crisis)

The phrase the *Last Mile Crisis* refers to the inconvenience and lack of ease in getting broad bandwidth to the "last mile" between the phone companies and the individual homes and businesses.[§] (Calling the problem the "last mile" illustrates how technologists tend to put the technology first instead of the customer. It's only the **last** mile if you think of the technology as the center. It's the **first** mile if you think of the customer as the center.) Early on there was a movement to change the terminology from the "Last Mile Crisis" to the "First Mile Crisis" to indicate the switch in focus from the technology to the customer, but the more common name *Last Mile Crisis* stuck.

Whether it is called the last mile or the first mile, in 2002 it was a real crisis. Most people and businesses connected to the Internet through a slow dialup connection. The cost to install high-speed cable or fiber to homes and offices throughout the United States was estimated to be 365 billion dollars. The U.S. was not alone; at the time the problem was even worse in other countries.

There were three paths to a solution, and in 2002 no one knew which path would be the one which would work the best:

1. install technology that provided broadband over existing wiring to the home,
2. install fiber cable to homes and businesses, or
3. provide wireless broadband access.

At the time, companies and individuals were still paying $300 - 1000 a month for point to point leased-line connections to the Internet to get broadband speeds (or suffering with slow dial-up connections over telephone lines). In the United States, all three solutions have been implemented and except for some rural areas, most people have at least one option for broadband connection.

---

[*] Fiber Optics refers to the type of cabling that transmits data using a light pulse instead of a frequency on a copper wire. Fiber optic cables can carry much more traffic than wire cables.

[†] Before the Internet, if a business wanted to connect one computer to another computer, it had to lease a direct phone line between the two computers. This was known as a leased data line, and was sold, usually, as a T1 line or a Frame Relay. More details can be found in Chapter 2.

[‡] An excellent description by a non-technologist of the impact of globalization on business can be found in Thomas L. Friedman's *The World Is Flat* published by Farrar, Straus, and Giroux in 2005

[§] Bandwidth is the term we use to describe the speed and or size of a network connection. Broadband refers to having a faster and/or bigger connection. More technical details can be found in Chapter 2.

Cable Modems are the broadband option of choice for most people in order to connect to the Internet for the first mile, with Digital Subscriber Lines (DSL) and Fiber Optic Service [FiOS]) as a second and third choice.

But why not everywhere? Businesses and residents in rural areas are not profitable markets for telecommunication companies such as Verizon and AT&T or cable companies such as Comcast (Xfinity). For twenty five percent of the population in the United States, the third option (wireless broadband) is the only possibility because many people don't live or work near enough to the switching equipment to get broadband over either copper wires or fiber optic connections, and even that requires either cell phone towers or expensive satellite service. Furthermore, satellite services introduce a delay that makes it impossible to videoconference or play games on the connection, so some would argue that those rural areas relying upon satellites are not providing true broadband. It may be decades before all areas of the United States have basic high-speed Internet access.

In other countries (especially Asia and Europe), wireless broadband was the more common solution – especially since they did not already have an installed infrastructure of copper and fiber. Additionally, wireless broadband is easier to upgrade, requiring only the receiving equipment to change rather than re-cabling. Some of the newest wireless protocols (such as 802.11ah) rival wired speeds for data transfer.

### *Switch to Digital*

As the First Mile Crisis has mostly been solved, the next revolution completed: the switch to digital everything. Instead of analog cameras where the negatives must be developed, we use digital cameras and store the pictures on memory sticks or hard drives. Cell phones were originally analog like their wired counterparts, but now are entirely digital. Even at home, instead of using analog phones, many people use Voice Over Internet Protocol (VOIP) phones*.

Films and movies are released on streaming channels or on Digital Video Disc (DVD) instead of on reels or videotape. Companies are beginning to charge their customers extra in order to receive paper bills or statements, and most invoices are paid digitally online or through a bank payment application or a credit card rather than sending a paper check in the postal mail. Even if you do pay with a paper check, instead of receiving the paper check with your statement as most people have for the past hundred years, a copy of your check has been digitized and can be viewed online.

Research is hardly ever printed in paper journals anymore; digital access makes it ever-so-much easier to learn what other experts are doing and add onto their findings rather than reinventing the wheel. When you go to the library, you no longer have to search through paper index cards or stacks of paper books to find the information you seek; you use the standing rows of computers to search for what can be found either in the library or within their digital collections.

The advantages to digital data are many.

1. Digital storage takes up much less room than paper.
2. Digital connections are cheaper than analog connections.
3. Digital data can be easily searched and indexed.
4. Digital data can be copied and shared much more easily and cheaply than paper.
5. Digital data is easier to manipulate and edit than non-digital data.
6. Digital data can be viewed and/or downloaded online.

The impact of this switch to digital for businesses cannot be over-emphasized. Many business processes describe the movement of forms (often in multi-part copies of white, pink, green, and goldenrod where each department gets their copy and passes the rest on to another department). If the form is digitized instead, the form doesn't physically have to move from place to place. If placed on an intranet or extranet, the form can be accessed by anyone in the company in any office

* In 2021 Kutztown University finally retired its PBX and installed a completely VOIP telephone system on campus.

anywhere - all at the same time. If we separate the data itself from the form, we can digitize the information without creating a graphic view of the form, enabling a much more flexible process.

For example, imagine a business process for requesting and approving funds for a capital project within a business. In the non-digital manual process, the physical form is sent through postal mail to four different offices, in sequence, to get the proper authorizing signatures (often being mailed or FedEx'd or UPS'd from Wilmington to Texas to Boston and back to Wilmington). The average time for the request to be approved might be 40 days (not very fast).

In the digital scenario, the form would become a series of fill-in blanks on a web page. Instead of postal mail (known as snail mail in technology circles) or FedEx, the information would be emailed to all four senior leaders who need to authorize the expense. They could simply reply with "approved" or "rejected." The average time for approval might get shortened from 41 days to 2 days. Imagine the cost savings in personnel and postage (especially since the speed of business commonly compelled people to FedEx the forms)*. Imagine the change in the speed of new projects getting started, new client acquisition, and new sources of revenues to start. For most industries, businesses began to understand the switch to digital and now they take advantage of it – or lose their ability to remain competitive.

### *Going Mobile*

It is difficult to express the speed with which mobile technology has taken hold. In the chapter on hardware, there is mention of the Newton (the first Personal Digital Assistant launched by Apple in 1992 which failed utterly in the marketplace). This section contains a description of how the PDA gave way to the smartphones, tablets, and phablets of today which have spawned an entire industry and changed the direction of technology.

In January of 1999, Research In Motion (RIM), a Canadian company, introduced a new device called the Blackberry. Although there were several personal digital assistants (PDAs) on the market at the time†, the Blackberry differed in that it included a miniature keyboard. Previous PDAs required the use of a stylus (a small instrument that looks like a pen but designed to write on a digital surface like a touchscreen or digitizing pad). Furthermore, instead of just synchronizing with a PC, the Blackberry could send and receive email from anywhere. Using wireless connection technology, the Blackberry downloaded email continuously, so busy executives were never out of touch. In March 2002, the Blackberry 5810 was released, the first Blackberry to include phone capabilities. RIM became the first company to combine the cell phone and a PDA in the same device.

In 2002, Apple release the first iPod. The iPod was a portable media player coupled with iTunes, a jukebox application that allows people to quickly and easily purchase, download, and manage the music on their device. Competitors arose, and the new devices became known as MP3 players, for the digital format of the music they play‡.

PDAs and MP3 players, such as the iPod, morphed into what is now known as smartphones§. Apple introduced the first iPhone in 2007 and has continued to add functionality since. The popular device combines a cell phone, a music player, a camera, a video camera, a web browser,

---

* If you look on my resume, you will see that this project was one of my achievements at MBNA when I was a Vice President there.

† Unfortunately for RIM, the company did not keep up with its competitors, and despite rebranding themselves as Blackberry and undergoing several changes, the company never returned to its pre-2010 glory. While still in business, it has dropped development of its own operating system and now just manufactures phones.

‡ MP3 is not really an acronym. The MP stands for Moving Pictures Expert Group, who published several standards for music and video encoding called MPEG-1, MPEG 2, and MPEG3 – which was shortened to just MP3.

§ The next commercially produced PDA was the Palm Pilot, which was followed by many others, including the Handspring Visor and its successor, the popular Treo. In 2003, Palm bought Handspring, and in 2005 Palm was bought by Robotics, Inc. PDA as a product disappeared after smartphones became popular.

email, and texting. The iPhone also revolutionized the interface by using a multi-gesture touch screen.

Since then, cell phone technology has been getting smarter and smarter. Cell phones now typically include built-in digital cameras (both video and still picture), Global Positioning Systems (GPS) navigation, and browsers to search the World Wide Web. Cell phones can send and receive emails and instant text messages, play music, maintain calendars, do word processing, run spreadsheets, edit video, organize photos, stream movies and shows, and synchronize with more powerful computing devices either wirelessly (using Bluetooth personal network technology or Wi-fi*) or through more traditional USB connections. In essence, our smart phones are almost as smart as our computers, and can accomplish almost (but not quite) as much. In 2010 Apple upped the game by inventing a completely new almost-computer-that-was-mobile, the iPad.

Most of the development going on today is taking previously-large screen information and making it available on small-screen mobile. Five years ago, laptops were decidedly inferior to desktop computers in power, speed, and storage. Now, laptops meet (and sometimes beat) the available power, speed and storage of desktops available today. Five years ago Wi-fi hotspots were few and far between; now whole cities are providing metropolitan LANs (MANs); Wi-fi throughout the entire city, from anywhere. Along with the growth of social media which runs on apps on smartphones and the convenience and ease of texting (which is a communication method solely used on smartphones), there is no doubt that this technology is here to stay.

### *Videoconferencing*

The most recent sub-revolution is around video technologies. Low-quality videoconferencing has been around for 35 years. Webconferencing has been around for about fifteen years. Skype, LiveMeeting, Webex, and countless other websites and browser add-ons can provide the capability for anyone with a camera and a computer to videoconference with others over the Internet.

For many years, connectivity problems at times cause the quality of the videoconference to be poor. Poor videoconferencing is not a minor problem; our brains are conditioned to expect that one's words match one's mouth and that movements are smooth instead of jerky. Low quality videoconferencing interferes with communication so much so that research shows it would be better to simply talk over a telephone line and not see the other individual. Furthermore, until recently, most webconferencing systems could only handle a tiny 2- or 3-inch videoconferencing window on the PC, which limits our ability to see emotions and communicate effectively. While television dramas seem to show people talking to each other, face to face, flawlessly, on mobile devices everywhere, the reality took quite a bit of time and technology magic to get to that point. Getting a stable connection used to be difficult, having enough bandwidth available for live videoconferencing used to be a rarity.

Note that there is a big difference in live videoconferencing with video streaming. For the most part, today people can watch full-size videos on their PCs after they have been downloaded (or "buffered") across the Internet. That is called video streaming. For videoconferencing, however, the conversation should be taking place in real time and cannot be buffered, which places a much larger strain on the system.

Until recently, there was still a great deal of resistance to utilizing web conferencing technologies. So what changed? The COVID-19 pandemic. For more than a year, traveling to work or school was a dangerous adventure due to a difficult-to-control virus that had governments shutting down industries and telling everyone to wear masks or stay home. Resistance to the new technology disappeared. Zoom, founded in 2011 by Eric Yuan, a former Cisco engineer and executive, utilized the new technologies (formerly called **telepresence)** to develop a system that could be utilized by anyone with a PC, tablet, or smartphone. Within one year, Zoom went from a respectable 25 billion dollar company to a 129 billion dollar company, breaking all sorts of

* Networking types are covered in *Chapter 2: Networking.*

records for profitability. The COVID-19 pandemic put a charge into videoconferencing like no other just as Zoom's easier-to-use and higher-quality videoconferencing software was reaching the consumer market.

**Figure 108. Telepresence system, the precursor for Zoom.**

Again we observe the speed of technology. In the year 2000, videoconferencing systems were extremely expensive, usually costing millions of dollars to implement just a few sites. By 2015 the cost had gone down to $400 per hour, or a few hundred thousand per site. And now, very high-quality videoconferencing is virtually free to anyone who already has a mobile device and a high-speed connection. Now that the infrastructure to handle the traffic is in place, high-quality videoconferencing is almost as ubiquitous as cell phones are today.

There are still a few unresolved technical problems. The Internet is "bursty" (transmits packets of data on different routes in bursts) and therefore sometimes have a hard time with the type of traffic generated by a videoconferencing system. Satellite service, which is often the only option in rural areas, have a delay caused by the lag of traffic from the satellite and back. This lag makes the videoconference connection virtually impossible.

Nonetheless, videoconferencing will continue to improve until it is very high quality and available everywhere. Wall-sized videoconferencing systems will have a major impact on businesses and education everywhere, especially when used in conjunction with mobile technology. Consider the impact: why worry about getting a corner office with a window when you can display beautiful live views of the Rocky Mountains on your entire wall? Why worry about traveling when you can see your friends and neighbors sitting at your table as clearly as you see them now? Add a Wii interface* and you can golf and fish with your buddies without ever leaving the room.

Imagine a classroom where students are shown in boxes on a large wall-sized screen, and all are watching the teacher on their own mobile tablets. The same face-to-face capabilities as today's in-person classroom, but without the travel or cost of a building. Why even travel to a building in a city to work when the entire team can meet more efficiently virtually from home? What reason would there be for working in cities and polluting our air with cars once people can connect, in full size, in person, without traveling? Working from home will become so commonplace, it won't even be called working from home – it will just be working.

Back in the sixties, famous author Isaac Asimov wrote a science fiction story about a universe where people never met face to face because they could conduct all their business and do all their socializing through room-sized videoconferencing†. While life may not become that extreme,

* Or more recently, the Switch. Wii is a hands-on video game controller that enables you to play physical games like golf and tennis virtually. Because of its large installed base, it may be here for a long time, but the newer Switch is being sold to most children these days.

† The only reason people got together physically was to procreate. Wanting to meet people outside of sex was considered a mental health problem, which formed the basis for the mystery of the murder which the protagonist had to solve in the story.

once the quality and the infrastructure are there, videoconferencing will be everywhere. And it will is changing the world, again.

# Obeying the Law

Within the world of business, there are several people who made statements that became "laws" that are familiar to everyone. For example, Murphy's law (proposed by Capt. Edward A. Murphy in 1949): *If anything can go wrong, it will.* The Peter Principle (proposed by Dr. Lawrence Peter and Raymond Hull in 1968): *In a hierarchy, every employee tends to rise to his (or her) level of incompetence.* Of course, these are not really laws, but rather *rules of thumb* - commonsense understandings that cover many cases, sometimes tongue-in-cheek. Nonetheless, recognition of the kernel of truth in these laws can help avoid problems in the future.

Within the world of information technology, various people have proposed "laws" that have proven themselves useful. These laws can help people understand just what and why certain technology changes occur (and why other aspects of technology never seem to progress). Knowing the laws can help business people make better technology decisions. We will explain seven laws, all named after those who proposed them.

## Loewy's Law

Raymond Loewy, the father of modern industrial design, coined the phrase MAYA - *Most Advanced Yet Acceptable* - to describe public acceptance of technology. It served as a guiding law for Loewy and those in his employ; reminding them not to push a design, however excellent, beyond the threshold of acceptability to consumers and manufacturers.*

## Moore's Law

Gordon Moore (Intel co-founder) proposed in 1965 what is now known as Moore's Law; *The number of transistors incorporated in a chip will approximately double every 24 months.* In other words the power of a computer chip will increase exponentially, approximately doubling every two years. This has resulted in a drop in price and an increase in power for computers every eighteen months or so.

## Metcalfe's Law

In the 1980s, Bob Metcalfe (founder of 3COM and inventor of Ethernet) proposed what is now known as Metcalfe's law: *The value of a network grows as a square of the number of nodes (i.e. computers) on the network.* Many people have attributed this underlying truth as a reason for the explosive growth of the Internet. Think about it; how useful would the Internet be if there were only a handful of people on it?

## Gilder's Law

In 2000, George Gilder, an author and technology pundit, expanded and connected the first two laws by proposing: *Bandwidth grows at least three times faster than computing power.* Considering the issue of the Fiber Glut and the increased demand for faster connections, this law

* **The MAYA Principle: Design for the Future, but Balance it with Your Users' Present** by Rikke Friis Dam posted April, 2020. https://www.interaction-design.org/literature/article/design-for-the-future-but-balance-it-with-your-users-present Last accessed October 30, 2020.

has proven pretty useful. (Gilder also predicted that broadcast television would become obsolete because of "telecomputers" connected to broadband fiber cable*, which is fast becoming true.)

### Jakob's Law

Jakob's Law (named after Jakob Nielsen, the father of web interface design) noted: *Users spend most of their time on other sites*. This means that users prefer all sites to work the same way as all the other sites they already know. Thus, anything that is a convention and used on the majority of other sites will be more familiar and desirable to the end user navigating the site. Web developers should avoid flashy new designs because if they don't give the user what they expect where they expect it, the users will not frequently visit the site.†

### Wirth's Law

This law is named after Niklaus Wirth since he wrote an article about it in 1995, but he attributed the statement to Martin Reiser. *Software is getting slower more rapidly than hardware is becoming faster*. Generally referred to as "feature bloat", it refers to the fact that as hardware gets faster and faster, the software programs increase in complexity and therefore make actual productivity slower. It has also been referred to as "What Andy givith, Bill taketh away" (meaning that as circuitry from Andrew Moore's Intel gets faster, software from Bill Gate's Microsoft gets slower).‡

### McGuire's Law

Russ McGuire is the Director of Strategy at Sprint/Nextel. He wrote a whitepaper in 2005 describing the law of Mobility: *Mobility exponentially increases the value of any product*. Thus, any technology that is developed as a fixed in space geography is of increased value to the consumer when mobility is added to the capability. Until recently, the cost of adding mobility outstripped the value of the mobility, but that changed when the other laws combined to make mobile technology cheap and available anywhere.§

## Summary of People Foundations

This chapter provided a foundation for understanding technology, including the historical people in the development of today's technology as well as the subrevolutions that formed the technology we are using today. We also discussed the current and future revolutions, as well as many of the rules of thumbs (i.e. the laws) that often explain how/why the technology we have today is the way it is.

---

* Gilder, George. (1985) *Life After Television* by W. W. Norton & Company, New York.

† Note that Jakob's Law is the only one named after the FIRST name of the proposer, rather than the last name. A video of his reciting the law can be found at https://www.nngroup.com/videos/jakobs-law-internet-ux/ Last accessed October 30, 2020.

‡ **A Plea for Lean Software** by Niklaus Wirth, posted February, 1995. https://www.computer.org/csdl/magazine/co/1995/02/r2064/13rRUwInv7E Last accessed October 30,2020.

§ **McGuire's Law of Mobility** by Russell McGuire posted June 2005. https://medium.com/clearpurpose/mcguires-law-of-mobility-32ebf919ad7d Last accessed October 30, 2020.

## 11-1. Interactive

1. **Discussion:** What videoconferencing experiences have you had? What kind of hardware and software was used? Has your videoconferencing frequency changed in the last two years? Why or why not?
2. **Discussion:** What do you think school will look like after so many schools were forced to go online in 2020? Can you see any advantages or disadvantages? What problems or issues might occur?
3. **Independent Activity:** Interview the owner or manager of a business.

   How do they use computers?

   How many employees use computers at part of their job? How many use mobile devices?

   What types of computers/devices are used by employees for business?

   What types of software do the employees use?

   What kind of connection do they have to the Internet?

   Do they also have a private network?

   Who is the top technology decision maker?
4. **Class Discussion:** This chapter only names a handful of the thousands of people who pioneered the foundations of IT. Can you name five other pioneers and what each of them did?
5. **Class Discussion:** What else do you think will change in the IT world? What will PCs look like 10 years from now? Will we still be using QWERTY keyboards? What will phones look like? What will videoconferencing look like?
6. **Class Discussion:** What other laws or rules of thumb impacting technology have you heard? Can you think of any yourself that might hold true?

## 11-2. Self Study Test Questions:

1. Who were some of the most famous people involved in the foundational history of Information Technology?
2. How has storage media changed over the years?
3. How has networking changed over the years?
4. What was the speed and penetration of the Internet before the World Wide Web? What about afterwards? How did it vary by country? Why?
5. What are the seven sub revolutions that led to the current status of technology? Which was the most recent?
6. What are some of the well-known technology laws (i.e. rules of thumb). How can knowing each one impact business decisions?

12

# Chapter 12: Information Technology Strategy

Overview: In this chapter, readers will learn the following topics:

- ⇨ Why We Want To Pounce Like a Panther
- ⇨ How to Have a Vision; A Hedgehog Concept.
- ⇨ When to Ignore Technologies That Don't Align.
- ⇨ How to Keep Track Of Standards Bodies.
- ⇨ Why It Is Important to Keep Track Of Vendor Success.
- ⇨ Why It Is Important to Keep Track Of Competitors' Technology Level.
- ⇨ Why It Is Important to Know Where All The Holes Are.
- ⇨ Why It Is Important to Play with New Technologies.
- ⇨ How to Be Patient For The Right Technology.

Throughout this book, we discussed various rules and guidelines for business leaders to follow in making decisions about business strategy that involves information technology. As you may have recognized, in many ways this book advises a conservative approach; don't buy until you can be sure, don't believe it unless it's proven, and so on.

This chapter, however, is both a contradiction for everything else in the book and its most important guideline. This chapter outlines the specific situations where the conservative advice in the rest of the book should be thrown out the window. This chapter will provide the culminating vision of the technology winner, so it will help put all the other chapters in perspective.

So here it is—the real secret to successfully using information technology for a competitive advantage: Pounce like a Panther.

## Pounce Like a Panther

Panthers have a very specific method for hunting. Not being as big or as swift as some of their competitors like elephants or cheetahs, panthers like to climb trees overlooking a path where their prey tends to pass. Then, at just the right moment, they pounce down on the prey. Similarly, we can pounce on new technologies, like a panther, but only in that very specific situation where the right technology is there at the right time.

When we look for certain instances for specific technologies that, when used at exactly the right time with full alignment of the business processes, we can transform a business and rocket it into stratospheric growth and profitability. This chapter is about how and when to recognize those truly strategic moments when inching out on a limb of a tree—even if it is a bleeding edge—is the right thing to do.

There are many steps to pouncing like a panther:

- Have a vision; a Hedgehog Concept.
- Ignore technologies that don't align.
- Keep track of standards bodies.
- Keep track of vendor success.
- Keep track of competitors' technology level.
- Know where all the holes are.
- Play with new technologies.
- Be patient for the right technology.
- When you see it, pounce.

## Have a Vision, a Hedgehog Concept

A vision is essential. As Stephen Covey makes clear:

> *More than any other factor, vision affects the choices we make and the way we spend our time. If our vision is limited—if it doesn't extend beyond the Friday night ball game or the next TV show—we tend to make choices based on what's right in front of us . . . If our vision is based on illusion . . . our choices [will] fail to create the . . . quality results we expect. If our vision is partial . . . we make choices that lead to imbalance . . . [A vision] clarifies purpose, gives direction, and empowers us to perform beyond our resources.*[*]

Fred Smith, founder and CEO, and Rob Carter, CIO, grew Federal Express from 186 packages a day (the average in 1973) to over 6 million packages a day. In the early 1990s, in both academic and industry writings, Federal Express was used so much as a technology success story that by 1999 I was sick of hearing about it. I figured that the magic of FedEx was probably more due to the needs of the market than to any particular technology strategy.

Then I met Fred and Rob at a conference, and I understood why they were continually used as *the* success story. Not only was their success clearly a case of correctly applied technology, but they are both very good at explaining exactly what they were thinking at the time and describing how things evolved. Their success was not an accident—and it behooves us to learn from them.

Their success started with a vision: to deliver packages overnight from anywhere to anywhere. Back in 1981, long before the Internet, and without any existing technology to rely upon, Rob Carter and Fred Smith laid out the technology needs for their vision of on-line everything. Not in general terms—in specific terms. The technology they needed to make the vision a reality wasn't feasible when they developed the vision, but over the next two decades it became feasible.

Fred Smith described his business idea in a paper as a student in an economics class in 1965[†]:

> *For instance, if you were a computer manufacturer like Burroughs or Sperry or IBM or Univac—all the people in those days who were competing for bank business—you'd go in and talk to, say, a banker in Amarillo and tell him he really ought to get rid of all his clerks and replace them with computers, which would be able to do the work much more cheaply, quickly, accurately, and so forth. And the argument was totally compelling except for one fact: The minute that computer went dark, the bank*

---

[*] Covey, S. R., A. R. Merrill, and R. R. Merrill. 1994. *First Things First.* New York: Simon & Schuster.

[†] If you've heard the story that Fred wrote this paper as a graduate student, or that it was a business plan for FedEx, or that he got a C in it, you've heard an urban legend. Fred doesn't remember what he got on the paper, and it wasn't a business plan, just the spark of an idea.

> *couldn't function anymore. When you automate a human function, either that device has to work all the time, or you have to be able to fix it rapidly. It was that simple an observation . . . You'd need a faster, more dependable, and more far-reaching kind of delivery system.*

We can learn two lessons from this insightful paragraph. The first lesson is the direct lesson that unreliable information technology is worse than no information technology. Once computers enter the picture, they can't go down without disrupting the entire business. Smooth computing means profitable computing.

The second lesson is more indirect, but the point of this section. To adopt the innovations that would help their business, FedEx needed to wait for the technology to become available, and feasible. In 1965 logistical information technology did not exist. But Fred Smith and Rob Carter kept the vision in mind throughout the years. They watched and they waited. When the time was right, they pounced on the technology—after it had been proven by others, after it became financially feasible, but before their competitors had adopted it.

Fred and Rob utilized technology to plan and maintain a series of systems that could fulfill their agreed-upon vision. In 1994, they introduced on-line tracking of packages (long before the Internet was the ubiquitous network it is today). By 1996 they were providing shipping information on-line, and by 1997 they were also providing return information. In 2000 they enabled customers to see the proof of delivery signature—immediately—on-line.

Note that Federal Express didn't always get it exactly right, either. In 1980 they launched Zapmail, a product that electronically transmitted documents from computer to computer. It was ahead of its time, and before the fax machine matured and then e-mail took over those functions. Zapmail flopped—big time. But Federal Express just picked itself up, brushed itself off, and tried again with a more successful service.

Not everyone can be a Federal Express. And not every business is served by the same technologies from which Federal Express benefited. But every business can improve by pouncing on the right technology at the right time.

In the late 1990s, Larry Bergman (COO) and his team at Boscov's, a large regional department store on the East Coast, had a vision that can be summed up in the headline of an article about it: *Out the Door in 7 Minutes.*[*]

Fast growth at Boscov's left it with a challenge: bottlenecks and slow throughput in the distribution warehouse. Sometimes it would take months for items to get to the selling store. For some items a delay wasn't a big deal, but for some items, especially items in the fashion world, Larry figured they were losing millions in sales.

At the time of the vision, there were no automated sorting conveyer systems that could handle the varied items sold by Boscov's. There were no standards for electronic invoicing. But Larry watched, and waited. As various technologies became available that would enable the vision, they were implemented so that over the course of three years, Boscov's succeeded in getting items to the stores an average of 170 days faster than before.

First they found ways to encourage more of their suppliers to use advance shipping notices (ASN). ASN is an e-mail that says "we just sent you a delivery of these items and we estimate it will arrive in four hours." The advanced notice enabled Boscov's to double-check the order and prepare to receive it. Next they increased the use of bar code scanning, chose a better monitoring system, and began to deal with errors differently. Finally, they installed a new sorting conveyor belt—a superhighway compared to their old one-lane road. The vision "out the door in 7 minutes"

[*] Maloney, David. February 2004. Out the Door In 7 Minutes. *Modern Materials Handling*. Volume 59 Number 2 pages 18 – 23.

had not changed, even while newer technologies were being employed to support the vision. Everyone—from Max Devlin, warehouse worker, to Olen Shaffer, shipping manager, to Bob Goonan, director of logistics—knew the vision was getting the items through the system quickly and accurately with minimum wait time.

The vision comes first and has to do with the business goals or desire to fill a market need. The vision is essential to guiding adoption of future technology, not the other way around. In other words, if we see a new technology and think "here's a good use for that technology," we are working backwards.*

Furthermore, the vision must be known to everyone in the company. We discuss ways to communicate the vision in more detail in *Chapter 8: Process of Organization*, so you already know that the vision can't be hidden away in the corner like a business plan that an entrepreneur is afraid to show people because they might "steal" the idea. The vision must be communicated again and again and again to everyone.

Jim Collins described the importance of developing and communicating the vision by using another wild animal, the hedgehog. Collins researched companies that were *Built to Last* and companies that went from *Good to Great* in an attempt to differentiate the key elements that made those companies successful. He determined that each successful business has a core strength, a core principle, a value proposition, a vision. Collins called this central vision the *Hedgehog Concept*, based on the idea that a hedgehog can do one thing (roll itself into a ball) very, very well, and is highly successful using that one well-known technique of defense. Collins notes that companies that have gone from good to great know what they do well and focus only on that capability, selling off divisions (even profitable ones) that don't directly support their Hedgehog Concept. Great companies pay attention to that one strength. Great companies don't waste time or energy trying to pay attention to everything.

Collins gives numerous examples of companies that are highly profitable while they stay focused on their Hedgehog Concept but begin to decline the moment they start to focus on something else. He also describes the difficulty people have in recognizing Hedgehog Concept but provides some guidelines for identifying and communicating it so that everyone in the company buys in to the same vision. For a Hedgehog Concept to work, everyone in an organization must be rowing in the same direction.

Having a Hedgehog Concept doesn't necessarily mean that companies can't diversify. Diversifying a product line does not mean that a company dilutes its Hedgehog Concept. The Hedgehog Concept has more to do with meeting a need in the marketplace (which can be met in a variety of ways) than it does with a specific product or service. But in all cases, the Hedgehog Concept was the guiding principle used by the leaders of these successful organizations in designing their strategy.

Technologies, especially potentially strategic technologies related to the Hedgehog Concept, can help forward the core strengths of a business. Any technology that can do that is worth some investigation, some research, some added attention and capital dollars. So the only time it makes sense for a company to be on the bleeding edge is when the company is investigating possible technologies that would launch new capabilities in the company's Hedgehog Concept.

---

* I believe that was one of the problems I had when I invested my heart and soul (along with significant funds) into Commercelinks.net to provide live videoconferencing systems for customer service. We saw the technology, and we were trying to find a good use for it. Not a good idea.

## *Ignore Technologies That Don't Align*

Have you ever gone fishing? Did you memorize all the names of all the fish in the ocean first? Of course not! We don't need to memorize the names of all the fish in the ocean before choosing the bait in order to start fishing. Similarly, we don't need to learn about all the technologies—only the relevant ones.

The biggest problem most leaders have is not that they are not aware of, or watching, the technologies that would align with their business. The biggest problem is that they are also watching all the technologies that don't. That is a mistake because it will distract them.

A company priding itself on its wonderful customer service should know all about the customer relationship management (CRM) programs on the market, although that doesn't mean, necessarily, that they've already implemented a CRM. But a business focused on the customer needn't spend the same amount of time and effort investigating back-end general ledger or financial systems.

If a company has a Hedgehog Concept of being the "most accurate and fastest shipper," implementing wireless radio-frequency identification (RFID) tags might be a good investment of time. But the company shouldn't spend the same amount of time investigating which is the best mobile device to purchase, or which PC manufacturer has the fastest processor—because neither of these technologies are related to that particular company's Hedgehog Concept.

The same would not be true for a company that provided medical transcription services. For that company, spending time looking at RFID tags would be a complete waste of time. A state-of-the-art mobile device and document management system investment, on the other hand, would be a strategic technology aligning with the Hedgehog Concept.

As Chad Dickerson, a noted columnist and CIO, said: "Just as one man's trash is another's treasure, one man's commodity is another's competitive edge."* As has been explained in earlier chapters, except for the Hedgehog Concept technologies, the best strategy is to stick with what you already have, or follow the crowd for anything new you might need. No one ever got fired for buying Lenovo and Microsoft. We are best served by choosing whatever our competitors are using. Don't sweat it, or waste time. Our best strategy is to completely ignore product brochures and articles touting new information technology that doesn't directly relate to our Hedgehog Concept.

At the same time, we need to follow the progress on technologies related to our Hedgehog Concept with a keen eye and our ear to the ground.

That's not to say it is easy. To me, focusing on aligned technologies is such a commonsense bit of advice, I had trouble understanding why so few businesses do it. Why don't decision makers follow this strategy? The research I've done points to various reasons this issue is such a problem.

First, those who know and understand the strategic vision of the company (typically the business owners and/or entrepreneurs) are often not technically literate enough to understand the underlying infrastructure, which is necessary to know which technologies to watch. Their technology maps are incomplete, and they haven't spent enough time trying to complete them. Typically the technology leaders of the company understand the underlying infrastructure enough to know which technologies to watch in support of the vision because they have a complete technology map. But they either don't know, or don't care, what that strategic vision is.

As explained more in the chapter on managing people, technology people get into this field because they love technology, not because they care about making money or supporting someone else's vision. Successful companies can only pounce like a panther when the technology people are fully aware of, have bought into, and are willing to support the central vision or Hedgehog Concept of the company.

---

* Chad Dickerson. July 11, 2005. CTO connection. *Infoworld.*

Another related problem, also uncovered in my research, is that often the chief financial officer (CFO) is involved in the decision to the detriment of the decision-making process. Indeed, one of the most common mistakes in technology strategy in businesses over the past 10 years has been investing in back-end enterprise resource planning (ERP) systems without paying attention to the impact on the customer. CFOs, being adept with spreadsheets and return on investment (ROI) calculations, can often make the case sound good. And, as the people holding the purse strings, they can often smooth the approval process. But the fact is, without a complete alignment with sales and operations, a new back-end financial system can sometimes do more harm than good.*

Many companies discovered this as new financial systems failed to deliver on what was promised. The problem wasn't the new financial system, the problem was that the new financial system was selected and implemented by the financial division of the company, with little thought toward sales, marketing, or operations; the true blood of the profit-making corporation. In the cases where the new financial system was chosen, customized, and implemented by the entire company rather than a single division, there was much more success. (Additionally, as discussed in *Chapter 10: People of Organization*, studies show that the amount spent on training had a significant impact on success).†‡

Furthermore, focusing too narrowly on the ROI calculation can lead to the exact wrong decision. Since the ROI calculation is dependent upon esoteric rules of capitalization rather than real dollars, it can often make decision makers think that a technology will pay off in a short period of time when it has no chance of doing that. Pouncing like a panther on the right technology at the right time has to do with strategy, not financial calculations. It must be customer driven, not financially driven.

## Keep Track of Standards Bodies

There is an old joke in the technology industry: The wonderful thing about standards is that there are so many to choose from.

A *standard*, literally, is an approved way of accomplishing a technical goal that is published by a standards body such as the IEEE (Institute for Electrical and Electronic Engineers); International Organization for Standardization (ISO); the World Wide Web Consortium (W3C); and several others (e.g., IEC, CEN, CENELEC, ETSI, ITU-T, and OASIS).[25] Some of those standards bodies are as old as the hills (such as IEEE§ which was established on May 13, 1884), and others are just babes in the woods (such as W3C which formed in October 1994).

Old or young, a standards organization is usually needed to broker negotiations for shared standards so that businesses can capitalize on interrelationships. Standards bodies are involved in the politics of a standard as well as the technical aspects of it.

The workings of standards bodies may be better seen with a nontechnical example. The size and shape of a freight container (one that can be moved from train to truck to ship to plane without emptying the contents) was one of the standards published by the ISO. Vince Grey, former

---

* Khosla, Vinod, and Murugan Pal. 2001. Real time enterprises: A continuous migration approach. *Information Knowledge Systems Management*. 3:53-80.

† Soja, P., July 2006. Success factors in ERP systems implementations: Lessons from practice. *Journal of Enterprise Information Management* 19:418-33.

‡ Gattiker, Thomas F., and Dale L. Goodhue. 2005. What happens after ERP implementation: Understanding the impact of inter-dependence and differentiation on plant-level outcomes." *MIS Quarterly* 29:559-85.

§ IEEE was originally called American Institute of Electrical Engineers. In 1963 it merged with the Institute of Radio Engineers (originally founded in 1912) to form the new organization Institute of Electrical and Electronics Engineers (IEEE).

secretary and chairman of ISO, talks about getting around the politics of establishing a standardized freight container size.*

> *One of the first problems we ran into was various countries trying to have the international standard reflect their own national practices. We really didn't want to do that. The trouble was, that when we decided to get into the middle type of container size, the ones that are like a truck body, the metric countries wanted to confirm the container sizes that had been in use in the UIC (International Union of Railroads). I'm bringing it up because I think how we approached that subject may have value. Instead of locking horns, and saying "no way!", we accepted these container sizes and called them Series 2 containers. . . . But when it came to the market place, no one bought [them]. . . . It's better to get on with the work as long as you can achieve the basic goal and let the merits of each series be judged by the users.*
>
> *Anyhow, once that last peg was put in place at that Moscow meeting, so that people could go out and procure standard containers, the industry just took off! Everybody started placing orders for containers. Until that time, the fear was that you would spend a lot of money buying the wrong containers.*

So why must we know about standards? Because in order to know when to pounce like a panther, we must know when standards are "real." It's not just a matter of seeing which standards are published, but in following the politics behind them and recognizing which standards are likely to be used by an industry, and which ones are likely to be phased out.

Once we've identified our Hedgehog Concept and have envisioned the technology that will improve and strengthen our ability to provide the product or service to our clients, we need to identify those bodies that manage the underlying standards supporting the technological devices involved in that technology.

Imagine that you are a movie distributor back in the 1980s. You would have been acutely aware of the impact of the competing VHS versus Beta tape format standard on the industry. If you had chosen to distribute all your movies on Beta only, you would today be out of business. If you had invested in both VHS and Beta, you would have lost your Beta investment, and perhaps because of overreaching, your business too.

The optimum path, back then, would have been to continue to focus on whatever you were doing before video players were widely and cheaply available. Then, the moment it was clear that VHS would win the tape format war, pounce on the VHS format. This strategy would have helped you avoid your competitors' mistakes of wasting time and money on the wrong format. Following the standards bodies would have let you clearly see that moment much earlier than waiting for the market to reveal the trend.

Let's work through an example involving a well-known technology standard—the network that connects your computers. Back in the 1980s, Ethernet (IEEE standard number 802.3) was just one of two popular competing standards (Token Ring, 802.5, was the other). Back in the early 1980s, if we didn't know that there were standards bodies defining wiring standards, we might not have known there were two potential solutions. (Actually there were many more, but by the middle of the 80s it was apparent that these were the two major local area networking standards.)

Let's imagine that we own a company that provides marketing research services to the transportation industry in 1985. A local area network is essential to our Hedgehog Concept because our employees need to communicate and collaborate using a network. Further imagine

---

* Grey, Vince. *Setting standards: A phenomenal success story.* https://www.iso.org/files/live/sites/isoorg/files/about%20ISO/docs/en/Friendship_among_equals.pdf Last accessed on October 31, 2020.

that we have just leased a building and must decide what kind of wiring to install. Our director of IT had previously worked for IBM, so she recommends Token Ring. If we didn't know any better, we might take her advice.

If we choose Token Ring for the wiring, however, we would be wasting a great deal of time and money. We would end up replacing the entire thing within a few years. The standards committee that defined Token Ring stopped actively working on it in the early 1990s, and it has since disbanded. Ethernet was cheaper, and it seemed to work just as well for most companies, so the better choice was Ethernet despite the fact that technically Token Ring was superior to Ethernet (just as Beta was superior to VHS).

Often, the technically inferior but cheaper technology wins. Technologists, however, often want to use the superior technology rather than the optimum technology. To give one example, my brother can be described as nearly legendary in some circles for his technical prowess. He held onto his Beta video system for years after everyone else had given up on Beta tapes. He still occasionally rants about how much better Beta was, and he would detail to anyone who would listen why VHS was a poor substitute. I'm sure there are a few die-hards who swear by Token Ring, and resent being forced to use the unstructured shout-on-a-wire Ethernet networking.

It makes sense to pay attention to the more enduring technology rather than the technically superior one. Pouncing on the wrong technology (even if it is the technically superior one) can be detrimental to our ability to pounce on the right technology. The cost to replace the entire Token Ring network would make the switch unfeasible for many years, but as time went on Ethernet became more and more essential to smoothly running networks. New devices such as switches and routers were released only for the more common network type, Ethernet, but not for the less common Token Ring.

Knowing the standards bodies, which standards they are working on, and which standards are competing with each other is an essential step to understanding the underlying technology. Generally, when there are competing standards we want to wait until there is a clear front-runner before pouncing. In the case of our network, by 1989 it was pretty obvious to anyone watching that Ethernet would beat out Token Ring as a standard even though it would be years before it stopped being supported by the industry.

Why are there competing standards? Once the standard is, well, standard, the technology equipment manufacturers have to do one better than the standard in order to have a competitive advantage. As a matter of fact, the standards of today are usually the innovations beyond the standard of yesterday. But innovation means that the equipment is no longer following just the standard. The Ethernet (802.3) standard, for example, is not what most people have today. The most common configuration is Fast Ethernet (10 times the speed, called 802.3u) or Gigabit Ethernet (100 times the speed, called 802.3az-2010). Of course, the next wiring standard will be a new Gigabit Ethernet protocols (802.3cg-2019). While knowing exactly what these standards are is not important, understanding that they exist and who publishes them is important. If we keep up with the activities of the committees, at some point a clear winner will emerge. That would be the right time to pounce.

Why is waiting for a winner important? Competing standards are often not interoperable. You can't have a network card "talking" at Gigabit Ethernet speeds (1000) while the hub port is "talking" at Fast Ethernet speeds (100). They must match, or the network doesn't work. Since you can't replace just one piece, switching to a different standard is a big deal and will cost a great deal of money.

Are you wondering which comes first, the product or the standard? Products using the standards are usually developed before the standard is published. Indeed, it is often the research and development arm of the vendors who submit the standard for publication. Token Ring, for example, was researched, developed, and submitted by IBM. Ethernet was researched, developed, and submitted by 3COM. So if a standard is published, there is already at least one vendor using it.

"Wait a minute," you might say. "Why would an organization pay for the research to develop a new way of doing things, only to turn around and submit that way for a standards body to publish? Wouldn't they want it to be proprietary, so only they have it and can charge more for it?" In a different industry, like drugs or chemicals, yes. However, information technology products are useless unless they interface with other information technology products. So a vendor can't keep its way of doing things proprietary and still interface with other pieces of the puzzle. If a vendor develops a way of doing things, but it is not adopted by the rest of the industry, that vendor goes out of business.

That's not to say some vendors haven't tried. Apple, for example, has a history of not submitting its standards for publishing by standards bodies. Instead it tries to maintain complete control over the entire system—hardware, operating system, software. They've even sued, and won, to prevent other vendors from using their methods and protocols. Some say that this is the reason Macintosh never grew to the level of PC-based industries. Other examples are the proprietary network protocols of the 1980s like Novell NetWare, IBM's SNA, and Microsoft's NetBIOS. Unlike the general TPC/IP protocols of the Internet, these networks could only interface with their own systems. All of them eventually disappeared, as proprietary technologies tend to do.

In summary, it is important to understand and follow the standards (or have a trusted adviser who is doing so on our behalf) that relate to the type of technologies that relate to our Hedgehog Concept, as well as the relationship between those standards and the products on the market, so that we can know when to pounce.

## Keep Track of Vendor Success

As noted previously, the products and services sold by IT vendors have a relationship with the standards published by standards bodies. But the success of the vendors themselves also has a profound effect on knowing when to pounce. As explained in the chapter about vendors, it is important to seek out vendors who have financial stability. We want to stay away from vendors who grow too quickly, whose growth is spurred more by venture capital than selling a product or service that meets a need. The smaller the company, the harder it is to have the necessary staying power to contribute significantly to the industry, a subset related only to the newest product or service.

Therefore, when the industry is full of new vendors in a product or service, it is probably too new for us to pounce on. To give some current examples; the vendors selling cloud technology have not yet had a shakeout. There are still 7 major cloud vendors, and 14 minor players. The government regulation of the industry has not quite stabilized, definitions aren't yet solid, and the business models of the vendors are still in flux. (Should they charge by the user? A monthly fee? Charge for each device? How is the income shared with application providers?) At the present moment, unless directly related to a Hedgehog Concept, cloud is not a technology anyone should be considering for replacing existing systems (although there may be a case made for new installations). But by the time you read this book, that may have changed.

At the point that consolidation has begun, we want to start watching to see whether the industry will disappear or mature. The signs that maturity is coming are subtle. The industry will have consolidated to three or four major players who are no longer dependent on venture capital but who are making a profit by selling their product or service. The regulatory environment has stabilized and there are no pending lawsuits that might change the business model. Definitions are solid, and business models have been tested.

The signs that the industry is not quite ready for prime time and has not yet matured (and may never mature) are similar but telling. The consolidation has resulted in a few puny or dying companies. The underlying technology upon which the entire industry is based has hit some problems that have not been overcome. Sometimes these problems take decades to solve. For example, the videoconferencing industry limped along for decades because inexpensive connections did not yet have the required quality. Virtual reality systems are still untested and largely unused. Other times the solution never comes. Blu-ray Laserdisks never hit their stride and is pretty much deader than a doornail now.

The moment to pounce, then, is just after consolidation, when there is a strong financially stable winner or two in the field.

## Keep Track of Competitors' Technology Level

In addition to watching the standards and watching the success and consolidation records of the vendors in the field, we also want to keep track of what our competitors are doing. One way to get this information is to use a competitive-intelligence consultant or send employees over to speak with their employees or observe their operation. We can purchase their products and look for clues in the invoice or paperwork. Often, though, all we need to do is ask our IT vendors; their salespeople will often know exactly what type of system all the major players in any industry are using.

Once we have the information, we use it to adjust our own strategy. This is a strange case of not being too early and not being too late. If the other companies in our industry are still on PC-based desktops, have no cloud services, and don't use mobile technology in the field, we don't want to be in a hurry to implement any of those technologies. Remember that as long as our competitors are using and maintaining the same technologies they've been using for dozens of years, their cost basis for technology is near zero. If they can lower the cost of their products and services below ours, and not give up too much on features and benefits, it is not to our advantage to invest in new technologies that will raise our cost basis for slightly incremental improvements to the products or services. In other words, it is too early to pounce.

At the same time, if we find out our competitors are investing in all new technology that will (supposedly) increase their market share or lower their costs or increase their quality, we don't necessarily want to jump on the bandwagon and invest in the same new technology. It might be that our competitors will implement it, and it will fail because it was too expensive, hyped, or not reliable enough. All we need to do to succeed is to sit back and wait for our competitors to do all the hard work to find the faults of the new technology.

If the technology our competitors invest in does not turn out to be a complete failure, that's when we should pounce! While our competitors are still struggling with getting their systems to work reliably, we can learn from their mistakes and implement the next generation of the information technology. Remember that vendors learn from clients' mistakes as well, so the vendor will come to you with a much deeper well of knowledge in how to implement the new systems. Even if our competitors are relatively successful in their new technology implementation, the second time a new technology gets implemented always is faster and cheaper than the first time, the third time cheaper yet, the fourth time even faster, etc. My experience says that by the time a technology has been implemented over a dozen times, the projects will start becoming smooth as silk—but not before then.

Some people might say, "But what about first-mover advantage? Once our competitors get their new technology implemented, they will have the advantage, and we won't be able to catch up." Well, the truth is; first-mover advantage is a myth.

"But what about the great examples, Amazon or eBay?" you say. Again, the truth is that there were many on-line booksellers and quite a few auction sites already on the Internet before Amazon.com or eBay.com. Their founders benefited by looking at why the others had failed. Cook and Proulx, the founders of Intuit and the highly successful QuickBooks software (often used as a first-mover example), joke that they had a forty-seventh–mover advantage.*

There are very few cases where the first company to implement a new technology actually succeeded. We think that the successful companies we hear about were the first ones because we didn't hear about the failures that came before them. As Mickey Rooney says, "You always pass failure on the way to success." The key to being the success and not the failure is to avoid being a front-runner until the perfect moment in the race (usually just before the end) when we can pull ahead.

## Know Where All the Holes Are

Innovation and technological advantage almost always come from plugging up a hole in the existing information technology. A *hole* is a function or feature that would be really great to have—but which doesn't exist in any of the current technologies. The technology map we've talked about can help us establish where the holes might be, but more often than not the hole must be defined by the need, not by the technology.

***The Search for Ubiquitous File Access***

An example revolving around a personal need that is still a hole might make this concept clearer. A few years ago, I realized that I really needed a storage place for my ever-growing data files. Between client files, photographs, videos, databases, my research, and my writing, I've aggregated quite a load of essential electronic files†—about 5 Terabytes. (In the past, every year I used to spend three days or more cleaning out and archiving files I no longer needed. Then when the size was about 10 gigabytes, I realized that three days of effort was much more valuable than the space I was saving since hard drive space was so cheap, so I now I don't bother deleting old file, but just keep buying larger hard drives. My most recent purchase was a 10 Terabyte drive.)

In past years, I would constantly send subsets of this data to my employees for various tasks they needed to accomplish — which was a pain. I longed for a password-protected virtual storage facility where I could store all my files. That way, I could just set up access authorization for my employees and I wouldn't have to waste my time finding and sending through e-mail or creating a flash drive for my employees when they need the files. Furthermore, someone else would be backing them up and maintaining them.

For years I investigated different options. I tried no less than five different companies that promised to provide exactly what I needed: a virtual storage area on the Internet where I could place my files and give access to the files to my employees.

* Taylor, Suzanne, Kathy Schroeder, and John Doerr. 2003. *Inside Intuit: How the makers of Quicken beat Microsoft and revolutionized an entire industry.* Cambridge, MA: Harvard Business School Press.

† It might be interesting to note that when I wrote the first edition of this book in 2002, I had amassed 10 Gigabytes of data, which was a huge amount at that time. Twenty years later that is hardly a pimple.

> So where is the hole? Until 2018 none of these services, *not one*, could give me instant access to all of my files across the Internet. Some promised to do so, but in every single case the time necessary to access the file exceeded the time-outs on the connection, and I rarely was able to work on the files virtually. In the end I gave up, purchased a portable hard drive and carried it with me when I saw my employees so that I could quickly download whatever files they needed into their systems. As of 2018 there were several reliable cloud services that provide me with **almost** what I need. But none of them work well enough for me to use them. Recently I tried DropBox Business, but even that didn't work. While it was close to what I needed, it wasn't an exact fit and was expensive to boot. The biggest issue was my video files which are huge, and the large number of pictures we need for the business. Dropbox Business was limited to 100 Gigabytes transfer at a time; still not large enough for my needs. And the videos took hours to upload. The hole still exists (but will probably be gone in a few years, so I'll keep watch).

It is hard to get vendors to talk about holes until they have a solution. They deal with the lack in one of two ways. (1) They ignore the problem, considering it unsolvable or outside of their purview. Alternatively, (2) they act as if they already have a solution, even when they don't, and will fail to understand your explanation of why their solution is not good enough. Once a vendor has a solution, however, they will tout it to the heavens for a while. Once every vendor has the solution and it is no longer a differentiator, they will go back to ignoring it.

For example, back in the 1980s, there was a major hole regarding managing the memory on personal computers due to a problem known as the 640K barrier (talked about in *Chapter 11: People - Foundations*). To solve this problem, virtual memory was developed, and dozens of memory management companies popped up so that people could use graphical programs, like Windows, which needed more than 640K of memory to run. But once memory management was built into the operating system of the personal computer itself, the hole disappeared, and so did all those memory management companies.

Another example: As recently as 2015 most antivirus packages did not do anything to prevent spyware. Antivirus programs were designed to eliminate computer viruses—little unwanted programs that actually damaged files or systems. Spyware is a little unwanted program that doesn't cause any direct damage, but watches what we do as we use our computer or surf the Internet and reports back to the vendor our behavior, ostensibly so that they can figure out what our needs are so that they can sell to us. It causes no direct damage, although it slows down our systems. A hole appeared—soon plugged by anti-spyware programs, which had to be purchased separately. It only made sense, however, that antivirus and anti-spyware programs would merge so that most systems are protected from both viruses and spyware by the same program. The hole disappeared.

To provide a business-oriented process example, there was, at one time, a hole when it came to calculating shipping costs. In order to calculate how much profit businesses receive on every sale, they have to know the shipping costs of the sale. But shipping costs are variable based upon the size, weight, source, and destination of the package as well as which shipping vendor was used. The costs were therefore unknown until the box had been weighed and stamped by FedEx or UPS or DHL or the post office. Because the actual costs could not be known at the time of the sale, most companies simply set shipping and handling charges to some number that they hoped was more than the cost, and went on their merry way.

But this could cause problems. In the early 1990s, I was responsible for the warehouse at a systems integrator firm. Every month, the warehouse manager was responsible for reconciling the estimated and actual shipping costs, and he came to me with a problem pattern he happened to notice. After investigation, we discovered that the salespeople were entering a low-ball figure for the shipping costs. The price to their customers was lowered without impacting the salespeople's net profit, on which their own commissions were based. But at the end of the month the company was losing thousands of dollars in shipping. Since this was before the Internet, we had no choice but to start charging the salespeople's commission the following month for overages from the previous month because there was a hole—no way to accurately calculate the shipping costs.

Nowadays the hole has been filled. Most financial packages can link automatically with the chosen shipping vendor who takes the size, weight, source, and destination of the package and accurately calculates the shipping cost. Once we know where the holes are, we can keep checking new technologies to see if the holes have been filled or not. To pounce optimally, we need to be aware of the hole, and the fact that it has been filled.

How do we find the holes if they are not readily apparent? We keep asking questions until we find something that a system can't do. It may help to know that a hole is usually related to one of six different requirements:

- Ease of use.
- Integration and connection with other needed function.
- High quality.
- Size.
- Security and privacy.
- Reasonable cost.

It may be that we can get the feature that we want, but it is very difficult to use and requires the skill of a technologist. One example of this is the Web browser. Before Web browsers such as Chrome, Safari, Edge or Foxfire were available, Internet surfers from 1973 to 1995 memorized and used esoteric commands to get around the Internet. Web browsers filled the hole.

Alternatively, the system might be easy to use, but it is not integrated with some other related device or software with which it needs to be integrated, such as navigation systems that aren't hooked up to Global Positioning System (GPS). Or, the feature or function is integrated, but we can't get the quality high enough. For example, we can get a grainy videoconference connection but not a television-quality videoconference connection.

On the other hand, we might be able to get the quality that we want, but not enough memory or space. We can feasibly get 25 gigabytes of on-line storage, but not 5 terabytes.

If you have too many contacts, you might bump into the number of contacts issue when trying to store them on your phone. Until recently Google contacts limited people to synchronizing only 2000 contacts with your phone. I have 6000 contacts and therefore could not use that service.*

We also might be able to get the quality, but not in the size we need. For example, in 2016 we could get a 3-inch videoconference window, but not a 17-inch monitor-sized conference window. Now we can get a large-monitor sized conference window, but are still limited to viewing only 49 other people at the same time on the same screen no matter how large our screen is.

Security and privacy are often a lagging hole because vendors don't work on these until there is a well-publicized problem. E-mail is not secure and cannot be used to send confidential information such as credit card data, health records, or personnel records. After 40 years, still a hole.

Finally, we may be able to get all of the requirements we want, but it costs an arm and a leg, outside of the reach of all but the most affluent. Large companies can afford all the most up-to-date secure Web site technology so that they can accept credit cards, but the small mom-and-pop

* I purchased a third-party package called CompanionLink to synchronize my contacts and calendar with my smartphone. Google now allows you to synchronize 25,000, but still limit you to import 3000. But I'm still waiting for another hole to be filled; I'd like to synchronize my phone with a mySQL database. Currently there are no software packages that do that.

shop still feels the hole. Quite a few Web sites still don't use SSL (secure socket layer) because SSL on a Web site is expensive*.

Of course, we also have to be careful that we don't lose existing capabilities when we plug up the holes. One of my clients purchased a new financial system to replace the green-screen mainframe system they had been using. The hole they were plugging up was the "easy to use" requirement because new employees found it very difficult to learn the old system. Until they implemented the system, however, they didn't realize that the accounts receivable on the new system was not directly integrated with the general ledger where deposited checks were recorded. They ended up entering each check twice: once for the deposit and once to record the payment against the invoice—something they didn't have to do on the old system. As mentioned earlier, it hadn't even occurred to the employees writing up the requirements for the system to include the question "When you enter a check for deposit, can you mark which invoice was being paid?" because their old system had that feature and they assumed any new system would also have that feature.

Another of my clients, Jim, decided to switch to a new Web site hosting service because it promised to provide a secure Web site with a shopping cart that would accept credit cards. Unfortunately, Jim hadn't consulted me before making the switch. He discovered the Web site had no way of syncing automatically with his inventory system as his previous hosting service could do. Jim had gained security but lost accuracy.

Every system has holes; the key to success is in knowing where the holes are and finding a way to work around them or live with them while keeping our eyes open for the solutions that give us everything we currently have, plus plugs up the hole.

## *Play with New Technologies*

How do we check to see if the holes have been filled? We play. We (or some technologists we trust) establish a test lab and install new systems to see if they fill the holes without losing existing capabilities. In the chapter on dealing with IT people, we discussed an excellent way to keep information technology gurus focused on improving our internal systems without tempting them to break the systems in order to play with new technology. We give them a budget to purchase, install, and play with any new technologies that support our Hedgehog Concept. But we can do the playing ourselves as well as long as we know what the existing capabilities are and where the holes are.

For example, for over 30 years I have been dreaming of a specific system that does not yet exist. I envision an Internet-based contact management/calendar/e-mail program, similar to the original Lotus Organizer, ACT!, GoldMine, or Outlook, that works seamlessly for a small group. This program would sync automatically with a wireless calendar/videoconference/camera/phone/Web browser/music device, but will be able to access many of the functions even when in an area that doesn't have connectivity to the Internet. Other people, like my employees, would have password-protected access to my calendar and direct access to the files on the corporate Web site (all 5 terabytes). The calendar portion of this system will provide me with a report each month that graphs different categories of activities. The contact manager will hold all 6,000 of my contacts, which are also available to my employees. Each contact's address will be automatically connected to a mapping program and will give me verbal directions from my GPS location at the touch of a button. All for less than $500 a year. And all very easy to use.

I already have all of these capabilities, but there is no single solution that provides all of them, and the combination is way beyond the $500 a year cost basis (and there are quite a few manual

* Though this is changing since the browser companies started warning users that any website without SSL is a danger, some refusing to even go to a website that doesn't have SSL installed.

processes that I programmed or designed for myself that are not generally available). Every time any software or hardware related to any of these capabilities comes out, I investigate it.

For example, when iPhone came out with much fanfare, I had a simple question: Can the iPhone maintain and sync with my calendar? If the answer was yes, I would buy one to play with to see if it could meet my other needs. I went to the Web site to get an answer to my question. I went to the Apple site and watched the video on the iPhone. Unfortunately, it did not answer my question. Nothing on the Web site, either. Eventually I went onto a series of on-line forums where people were discussing the iPhone and discovered that the initial IPhone *didn't even have a calendar program*. Quite a hole! Furthermore, there seemed no way to convert contact information from my existing contact manager. No calendar? Re-enter all my contacts? I don't think so.

In summary, if we know the capabilities of our own system, and we know where all the holes are, we can quickly look at each new technology to investigate if it fills the holes without losing existing capability.

## Be Patient for the Right Technology

It is hard to simply sit back and wait for the right combination of events to align so that we can pounce. Remember the client of mine, Jan, who called me up to find out what combination of software and hardware would allow her to start a transcription business. She'd seen a demonstration of a system that could automatically type up the spoken word. She envisioned a business where doctors and lawyers would speak into a phone and have their words automatically converted to text and e-mailed to them. I told her that what she was envisioning wouldn't work, because that the hardware and software that she wanted didn't exist yet. She was sure that the technology was just around the corner, and would be available within the next year or so.

Fifteen years and we are still waiting. Last month I was speaking with some other people, Bob and Mary, telling them about my friend Jan, only to have them inform me that they are actually in the medical transcription business. Furthermore, they had just purchased the hardware and software to do what I described. Their clients (mostly doctors) speak into a phone and their words get recorded. Until recently, a human being listened to the recordings and typed up the information. However, Bob was sure that the system they had just purchased and implemented would eliminate the need for typists. It would automatically generate the text.

Of course, after some in-depth questioning on my part, I clarified that currently the system only helped; it was still only 80 percent effective. Currently the new system was costing more to have the typists correct the mistakes than it used to cost to have them do the typing from the beginning. But Bob was sure that within a short time the system would get more accurate and it would stop costing more to correct than to type. If he'd been my client, I would have told him point-blank that he had wasted his money. These systems were 80 percent effective back in 1987, the height of the natural language processing research. In 30 years they have not improved their track record much. The necessary technology was beyond the capabilities of any but the most powerful supercomputer system today. The typists have been, are, and will be cheaper for years to come. Natural language is one of those things computers just aren't very good at.

***Being Patient is Easier to Say than to Do***

No one is immune from impatience. I told the story of how I allowed someone to convince me that videoconferencing over the Internet was not only possible, but currently being done. My impatience for that technology colored my views. If I had been more patient, I would have saved the hundreds of thousands of dollars that I lost when I invested in CommerceLinks.net and almost a year of my life working on the patent for videoconferencing technology before its time.

Trying to push the envelope are companies selling video e-mail services, end-to-end enterprise network management, natural-language processing, virtual gaming systems, voice-recognition services, mobile team contact management, radio-frequency identification, and many, many more "new" services. We all want these capabilities, and they seem so near to the capabilities we do have that it is hard to understand why we are still waiting. Remember that the panther may end up waiting for weeks on the limb of that tree until the prey comes walking down the path. Hunger pangs do not make the prey appear. Wishing does not make it so. There is no way to force the prey to appear.

Some business leaders try to force the needed information technology by investing in research and development. They hire programmers or engineers to plug the holes or develop the needed information technology. The difficulty with that path is the high cost. If one company pays millions to develop an easy-to-use shopping cart so that they can sell their items on-line, and another company just waits two years for easy-to-use shopping cart software to become available commercially, which one will win out in the end? The company who waited, because its cost basis will be many times lower than the first company. Once again, until the right time, we must be patient.

## When You See It, Pounce

After having watched for years and years, following the guidelines in this book, when the exact right technology comes along, we will recognize that critical moment in time. Our vision aligns with the people, the process, the hardware, and the software. We have only one thing left to do. Pounce.

When the just-right technology finally matures, we need to jump into it with every muscle, every brain cell. Great companies allow themselves to be completely transformed by new technology—every division, every department, every decision maker, every employee. Pouncing is not incremental. It is not inching out on the tree limb. Every person, every process, every system usually has to change to incorporate truly new technology. Only by being willing to transform the entire business to take advantage of the new technology—only then will we accomplish far more than any stuck-in-the-old-way competitor can keep up with.

The jump cannot be accomplished without the watchful waiting first—because hype and lack of knowledge will keep us from being sure when the transforming technology is ready. If we jump as soon as we wade in we will get fooled. We need the research lab, the careful testing, knowledge of the holes. We need to be absolutely sure.

Jim Collins noted that technology is an accelerator. If we choose the right technology, one that fits into our Hedgehog Concept, then it can accelerate the entire business to great new heights. But if we choose the wrong technology or the wrong time to implement the technology, then that very same technology **will accelerate our demise**.

***My First Business***

I have personal experience with the accelerating power of technology—both positive and negative. I started my very first company in 1986, Computer Educational Services (CES), a computer training and consulting firm. I had already had the vision of contact manager/calendar/file management during the infancy of CES. I wanted to be able to access my files from any of my three offices (home office, training room, and administrative office) and from either the PC or the Macintosh. The files were only a few dozen megs' worth back then, but no less important to me and, a decade before the Internet, much harder to access. In my quest for the technology to fulfill my vision I got to know certain people who ended up being essential to my success. One was Warren "Pete" Musser, the CEO of Safeguard Scientifics, a venture capitalist who invested in Novell NetWare in 1987. We appeared together in an entrepreneur's panel for alumni at Lehigh University. When I met him, being focused on making my vision a reality, I asked him what he knew about accessing files from either a PC or a Macintosh on a single server. Pete told me that I could do what I wanted with Novell NetWare, and he made arrangements for Novell to give me a copy of NetWare 286 (along with the shelf full of manuals) so that I could share files from the server from either the Macintoshes or the PCs in my training room. Of course, in the end the file sharing didn't work as seamlessly as I would have liked. And I'm still waiting for a contact manager/calendar/file management system. But my knowledge of this difficult-to-use network operating system led me to great success in the infant field of networking so the vision still helped me.

It was because of my knowledge of networking that I was able to impress Ray Melcher and Ed Barrett, the owners of Hi-TECH Connections, a systems integrator firm, by solving a problem on a multiuser version of WordPerfect that their network engineers were unable to solve. Eventually I transformed my company from an applications training and consulting organization to an enterprise network management training and consulting company, which turned out to be much more lucrative, and sold it to Hi-TECH Connections.

My knowledge of networking had accelerated my business to unbelievable heights. But I've also seen the downside. The third company in which I invested jumped on videoconferencing technology, as mentioned, before its time. Our focus on the technology accelerated us straight to bankruptcy within a few short years. So timing, as they say, is everything.

## Implementation: Pouncing Like A Panther

Business leaders who have followed all these steps in their own industry are more likely than anyone else to see the magic moment when the technology that fills the hole is feasible.

Let's summarize.

We start with a vision, a Hedgehog Concept. All of our employees know our Hedgehog Concept, what we are good at and how our clients benefit from working with our company. We know which technologies relate to our capabilities to meet the client needs.

We carefully watch and keep track of technologies that align with our vision. We avoid wasting our time on technologies that don't. We keep track of standards bodies, and know exactly which standards are current, which ones are likely to become real, and which ones will probably fall by the wayside.

We keep track of vendor success. We know which subindustries within the information technology world are blossoming and which ones are floundering.

We keep track of our competitors. We know what level of technology they are using. We know what the customer's expectations are, and whether we (or our competitors) are fulfilling those expectations with existing technology.

We talk with the people on the front lines, so we know where all the holes are. We know what kind of work-arounds we are currently using. We spend quite a bit of time playing with new technologies to see if they fill the holes. We extensively test to ensure that we won't lose existing capabilities when we implement the new technology. We remain patient for the right technology, and we avoid succumbing to the power of wishful thinking. The right technology comes along, and we recognize it. We are ready. Okay . . . **Pounce.**

## 12-1. Interactive

1. **Discussion:** Consider some of the companies you've worked for or had dealings with. Did they have a well-known and agreed-upon vision? What was it?
2. **Discussion:** Have you ever invested in a technology that disappeared after a few years? How much did you invest? How could you have figured out ahead of time that the technology was not going to be around for the long term?
3. **Discussion:** Which current companies have "hot" technology. Do you think it will be around for the long term? How do you know whether it will blossom or not?
4. **Independent Activity:** Look at the technology map of your personal technology. Write down the standards bodies that is involved with one of the items in each layer of technology.
5. **Independent Activity:** Identify some hardware or software that you are currently using that has a "hole". What is the hole? When do you think the hole will be plugged?

## 12-2. Self Study Test Questions:

1. How do you know which technologies relate to the vision of the company?
2. How do you keep track of the progress of technologies related to the vision?
3. Why should you ignore technologies that are not related to the vision?
4. What are some of the standards organizations, and why are they important?
5. How do you know if a vendor is going to be around in the future?
6. Why is it important to know what technology your competition is using?
7. Why should you invest in a research lab with the latest technology related to the vision of the company?

# Index

Made in United States
Orlando, FL
04 November 2026

90209482R00195